The Worldmakers

The Worldmakers

Global Imagining in Early Modern Europe

AYESHA RAMACHANDRAN

The University of Chicago Press

CHICAGO AND LONDON

AYESHA RAMACHANDRAN is assistant professor of
comparative literature at Yale University.

The University of Chicago Press, Chicago 60637
The University of Chicago Press, Ltd., London
© 2015 by The University of Chicago
All rights reserved. Published 2015.
Printed in the United States of America

24 23 22 21 20 19 18 17 16 15   1 2 3 4 5

ISBN-13: 978-0-226-28879-6 (cloth)
ISBN-13: 978-0-226-28882-6 (e-book)

DOI: 10.7208/chicago/9780226288826.001.0001

Published with the assistance of the Frederick W. Hilles
Publication Fund of Yale University.

Library of Congress Cataloging-in-Publication Data

Ramachandran, Ayesha.
   The worldmakers : global imagining in early modern Europe / Ayesha
Ramachandran.
      pages cm
   Includes bibliographical references and index.
   ISBN 978-0-226-28879-6 (cloth : alk. paper) — ISBN 978-0-226-28882-6
(ebook)   1. Geographical perception.   2. Human geography.   3. Historical
geography.   4. Cosmography.   5. Philosophy and science—Europe—History.
6. Europe—History.   7. Europe—Civilization—16th century.   8. Europe—
Civilization—17th century.   I. Title.
   G71.5.R35 2015
   910'.01—dc23

                                                    2015005557

   ∞ This paper meets the requirements of ANSI/NISO Z39.48-1992
               (Permanence of Paper).

mutat enim mundi naturam totius aetas
ex alioque alius status excipere omnia debet
nec manet ulla sui similis res: omnia migrant,
omnia commutat natura et vertere cogit.

[For time changes the nature of the whole world,
and one state of things must pass into another,
and nothing remains as it was: all things move,
all are changed by nature and compelled to alter.]

LUCRETIUS, *De rerum natura* 5.827–30

Contents

Figures

Introduction: Worldmaking and the Project of Modernity

Antwerp, 1579. A new pocket collection of maps is on sale from the printing house of Plantin. The first of its kind in French, it offers a scaled-down version of Abraham Ortelius's recent folio bestseller *Theatrum orbis terrarum* [*Theater of the World*], the first world atlas produced in modern times. This little volume bears a similarly imposing title: *Miroir du monde* [*Mirror of the World*].[1] A reader who opened the "epitome" expecting to find Ortelius's famous world map, however, would have been disconcerted. The opening pages feature an allegorical frontispiece (fig. 1) that illustrates the work's title and scope. A muscular figure, his face obscured by a giant globe topped by a cross—a *globus cruciger,* the ancient symbol of dominance over the world—looms over all. He is identified as "omnipotentia dei," the all-powerful God. Flanking him are two naked women, God's Prudence and God's Truth, holding mirrors to reflect his glory. Streams of light illuminate their bodies. But the center of the image remains dark and difficult to see. *Venite et videte opera Domini,* invites the caption, echoing the Psalms, "Come and see the works of the Lord."

Amsterdam, 1633. Fifty years later, another Dutch printing house brings out the latest collection of maps, the prized *Atlas ou representation du monde universel* [*Atlas or Representation of the Universal World*] with its iconic double hemispheric world map (fig. 2).[2] Here, the precise contours of continents embedded in geometric matrices of longitude and latitude take familiar, modern forms. Allegorical symbols of dominion are reduced to a cartouche in the center, while the map's richly decorated frame celebrates the achievements of human cartographers in four medallions. In a proto-scientific gloss

FIGURE 1. "Speculum mundi" from *Miroir du monde* (Antwerp, 1579). Courtesy of the John Carter Brown Library at Brown University.

on the nature of the planet, the hemispheres are surrounded by emblems of the four elements. This world emerges into view not through divine revelation but by dint of human effort. It is the outcome of a long quest to make visible the global whole—now understood as the "universal world," a fusion of the earth and the heavens—that could never be seen at once through the naked eye.

Apollo 8, 1968. Three hundred years later, the dream of encompassing the world in a single glance would be fulfilled when the first manned NASA mission to the moon photographed the earth from lunar orbit. "Earthrise at Christmas" (fig. 3) finally confirmed the picture of the world that the early modern mapmakers could only construe through the imagination. In the multicolor interplay of land and sea across the earth's surface as the orb rises above the horizon, the photograph provides a god's-eye view of the world—a view previously reserved for the deity and only partially revealed to a curious human gaze.

The Worldmakers reconstructs this imaginative struggle to capture the

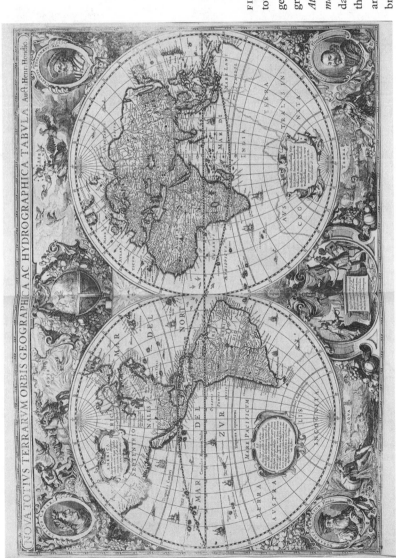

FIGURE 2. "Nova totius terrarum orbis geographica ac hydrographica tabula" from *Atlas ou representation du monde universel* (Amsterdam, 1633). Courtesy of the Beinecke Rare Book and Manuscript Library, Yale University.

FIGURE 3. "Earthrise at Christmas" (1968). Courtesy of NASA.

world's entirety through the self-conscious efforts of particular human makers. It tells the story behind these changing images, tracing the transformation of the world from an expression of a creative, omniscient deity to a modern conception of cosmic totality—from a world revealed to a world made up. Looking back at the long history of the desire to see the world whole, a desire that culminates in "Earthrise," *The Worldmakers* asks how it became possible to capture such a vision of the world in the sixteenth and seventeenth centuries when space travel was a metaphor confined to fantasy. And it investigates why all aspects of early modern culture were fueled by this desire to comprehend the world, to organize and capture its variety in a single, harmonious frame.

*   *   *

Traces of a resurgent interest in "the world" as a whole appear everywhere in the early modern period. The boundaries of the world slowly expand on planispheres and across the love-lyrics of Petrarch, Scève, Bruno, and

Donne. Ptolemy's *Geography*, the classic work on how to draw a map of the world, becomes a sixteenth-century bestseller.[3] Early advertisements for navigational tracts, scientific instruments, and maps promise untold riches in lands yet-to-be-discovered.[4] Political treatises dream of empires vaster than any classical civilization. The new, global scale of such dreams is also indexed by their immense cost: the slaughter of the Amerindians at Cuzco and Tenochtitlán, the flames which burned heretics at the stake for daring to think of plural worlds or different origins for the cosmos.

The verbal omnipresence of "the world," a familiar refrain in various texts of the period, thus signals a brave new intellectual conundrum. The intelligibility and scope of the known world had been called into question over the course of almost two centuries, ever since the early Spanish and Portuguese voyages of exploration. The pursuit of colonial and commercial exploration, the growing intellectual trends of skeptical thought, theological questioning, astronomical speculation, and the emergence of a new "historical consciousness" all raised the urgent question of how the extent of the known world—whose boundaries were not immediately visible or tangible—was to be described.[5]

When in 1651 Andrew Marvell mused, "'Tis not, what once it was, the world," he was speaking for at least two generations of Europeans who had experienced at first hand the effects of an expanding world, transformed by the discovery of a new continent on the other side of the Atlantic and of new planetary bodies circulating in space. No longer the divinely ordered terrain familiar to classical antiquity or the Middle Ages, "the world" now seemed, in Marvell's words, "but a rude heap together hurled." With a mixture of elegiac solemnity and wonder, the poet articulates one of the most profound intellectual shifts of early modern Europe: the definition of "the world" as a new category encompassing a previously unknown intellectual expanse and holding new imaginative power. For the poet and his contemporaries, the crumbling of old systems of explanation had left the concept vague and undefined. No longer did a golden chain connect "this pendant world" to Heaven. The human and natural world seemed decentered and disconnected, leaving the idea of "the world" itself desperately in need of redefinition, re-imagination, and renewal.

Early modern Europe responded to this quest with an explosion of images, descriptions, measurements, hypotheses, and debates about the nature of the world. From the Dutch print of a world map in a jester's cap to the mammoth Coronelli globes, from vast Flemish tapestries showing *The Spheres* to small octavo epitomes of compendious cosmographies such as the *Miroir du monde,* from global trade networks that brought pineapples to England and Chinese slaves to Mexico, to furious local debates over cosmic theories, such as Galileo's *Dialogue Concerning the Two Chief World Systems*— the very idea of "the world" becomes a foundational but fluid and fiercely contested category.[6] It would be no exaggeration to identify the central intellectual task of the late Renaissance, which affected all aspects of early modern life and thought, as the problem of "the world" itself.

Writers from More to Leibniz make the collision between worlds—old and new, ancient and modern, imagined and real—central to their depiction of what has since been called the "epistemological crisis" of the period, that increasing emphasis in the late sixteenth and seventeenth centuries on worldly plurality, contingency, and the limitations of human perception and knowledge. The last such comparable effort dated back to Roman Egypt of the second century, where Claudius Ptolemy had established the boundaries of the *oikoumene.* For almost two thousand years then, until the Columbian age of exploration, the world had remained a stable concept. In the age of the Renaissance worldmakers, it had to be rethought and reshaped once more.

WORLDMAKING

At first, the intimation of a world beyond could only be sensed in slivers of new knowledge, in local details, anecdotes, "singularities." A feather headdress might stand for America; a piece of coral for the beauty of the Pacific islands; the sketch of an unseen coastline might promise a sea route to China, or prove to be a new continent. But if "the world" stood for some idea of unity—an ordered system—then these fragments had to be synthesized into an intelligible conceptual framework, a coherent world picture. How was such a synthesis achieved? What tools helped navigate the passage from an old order to a new one?

This book follows the hard-won renovation of the world across the sixteenth and seventeenth centuries, tracing the story of its emergence as a foundational category of modernity. "Worldmaking" thus describes the methods by which early modern thinkers sought to imagine, shape, revise, control, and articulate the dimensions of the world.[7] It captures the relentless intellectual and cultural drive to uncover a comprehensive vision of the whole—global and eventually cosmic—by attending not only to large-scale macro-historical processes, but to the conceptual, imaginative, and metaphysical challenges posed by the task of envisioning an abstract totality.

The modern world comes into view as it is measured against its various parts—from the microcosmic self, to national and imperial communities, to the sweep of the cosmos. Worldmaking was a ubiquitous cultural practice in the early modern period. It informed the commerce of sailors and merchants, the battles fought across continents for global imperial dominion, the crafting of precision instruments and the printing of books in the workshops of European capitals. It colors the work of the land surveyor in Peru as it does the rhetorical bombast of Marlowe's Tamburlaine who demands, "Give me a map; then let me see how much / Is left to conquer all the world."[8] It fuels voyages of exploration, habits of collection, and the rise of the "New Science." And it demands the interrogation of traditional forms of religious belief and faith in the divine.

Underlying all these activities is a need to synthesize new global experiences into a structure that would bind individual fragments into a collective unity. To comprehend the world thus required deft oscillation between local details and global frameworks and a reconfiguration of the particular against the universal. It was a task of metaphysical, and not just practical, dimensions. These abstract questions were familiar from a long tradition of medieval mereology and its classical antecedents in Plato, Aristotle, and Cicero. But they took on a new urgency when "world," that all-encompassing but all-too-nebulous category, itself was in the balance. To redefine the whole demanded a new consensus for determining the relative autonomy of the individual vis-à-vis a collective whole. While measurement and observation—soon to become the standards of empirical science—could provide a record of local details, synthesis into a global whole required an act of imagination, a leap of theoretical speculation that left the precision of the example for the abstraction of totality. A concept, a category, and a system of order, the world thus had to be self-consciously refashioned by individual human makers. But this was a gradual and difficult recognition.

For much of European history, worldmaking had remained tied to the idea of creation, an original divine act that had prescribed an absolute order to all things. Even the frontispiece to Pieter Heyns's 1579 *Miroir du monde* reflects this view: the globe *is* the face of the Deity, simultaneously covering and revealing it; it is both subject to divine dominion and a privileged expression of divine creative power. That identification provided the foundation for a vision of world order whose intellectual contours had crystallized into the Thomist-Aristotelian synthesis of Christian religion and classical science. But by the late sixteenth century, this *summa* lay in shreds. Even though many dreamed of uncovering a perfect, perhaps divine, system of world order that would heal the damage, the impossibility of that aspiration was soon evident: when Montaigne writes movingly of a world in flux at the end of the "Apologie de Raimond Sebond," we sense the emergence of a terrible skepticism about our ability to apprehend the order of the universe. Early modern worldmaking, as it is chronicled in these pages, begins in response to this dilemma. It reflects a new recognition of our existence in a radically uncertain world where we must create our own order. And it therefore emphasizes the importance of *poiesis*—artful making—as a means of eliminating contingency and making sense of the pieces.

Despite its recent association with the anti-realist, neo-Kantian philosophy of Nelson Goodman, worldmaking has a complex ideological history that derives from physico-theology and was only later exported into modern philosophical discourse.[9] At its core is the idea of creation—the belief that a world can be made and transformed, rather than being a preestablished entity awaiting discovery. When Nathaniel Fairfax first used the term "worldmaker" in the late seventeenth century, he was asserting the importance of metaphors of construction that were popularly used to describe divine creation. "I can't find in my heart to deny that skill to a World-maker, that I must needs give to a Watch-maker," he wrote, alluding to the mechanical philosophy with its vision of a perfectly ordered natural world that functioned according to preestablished laws.[10] His emphasis on materiality signals a literalist vision in which physical matter is carefully crafted and given a specific form. While this metaphor of construction was immensely popular throughout the Renaissance, it was only one of several possible models; others included theories of spontaneous generation, instantaneous creation called forth by the Word, or random evolution through the collision of material particles. To assert that the world was *made* implied staking a position in a charged debate on the existence of deity and the extent of God's involvement in human affairs.

Paradoxically, the constructivism implicit in Fairfax's version of worldmaking becomes staunchly realist because it is grounded in theistic belief. It conceives of the world as a discrete object given form by a single identifiable creator. The existence of a world-picture as a subjective human creation that might itself replace or construct a sense of the objective world is utterly absent; human representations of the world are always secondary, imitative shadows of the divine original. But while this remained the orthodox view, it was already under attack by the late seventeenth century, when Milton would explore the bounds between human and divine making in *Paradise Lost*.

By the early eighteenth century, there are signs that e term "world" had become detached from this literalist, theistic context to encompass more metaphorical meanings. Thus, Matthew Prior's use of a similar phrase "system-makers and world-wrights" in 1721 suggests that worldmaking could refer to the construction of competing models of world order rather than to the physical world itself.[11] "World-wrights" seems to be derived from such Anglo-Saxon compounds as "shipwright," "wainwright" or "playwright," which describe human artificers, specifically handicraftsmen. While it retains a trace of Fairfax's realism, the synonymous use of "system-makers" suggests that Prior's "world" is not that of physical matter but rather one of philosophical theory.

It is, however, not until Hume's *Dialogues concerning Natural Religion* (1779) that "worldmaking" receives its first philosophically deliberate use, though once again in a theological context, as the skeptic Philo questions the logic of the cosmological argument for the existence of God:

> But were this world ever so perfect a production, it must still remain uncertain, whether all the excellences of the work can justly be ascribed to the workman. . . . Many worlds might have been botched and bungled, throughout an eternity, ere this system was struck out: much labour lost: many fruitless trials made: and a slow, but continued improvement carried on during infinite ages in the *art of world-making*. In such subjects, who can determine, where the truth; nay, who can conjecture where the probability, lies; amidst a great number of hypotheses which may be proposed, and a still greater number which may be imagined?[12]

Hume explodes the materialist-realist vision of a divine world-maker articulated by Fairfax. The skeptic Philo sees the art of worldmaking as an extension of human artifice rather than an illustration of divinity: it is therefore subject to the vicissitudes of trial and error. Here, Hume gives the notion of worldmaking its current, double-edged meaning—it refers *both* to the actual origin and order of the physical world as well as to the theories that we invent to comprehend the vastness of that whole. In this, he anticipates both the realist and antirealist positions of recent philosophers while highlighting what was at stake in the struggle over the nature of worldmaking: the basis of religious belief, the possibility of scientific truth, and the nature of all systems of order as imagined representations rather than demonstrable facts.

The *Dialogue* thus illustrates the outcome of a long struggle to reconstruct a new world order, and Hume's specific use of the term "the art of worldmaking" marks the end of an intense phase of such activity rather than its beginning. It reflects a radical shift from a primarily realist-theistic view of the world as divinely created to a skeptical-constructivist view of the world as humanly fashioned. But as the controversial reception of the *Dialogues* suggests, this destabilization of the world was accompanied by a tremendous cultural anxiety that is already palpable in many early modern works.

Hume's skepticism was not itself new. It draws on well-known arguments by writers such as Lucretius, Montaigne, Descartes, and even Milton, and is similar to the position taken by the French encyclopedists. However, the *Dialogue* contains an important insight that was never explicitly articulated before: the world must be understood as no more (and no less) than a human representation because certain, complete knowledge of the objective world is ultimately impossible to achieve. Hume thus touches on the great secret of the early modern system-makers—worldmaking is possible, even necessary, because of the insurmountable gap between our fragmentary apprehension of the phenomenal world and our desire for complete knowledge of it.

Worldmaking is thus a creative process emerging from a renewed celebration of *homo faber*.[13] I use the term in the wake of Hume and Goodman to accent the processes by which the world is remade in the early modern period through a combination of rhetoric, aesthetics, *poiesis*, and the speculative imagination. This new belief in the world as an artifact also marks its modernity: the world comes into view as a thing made, shaped by human skill and ingenuity, and subject to historical transformation.

A SHORT HISTORY OF "THE WORLD"

One of the first signs of change is lexical: the words used to designate "world" in both classical and vernacular languages undergo significant reconfiguration over the sixteenth and seventeenth centuries. Derived from two related but distinct classical concepts—the *oikoumene* or *orbis terrarum* (the "circle of lands") and the *kosmos* or *mundus* ("the world" or more amply, universe)—the words for "world" in most European vernaculars (*world, welt, monde, mondo, mundo*) begin to combine both meanings into a single term in the early modern period. This gradual fusion is evident, for instance, in the difference between the expressions of the idea "world" in Cesare Ripa's late sixteenth-century *Iconologia* (1593) and Giambattista Vico's early eighteenth-century *Scienza nuova* (1725). Taken together, they measure the intellectual transformation witnessed by the early moderns; at the same time, they suggest models for the study of such cultural change.

Ripa's influential *Iconologia,* the Renaissance sourcebook of iconography, contains several emblematic representations of the world—or rather, of various aspects of it. Ripa includes detailed instructions on how to represent *Terra* (the element of earth), *Mundus* (the World), the four continents (America, Asia, Europe and Africa), as well as emblems for the disciplines of cosmography, chorography, and geography. Each represents a particular world-picture, and their cataloguing as distinct images and categories marks a process of fragmentation. But Ripa's divisions also reveal how a multifaceted conception of the world was being developed in symbolic form.

The *Iconologia* depicts a distinction that underpins words associated with "world" throughout the early modern period (*weorold, worold, world* in Old English; *kosmos* and *oikoumene* in Greek; *terra, orbis terrarum, mundus* in Latin). On the one hand was the natural world, the bounty of the earth and the glory of human culture and civilization: *terra* signifies the fertile land and all that it sustains, organically or architecturally. At the other extreme was the charged moral field of *mundus,* the world, which retained its medieval associations with vice, corruption, and metaphysical decay (worldliness) even as it came to signify the immensity and beauty of the cosmos.[14]

The emblems for *Terra* and *Mundus* thus present dramatically different, gendered versions of the world. Earth, grouped with the other elements early in the work, is described as "a matron sitting upon a globe, with a cornucopia in one hand, and a sceptre in the other."[15] She is said to wear a "mural crown" or a garland of flowers and fruits, and her typically green garments are decorated with floral motifs. This iconography links *Terra* to *Natura* and *Scientia,* and Ripa explains that her attributes transform her into a figure of both nature and culture: she is the mother of all animals; the globe denotes the sphere of the earth, while the cornucopia and foliage represent the products of the land; the crown "alludes to the buildings for the accommodation of the inhabitants." *Terra* thus signals the conjunction of human and natural worlds, an intersection that produces political and social action as well as scientific inquiry.

Mundus, however, is a pictured as an Atlas-figure, "a strong man, supporting a golden coloured globe on his shoulders with the constellations marked upon it." He is dressed in a garment of haircloth, covered with long bejeweled purple robes. Ripa's exposition moves away from the language of fruitfulness and civilization associated with *Terra* and instead enters the realm of natural and moral philosophy. Strength and support of the globe allude to "endurance of the evils, toils, and labours of this World"; this time, the globe denotes "the splendour, perfection, order, and harmony of the

Universe, and the amazing works of Creation and Providence." The hair-cloth, however, is a reminder of "the miseries, misfortunes and difficulties of this present state," while the pomp of his robes "signify that the pursuit of riches and worldly grandeur is vain and transitory" (2.160). *Mundus,* the World, figures the lure of knowledge and the transfiguring beauty of universal creation; it also reminds us of the need for metaphysical reflection.

While Ripa's emblems synthesize these differences into intelligible visual wholes, it is only in Vico's mammoth *Scienza nuova* that we get an etymological history of the concept "world" that reflects back on the transformations of the two previous centuries:

> The theological poets felt the earth to be the guardian of boundaries, which is why it was called *terra*. The heroic origin of the word is preserved in the Latin noun *territorium*, territory, meaning a district over which dominion is exercised. . . . The Latin grammarians mistakenly derived territory from *terrere*, to frighten, because the lictors used the terror of the fasces to disperse crowds and make way for the Roman magistrates . . . [but] the true origin of the verb *terrere*, to frighten, derives from the bloody rites by which Vesta guarded the boundaries of the cultivated fields, in which civil dominions were to arise. The Latin goddess Vesta is the same as the Greek Cybele or Berecynthia, who is crowned with towers, *torres*, or strong situated lands, *terrae*. From her crown there began to take shape the so-called *orbis terrarum*, or world of nations, which cosmographers later expanded and called the *orbis mundanus*, mundane world, or simply *mundus*, world, which is the world of nature. . . .
>
> [In] early Latin *mundus* meant a slight slope. . . . Later, everything that trims (*monda*), cleans, and adorns a woman was called *mundus muliebris*, feminine ornament. Eventually, the poets understood that heaven and earth are spherical; that each point of their circumference slopes in all directions; and that the ocean washes the earth on all sides. So when they saw that the whole is adorned with countless various and diverse sensible forms, the poets called the universe *mundus* as a beautiful and sublime metaphor for the ornament with which nature adorns herself.[16]

Vico's creative reconstruction of the concept's evolution from the specificity of the land (*terra*), to civil dominion over a wider region, and eventually to a universal ideal of beautiful order (*mundus*) parallels Ripa's differentiation, and like the iconographer, owes much to a long literary and philosophical tradition. But the categorical differences in Ripa are, in Vico, part of an intellectual-historical continuum.

Isidore of Seville's seventh-century *Etymologiae* had already divided the study of the world into sections on the cosmos ("De mundo et partibus") and on the earth ("De terra et partibus"), noting that the Latin *mundus* was an attempt to translate the meanings of the Greek *kosmos,* a word that presented a distinctly aesthetic understanding of the universe, since it signified order, beauty, form, fashion, and ornament.[17] On the political plane, the Roman historians Livy and Herodian had suggested links between Roman territorial concepts and religious ritual—accounts that were then faithfully reproduced by Renaissance cosmographers and lexicographers. Early modern thinkers, however, added a new term—"the universal world"—a hybrid that marked the integration of land and sea into a single terraqueous planet.[18] By the early eighteenth century, the geographical contours of the world had been reconceived by cartographers such as Ortelius, Mercator, Hondius, and Blaeu; the "world of nations" had been brought into political existence by conquests in the Americas and the Peace of Westphalia; and the "world of nature" seemed continuously to expand as scientific study probed both infinite space and the infinitesimal microbe. Vico's attempt to integrate classical origins and modern reconfigurations into a single seamless narrative reflects a point of culmination, the description of a newly completed event.

The Italian philosopher's "poetic cosmography" differs sharply from earlier compilations in its focus on the unexpected ways in which cultures synthesize meaning and create new conceptual categories. Vico is less interested in establishing what the concept "world" means than in how it comes to have multiple meanings and why it means in these particular ways. Here, as elsewhere, Vico emphasizes the intersection between poetic mythmaking and historical contingency: the *orbis terrarum* (circle of the earth), once derived from the crown of towers worn by the earth goddess Cybele, is now simply a collective term for the world of nations; the beauty of feminine ornament has, over time, become a "sublime metaphor" for the ordered universe.

If such etymologies and emblems are the fragments through which Vico traces a culture's transformation, his interest in the assimilation of poetic naming and narrative into cultural memory offers an unusual model for this book as well. The *Scienza nuova*'s recovery of long-forgotten acts of *poiesis* invites us to reexamine them too from a postmodern stance. *Poiesis,* the act of making, is an epistemological practice for Vico, the only mode of knowing with certainty. If, as he had famously argued, "verum et factum convertuntur" (the true and the made are interchangeable), we can only truly understand what we have made.[19] Full knowledge of any thing involves discovering how it came to be what it is as a product of human ac-

m this perspective, Vico's discussion of the world suggests that it
anly made through constructive acts of naming. The centrality of
l poets" to Vico's method signifies a crucial link between *poiesis*
aology, making and knowing, and thereby lays the philosophic
foundation for understanding how a plethora of local details may be trans-
muted into encyclopedic knowledge of the whole. Vico's vision of a poetic
epistemology and his history of the term "world" provide inspiration for
this project, which tells a previously unexamined cultural and intellectual
history of "the world" by excavating its symbolic, ideological, and metaphysical freight.

A PROJECT OF MODERNITY

Few ideas have become so thoroughly associated with the emergence of
modernity in Europe as that of a globalized, interconnected, secular world.
The phrase "modern world" has in fact become a shorthand for a global environment characterized by scientific rationalism, large-scale economic networks, international realpolitik, and agnostic skepticism. Not surprisingly,
then, recent scholarship on globalization and world-systems has emerged
primarily from the social sciences, particularly economics, historical sociol-
ogy, and cultural anthropology, thereby reiterating the basic elements of a
familiar historiography despite overt gestures of critique. But to recognize
the world as a subject in its own right—rather than as a background for or
byproduct of large-scale historical processes—is to rethink traditional narratives about the genesis of the Copernican universe and the making of the
modern world.

In its emphasis on human making, *The Worldmakers* tests one of the key
shibboleths of modernity: the entwined rise of secularism and scientific
empiricism. Contrary to the now-classic Weberian narrative of modernity
and disenchantment, I argue that the invention of the modern world owed

much to theology and the spiritual practices of imaginative identification; it
remained enmeshed in metaphysics and the creative faculties of the "intellectual imagination" even as it drew on the tools of empiricism, mathematics, and the new science. Central to this story is not only a new technological facility and belief in human reason but also an integration of earlier
forms of magical thinking—hypothesis, metaphoric association, symbolic
correlation, aesthetic formalism—into scientific practice.

The Worldmakers thus seeks to move conversations about globalization
and modernity beyond the events and material processes that were its cata-

lysts to the imaginative responses that sought to comprehend them. Philosophical critiques of modernity in the twentieth century from Heidegger to Habermas and Foucault have argued that the modern world was founded upon a rationality that stripped away alternate forms of knowing—speculation, meditation, intellectual intuition—in order to establish the hegemonic universalisms of the Enlightenment. But my inquiry into early modern worldmaking raises fundamental questions about such accounts as it reveals the persistence of those earlier modes of thought. I argue instead for an alternate genealogy for modernity, one that emphasizes the collusion of empiricism and the poetic imagination and highlights the continued significance of metaphysics alongside a supposed "epistemological rupture."[21] If the modern age, for Heidegger, begins when we no longer seek a picture of the world but rather when the world comes to be "conceived and grasped as a picture," the early modern project of worldmaking illustrates how this inversion came about.[22] The making of the modern world, in this book, depends finally on the synoptic energies of the imagination even as its individual elements are produced through rational inquiry and action.

Recognizing modernity's debt to self-conscious worldmakers brings a new perspective to two distinct matters: the question of religion in modern life and the much-debated connection between modernity and empire. Attention to the spiritual and theological roots of worldmaking reminds us that the world's creation and its domination were traditionally the provenance of the deity.[23] The transfer of worldly authority from divine to human hands provided the legitimation for early European imperial ambitions (the title *dominus totius mundi*, once reserved for God, was later appropriated by individual monarchs). It also underwrites a now conventional narrative about the rise of secularism as a condition of post-Enlightenment modernity. And yet, the persistence of theological rhetoric in worldmaking accounts suggests how the skeptical crisis of modernity could also engender a new, more robust faith—a historical insight that is in keeping with Charles Taylor's recent analysis of the persistent place of religion in the modern world.[24]

Consequently, this book argues for the importance of reevaluating the metaphysical foundations of the modern world. These are discernible in the early modern competition between different philosophical systems, particularly the repeated confrontation between Platonic and Epicurean philosophy which epitomized a wider struggle between two kinds of metaphysics: one founded on the (theistic) principle of divine creation and cosmic order, the other based on an (atheistic) belief in worldly contingency, mutability, and evolution. Historians of philosophy have long acknowledged the significance of this opposition: it informs the emblems and images which contrast the eternal and the mutable; it underlies clashes over the closed, Ptolemaic system with its unchanging celestial spheres and the infinite, Copernican universe composed of mutable matter; it infects arguments over the status of scripture as unchanging, literal truth or as allegorical narrative open to changing interpretation. But the difference between the Platonic emphasis on the primacy of form and the Lucretian insistence on the centrality of matter precipitated a cultural debate on the nature of world order and its relation to God that continues even today: was the world preestablished by divine sanction or is it unstable, ever-evolving, and open to human intervention? *The Worldmakers* charts the oscillations between these positions, connecting such debates to contemporary reflections on secularization and faith.

This book, however, does not neglect urgent political and ethical concerns. Laura Doyle speaks for many scholars when she argues that "modernities are often organized and motivated by the will to empire."[25] And indeed, worldmaking has frequently been regarded as a euphemism for the empire-building aims pursued by European states across the globe, both in accounts from the sixteenth and seventeenth centuries and in contemporary historical and theoretical analyses. This identification between modernity, empire, and worldmaking seems almost self-evident: Charles V's emblem and motto, "plus ultra," alluded to the vast world beyond the Pillars of Hercules, while his successor Philip II's motto, "non suff101t orbis," would proclaim that even the world was not enough for his ambitions.

It is, in fact, impossible to speak of an emerging "modern world" in Europe in the sixteenth and seventeenth centuries without grappling with the impact of European colonial and imperial projects across the globe. For the language of worldmaking, universalism, and cosmic comprehension is frequently invoked in the context of political desire; the triumphant narratives of global expansion remain tied to the rapacity and destruction that accompanied cross-cultural encounters and which were their bitter, lingering aftermath. As Walter Mignolo has powerfully argued, imperial annihilation and colonial oppression are the "the darker side of the Renaissance," the corollary to the anthropocentric humanism celebrated by writers such as Alberti, Pico, and Leonardo.[26] Contemporary postcolonial critiques of empire and Enlightenment insist on this conceptual overlap between empire and world, universal humanism and European hegemony.[27] But were the two terms—empire and world—indeed synonymous and indistinguishable in the early modern period? How, in fact, does the totalizing concept of "world" relate to other categories of belonging, such as the nation or the state? The conceptual challenges inherent in these questions motivate this book's quest to distinguish between the nebulous, always present political dimensions of the world and of world order that lurk beyond early modern empires and nation-states.

Despite the enthusiasm for global approaches in a variety of disciplines, and the apparently relentless push toward the transatlantic, the transnational, the multicultural, and the diverse, there remains a significant lacuna: the emphasis on globalism has actually produced a plethora of local narratives and detailed micro-histories, but the intellectual problem of understanding the "world" and how it is imagined *as a totality* still demands critical attention. Sanjay Subrahmanyam has observed, "The early modern world was for the most part a patchwork of competing and intertwined empires, punctuated by the odd interloper in the form of a nascent 'nation-state.'"[28] This picture of inter-imperial rivalry punctuated by gradual consolidation points to a sliding scale of community and political belonging within which "world" represents an intangible, theoretical amalgam of all polities, or of all peoples. At a literal level, such as that evoked by Tamburlaine and Philip II, "world" signifies all the territory available for conquest, the desire for universal dominion. But "world" also opposes "empire." It points to the territories and peoples *not* part of a state's control. It thus offers an alternative order and an alternate means of identification that both resists and transcends the hegemonic energies of empire — even though it threatens to be co-opted into displays of imperial ambition.

In the wake of recent calls to reevaluate cosmopolitan notions of global belonging, as well as their critique, it is particularly urgent to recover the pre-Enlightenment vision of "world" as a politically potent, morally compelling category.[29] It is now a commonplace to argue that the rhetoric of the universal and the global merely masks the imposition of European norms onto 'the rest of the world,' often through the violent assimilation of the colonizing process. But the elision of earlier worldmaking conceptions with individual national imperialisms reiterates a post-Enlightenment confusion of scale. For the early modern period sees the emergence of the modern world through the oppositional energies of nation- and empire-building on the one hand and worldmaking internationalism on the other. The word "cosmopolite," itself a sixteenth-century invention, carried with it from these early modern origins the conflicted freight of imperial ambition and transcendent global aspiration.[30] In these hesitant movements toward a political world beyond the interests of the individual state and its imperial ambitions, a world existing prior to and greater in scope than the sum of singular nations, we can see the beginnings—and also perhaps the future—of the modern global order in the political sphere. Tracing the seeds of a universal desire that opposes imperial aggrandizement, this book finally suggests how worldmaking in its early modern iteration may offer insights for contemporary theorists in a post-imperial world.

INTRODUCING THE WORLDMAKERS

Emphasizing connections between classical, medieval, and modern desires for comprehensive knowledge of the world, *The Worldmakers* traces the rough chronological arc of the early modern period, from the late fifteenth-century voyages of exploration to the late seventeenth-century debates over religion and science. It moves across increasingly expanding scales of analysis that connect part and whole: from the microcosmic relation of self and world, the communal affiliations promised by the nation and the lure of imperial dominion, to the vast expanse of cosmic space, where the world itself is subsumed into the universe.

Built around five case studies, the book explores varied terrain: the maps of Gerhard Mercator, the essays of Michel de Montaigne, the imperial epics of Luís Vaz de Camões and Edmund Spenser, the philosophical meditations of René Descartes, and the cosmic poetry of John Milton. These encounters with particular intellectual topographies highlight central historical cruces and suggest how new genres and styles of representation emerged in response to the problem posed by a world in flux. Thus, my consideration of Mercator's *Atlas* interrogates the intellectual strategies implicit in the development of the world atlas as a genre and as a tool for envisioning the world, while the companion chapter on Montaigne's *Essais* examines how the rhetorical conjunction between cosmography and autobiography was instrumental to a skeptical reappraisal of the world's order. Turning to the relations between political and cosmic order, the pairing of *Os Lusíadas* with *The Faerie Queene* reveals how the Renaissance epic showcased a contemporary tension between national-imperial desire and cosmic-universal claims, anticipating the dilemmas of imperial expansion even before the era of postcolonial critique. The final set of chapters turn to the question of Genesis and ask how this foundational worldmaking text was reimagined in the wake of the New Science: it follows Descartes and Milton as they rewrite and rehabilitate the subject of creation itself.

I turn to such familiar texts precisely because they are so thoroughly assimilated into a modern intellectual tradition that we have forgotten the radical charge that they carried. These texts established habits of thought that formed the bases of worldmaking across a variety of disciplines— cartography, geography, political theory, literature, natural and moral philosophy—that continue to inform ways of encountering the world to-day. Each of these writers invents new genres and modes of expression in response to the challenge of a new world; they cross linguistic and national boundaries as well as the battle lines of class and confession. To return to their work with fresh eyes is to look past the disciplinary frontiers erected by the Enlightenment to a time when the philosophers' most cherished goal was the union of all forms of knowledge.

With its immersion in multiple disciplines and languages, this book— perhaps unsurprisingly for an account of efforts to comprehend the world —speaks to a range of diverse interests, engaging with scholarship in the history of philosophy, early modern literature, studies of comparative empires, the history of cartography, cultural geography, and the scientific revolution. It tells of exploration and mapmaking, of the struggle to shape a nascent selfhood against an engulfing world, of the march to national self-definition, of quests for scientific truth, and of the perennial desire to contemplate the cosmos. But while it seeks to join together philosophical, literary, and historical approaches in rethinking the making of the world, it inevitably leaves out many details—a problem that early modern world-makers themselves would recognize. The story of worldmaking could take in many other figures, texts, and objects; it could include other genres, dis-ciplines, languages, for the scope of the world is vast and finally cannot be contained within the covers of a book. In writing of the early moderns' de-sire to comprehend the world, one must inevitably succumb to their pitfalls as well: like the texts it engages, this book too suffers from a partiality of perspective, a struggle to surmount the intellectual abyss that separates an individual's limited gaze from a yearning to acquire an all-encompassing view of the whole.

A final word is in order about the book's scope. This project presents an unmistakably Eurocentric approach to the question of the world—a ques-tion that justly demands to be addressed from a variety of cultural and geo-graphic vantage points to be truly global. The critique of Eurocentrism, however, is also bound up with the vision of a modern world whose theo-retical coordinates have been fixed by Europeans and exported to the rest of the world. Is a sustained excavation of European forms of worldmaking merely a reiteration of post-Enlightenment global hegemonies, or can it move beyond them? Paradoxically, perhaps, this book emerged from a post-modern, postcolonial suspicion of ideas of unity and totality—concepts that have often been used to erase local difference and impose imperial struc-tures on subject peoples. But as Martin Lewis and Karen Wigen suggest, "to transcend Eurocentrism initially requires a very close engagement with European thought."[31] This study consequently aims to deconstruct critical orthodoxies and seeks a subtler understanding of the stakes of universalist dreams. *The Worldmakers* thus joins in a project of provincializing Europe at the very moment it laid claim to the world; in this, it suggests avenues for research in a similar vein in other, non-European, contexts.[32] Through the book's analyses, worldmaking appears as a courageous, ethically em-powering, imaginatively ambitious response to forces of entropy and disor-der. It may be this response—and not the Weberian analysis of institutional change—that can define a modernity capable of transcending the particu-larity of the European model.

*   *   *

Such an ample vision already surfaces in the sixteenth century. In 1583, Justus Lipsius's *De constantia* urged readers to imagine themselves as part of one world:

> O fool: Are not they men, sprung first out of the same stock with thee? Living under the same globe of heaven? Upon the same mold of earth? Thinks thou that this little plot of ground . . . is thy country? Thou art deceived. The whole world is our country, wheresoever is the race of mankind sprung of that celes-tial seed. Socrates being asked of what country he was, answered: Of the world. For a high and lofty mind will not suffer itself to be penned by opinion within such narrow bounds but conceives and knows the whole world to be his own.[33]

This exhortation to cosmopolitan connection carries with it a new, fragile intuition of global unity, one nevertheless charged with a recognition of human divisions and differences. Its turn back to Socrates, however, may have been inspired by new cartographic vistas: Ortelius's *Theatrum orbis ter-rarum* contains a similar neostoic sentiment juxtaposed alongside a world map. And Montaigne, too, would return to Socrates' assertion in the essay on vanity as he contemplated Rome, *caput mundi*, the head of the world.

Almost four hundred years later, responding to television broadcasts showing the Earth rise over the lunar horizon on Christmas Day 1968, the poet Archibald MacLeish echoed this rhetoric when he spoke for the first generation to see photographs of the world from space:

> To see the Earth as it truly is, small and blue and beautiful in that eternal silence where it floats, is to see ourselves as riders on the Earth together, brothers on the bright loveliness in that eternal cold—brothers who know they are truly brothers.[34]

For MacLeish, as for a range of commentators, the shock of the first literal moon voyage was not the experience of space or the materiality of the lunar surface. It was that unexpected first long look at the Earth floating in the dark expanse of the universe. It was a sight that evoked a renewed sense of cosmic loneliness, first experienced in the seventeenth century, as well as a powerful celebration of global human community.

When the astronauts' photographs were later published, "Earthrise at Christmas" (fig. 3) would be met with awestruck wonder—a human response to a panorama reserved for God. Robert Poole does not exaggerate when he notes that "the sight of the Earth came with force of a revelation."[35] It was a revelation. If the early moderns could never achieve such a direct view of the planet from a point outside it, *The Worldmakers* shows how they anticipated its metaphorical and metaphysical impact—it was a view they had repeatedly tried to imagine and recreate in texts, images, and objects. They had prepared the way for Earthrise. It is no accident that the astronauts themselves, invoking a time when such a vision remained suspended between deistic and artistic creation, prepared to introduce the moment with an ancient text. They read from the opening verses of the Bible, speaking aloud the creation story of Genesis, as they looked at the world coming into view.

CHAPTER ONE

Mapping the Body, Mapping the World: Mercator's *Atlas*

The 1595 Duisburg edition of Mercator's *Atlas* opens with a now iconic engraving (fig. 4). A muscular man with a white beard and flowing red mantle scrutinizes a globe and a pair of compasses. Another globe rolls between his legs as the sky unfolds behind him. The entire tableau is encased within a classical architectural façade whose Corinthian columns may gesture symbolically toward the Pillars of Hercules. Above, two *putti* support an enormous astrolabe as though imitating the figure below them. *Atlas sive Cosmographicae Meditationes de Fabrica Mundi et Fabricati Figura*, reads the title: "Atlas or Cosmographical Meditations on the Making of the World and the Image of the Made [World]."[1]

The iconographic concision of the image combines technical illustration with the visual rhetoric of contemporary allegorical prints and emblem books, offering a counterpart to the double title that explores representational and conceptual boundaries.[2] For this text simultaneously names a person (Atlas) and a set of spiritual and aesthetic practices (cosmographic meditations). It reflects both on the making of the world itself (*fabrica mundi*) and on the figuration (in maps) of that world understood as an artifact (*figura fabricati*). Presenting a particular individual perspective as well as a universal gaze over global spaces, the frontispiece seeks to bring these opposing scales into relation. But as an allegory of making, it announces a bold new argument: the mapmaker *embodies* the world. In material and metaphorical ways, the human body and the global body become one.

ATLAS
SIVE
COSMOGRAPHICÆ
MEDITATIONES
DE
FABRICA MVNDI ET
FABRICATI FIGVRA.

Gerardo Mercatore Rupelmundano,
Illustrißimi Ducis Iuliæ Cliviæ et Mõ
tis &c. Cosmographo Autore
Cum Privilegio

FIGURE 4. Frontispiece from Gerhard Mercator, *Atlas* (Duisburg, 1595). Courtesy of the Library of Congress; reproduced from an original in the Geography and Map Division.

COSMOGRAPHIC DESIRE

This striking visual proclamation identifies the *Atlas*, and its author, Gerhard Mercator, as icons of worldmaking. An innovator in the development of world mapping, Mercator, perhaps more than any other historical figure, is closely associated with a paradigmatic shift in the image of the world. His 1569 navigational projection—the so-called Mercator Projection—produced a vision of global space that remains familiar; it is even the basis of the Web Mercator platform used by Google Maps and in ArcGIS systems today.[3] Beyond this technical breakthrough, his influential *Atlas*, whose title now names the genre, established the form, structure, and organization of world atlases for over two centuries, shaping cultural perceptions about the nature and order of the world itself. Thus, like the quincentennial commemorations of Columbus's first voyage to the Americas in 1992, the five hundredth anniversary of the Flemish mapmaker's birth in 2012 was also an occasion to reflect on the historical emergence of a modern world-picture. "He mapped the world, and we saw ourselves," began one popular tribute, honoring the transformative power of his cartographic achievements, which quite literally "changed the way we see the world."[4]

Endlessly subject to revision, the world and its recognizably "modern" visual representations in maps slowly came into focus over the course of the sixteenth century. If the world map emblematizes a changing world picture, it is also almost synonymous with worldmaking. The history of planispheric images—from Juan de la Cosa's *Mappa mundi* (1500), the earliest extant European map to incorporate the Americas, to the 1599 Wright-Molyneux chart, which popularized the Mercator projection—tells the story of a fluctuating idea of a global totality. Mercator's *Atlas* marks the culmination of this decades-long effort. Its innovation lies in recognizing that the cosmographic challenge was not merely technological or political but rather metaphysical and therefore radically conceptual in scope. Ironically for a work hailed as a cartographic point of origin, most of its maps had already been published in the 1570s and 1580s; even the plates used for the *Atlas* were engraved not by Mercator but mostly by his sons and grandsons. The importance of the *Atlas*, however, lies in its understanding that reimagining the world as a visual whole on a map necessarily demanded a philosophical, theoretical counterpart—a reevaluation of the world's structure and the individual human being's position in relation to it. Moving beyond traditional cosmography's compendious data collection and compilation of maps, the *Atlas* defines an intellectual watershed by seeking to envision the totality of the world.

Cosmopoesis

Capturing the world in a single glance on a map demanded unique skills and a vast network of informants and sources. Mercator's career-long quest to understand and represent the world in its entirety therefore participates in the complex web of relations joining various sixteenth-century communities and interests. Mapping the world depended on a synthesis of information drawn from a heterogeneous group—from cosmographers, craftsmen, and printers, to sailors, diplomats, political patrons, philosophers, and would-be theologians. The mapmaker acted as a synapse, gathering, consolidating, and relaying visions of the world through a mix of technical skill, scientific knowledge, humanistic learning, and a capacious, creative imagination. Eventually settling in Duisburg after leaving volatile Louvain in 1552, Mercator crafted this intermediate professional position—between the anonymous craftspeople in artisanal workshops, the gentlemen-virtuosi of the court, and the orthodox schoolmen of the universities—for himself. As a calligrapher, instrument maker, cartographer, cosmographer, mathematician, astrologer, printer, and would-be philosopher, he was thus uniquely positioned to confront the challenge of worldmaking. In his diverse attempts to map the world—from the *Orbis imago* (1538) to the posthumously published *Atlas*—we see one of most multifaceted early modern attempts to grapple with the integration of new global knowledge into a coherent system.

The development of this lifelong experiment emerges clearly in Mercator's writing, for unlike many contemporary cosmographers, he returned repeatedly to questions of cosmic scope, rarely engaging in regional or national mapping projects. When he did focus on regional maps—such as his detailed atlas of Europe (1585)—they were conceived as part of an *opus magnus* on the whole world. In the late 1530s and 1540s, while still in Louvain and part of the circle around the Flemish physician, cosmographer, and mathematician Gemma Frisius, Mercator worked on terrestrial and celestial globes and dabbled in astrological questions. A recently discovered astrological disc made by him attests to an early interest in astral and cosmic matters by 1551.[5] As early as 1563, his edited lecture notes on cosmography published by his son, Bartholomeo, gesture toward an ample understanding of the mapmaker's task: a marginal keyword describes his project as a "cosmopoeia"—literally, a world-making.[6]

By the late 1560s, Mercator seems to have already envisioned a place for the *Atlas* outside the bounds of geography and within the broader realms of philosophy and theology.[7] The introduction to the *Chronologia* (1569), his work on universal chronology published the same year as his famous map projection, describes what he intends to pursue:

> At first, I had intended a work in two volumes, the description of the heavens and that of the earth. As, however, in philosophic study history takes the first place, I recognized that to cosmography and geography also belong the origin and the history of the heavens and of the earth and of their parts.[8]

These lines connect the map, *Nova et Aucta Orbis Descriptio* (1569), to a larger philosophical project and also anticipate the *Atlas*'s grander frame by placing both works beyond the typical scope of cosmographies, such as those of Apian or Münster, or of a largely geographic atlas, such as Abraham Ortelius's *Theatrum orbis terrarum* (1570). Mercator notes that the task of description leads to philosophic and historical questions of origin and order, of the temporal as well as spatial connections between the earth, the heavens, and their parts. A similar sentiment, articulated as a biographical narrative, emerges in the dedicatory letter to his 1578 edition of Ptolemy:

> In my youth, geography first became my field of study, in which as I progressed, by using natural and geometrical speculation. . . . I found marvels, not only in geography, but in the constitution of the whole machine of the world, much of which has until now been explored by no one.[9]

Mercator traces the development of his own fascination with the "constitution of the machine of the world" (*mundanae machinae constitutione*), that is, the entire structure of the universe itself, to his early interest in geography. Pursuing interests in geometry and natural philosophy, he moves beyond the confines of geography and casts himself as an intellectual explorer who seeks what "has until now been explored by no one." Shifting from the earthly plane to a cosmic vision, these passages chart a movement from the act of representation to a reflection on the process and consequences of representation, from a technical mapping of the world to a theoretical meditation on cosmographic themes.

In his *Life* of Mercator (1595), included in all versions of the *Atlas*, Walter Ghim also traces these ideas to the dedicatory letter prefixed to Mercator's maps of France and Germany, published in 1585:

> The arrangement and order of the work demanded that I first treat of the making of the world and the arrangement of its parts generally; then of the order and motion of the celestial bodies; thirdly of their nature, radiation, and conflict in their workings, in order to inquire more truthfully into astrology;

fourthly, of the elements; fifthly, of the description of kingdoms and the whole earth; sixthly, of the genealogies of princes from the beginning of the world, in order to investigate the migration of peoples and the first inhabitants of the earth and the times of inventions and antiquities. For this is the natural order of things, which easily demonstrates their causes and origins and is the best guide to true science and knowledge.[10]

Almost twenty years after the *Chronologia,* Mercator's project has crystallized in form, but what he envisions bears little resemblance to contemporary map books. It might best be described as a world-systems theory, a comprehensive exposition of a vision of world order, a structuring of all knowledge that moves systematically from a macrocosmic, celestial, and planetary plane to a microcosmic, socio-political, and individual plane. While this vision was never completely executed, its shape seems clear. It was to be in five parts: a treatise on the creation of the world; a description of the heavens; the description of the earth; the genealogy and the history of states; and finally, a universal chronology, from the Bible to the present. When thinking of representing the world, what Mercator has in mind is a new *organon.*

Mapping and Mimesis

The posthumous publication of the *Atlas* marks the final form of Mercator's lifelong ambition, whose evolution spans most of the sixteenth century and bears its imprint. From a literary historical perspective, the volume's title is striking and strange. It introduces a new nomenclature for the map book, replacing the conventional "theater" or "mirror" metaphors with the materiality of a human body. It therefore effectively interrogates and re-imagines the mimetic claims of the mapping enterprise. The *Atlas* troubles the illusion of the map as a transparently mimetic object—a "mirror of the world"—by highlighting the function of the human mapmaker as a mediator; it is through his particular perspective and technical gaze that we see the world visualized on the page. The title's strategic balancing of two terms, *Atlas sive* ("or") *cosmographicae meditationes,* joins the external spatiality of the material world with the internal spaces of the corporeal body and the imagining self, exploring the favorite early modern trope of the "great body of the world."

Mercator's intuition that the task of mapping the world was intimately linked to knowledge of the self is clearly on display in the *Atlas.* It touches on a fundamental dilemma that underlies the cartographic enterprise: to represent the world, the mapmaker must first create it. At the opening of the *Geography*, Ptolemy describes the discipline as a *mimesis*: *"Geographia is an imitation [mimesis] through drawing of the entire known part of the world together with the things that are, broadly speaking, connected with it."*[11] But describing the practice of world mapping as *mimesis* is at once deeply intuitive and utterly counterintuitive. Though we imagine that the map represents the 'real' world with all its particularities, the world map has no external referent because a simultaneous view of the globe's convexity cannot be captured. Opening the three-dimensional sphere onto a two dimensional plane is a heuristic exercise forged through the union of art and number, the synthetic imagination of the mapmaker and his grasp of mathematical geometry. As Christian Jacob argues, the map is "the materialization of an abstract intellectual order extracted from the empirical universe" so that "without the map, the world has no contour, neither limit, form, nor dimension"; it creates "the world" as a clear concept and field of inquiry in the very act of representing it.[12] "The world" as a visual object comes into existence only *through* its representation on a map; it is always posterior to its supposed imitation. In world mapping, creation and imitation become indistinguishable.

This understanding of the cartographic task invests the individual mapmaker and atlas-compiler with tremendous power. He is no mere copyist who mechanically reproduces a finite object from life. Instead, he becomes a worldmaker—one who creates in the very act of representing. As the world is known and produced through the body of the artisan-cartographer, from the intellectual processes of abstraction to the manual skill of drawing and engraving, the particular and the universal interpenetrate. From this perspective, the mapmaker mimics the creative deity, activating a complex analogy between human and divine making. By reframing the map book as a "cosmographical meditation," Mercator emphasizes its place within a spiritual context of meditative contemplation even as he celebrates its scientific mastery. Eventually, the cartographer will threaten to displace the theologian: Mercator's final work, published as the first part of the *Atlas*, is a commentary on Genesis entitled "De mundi creatione ac fabrica" ("On the Creation and Making of the World").

The *Atlas* rests on a fusion of the technical, the theological, and the poetic, self-consciously asserted in its subtitle, "de fabrica mundi et fabricati figura." *Fabrica* and *figura* are key terms here, derived from quite different (though related) discourses. Associated with technical manuals in the early modern period, *fabrica*, from the Latin *fabricare* (to make), refers to the artisan's workshop, and in a transferred sense to his art, trade, or profession, or to the products of his craftsmanship. *Figura*, on the other hand, draws on a lexical history that derives primarily from rhetoric and aesthetics, and then becomes linked to biblical hermeneutics. Concerned with "plastic form," with artistic or poetic figuration, it confronts the problem of mimesis—the relation between original and copy. As Erich Auerbach has famously shown, *figura* also acquires a historical dimension, as a term that mediates different but related historical events.[13] Deploying these terms in tandem, Mercator suggests that worldmaking through a map book requires the union of technical craft and aesthetic criteria to make historical connections between past, present, and future. Worldmaking in this view demands the skills of artisan, poet, and prophet.

These analogies between making and representing in the *Atlas* explain why the growth of world atlases as a genre is so central to the world's emergence as a distinct object of inquiry in the early modern period. The term "world" acquired a stable meaning only when it had acquired visual form in images that were rapidly canonized as *typus orbis terrarum*—the ideal type or image of the world. If, as Lorraine Daston and Peter Galison have noted, scientific atlases register an outbreak of epistemic anxiety, then the signs of global redefinition should be legible in the outpouring of cosmographies and cosmographic atlases in the sixteenth century—a process recorded in the conceptual leap from Ortelius's *Theatrum orbis terrarum* (1570), considered the first world atlas, to Merctaor's *Atlas* (1595), the first map book to bear that name.[14] The crystallization of the geographic atlas as a genre occurs in tandem with the crystallization of "the world" as a visually recognizable "working object."[15]

The *Atlas* emphasizes the process and stakes of global representation as much as the maps that are its final products. In the symmetry of *fabrica* and *figura* lies a question about the world and its image. The body of Atlas on the frontispiece celebrates that interplay between a particular self and the world as a whole. The vision of a modern world produced by would-be worldmakers such as Mercator thus claims mimetic transparency even as it acknowledges ideological opacity. It is a world discovered and invented, a world charted and a world made-up.

FABRICATING THE WORLD

Mercator's *Atlas* is the only map book in the sixteenth century to claim that it is about the making of the world—*de fabrica mundi*. Of all the published books in Europe with the word "fabrica" in their titles, almost all are either technical manuals (with the generic title, "de fabrica et usu"—on the making and use) or anatomical treatises.[16] As Jackie Pigeaud notes, "fabrica" becomes an integral term in the technical lexicon of medicine in the Renaissance, particularly of human anatomy.[17] From Theophilus Protospatharius's epitome and commentary on Galen's *De usu partium*, which was entitled *De corporis humani fabrica*, to Vesalius's groundbreaking anatomical atlas, *De humani corporis fabrica* (1543; 1555) and its many imitators, by the late sixteenth century the term "fabrica" was almost synonymous with studies on the structure of the human body. Why then does Mercator use it in his *Atlas*, connecting his subject—the world—with a human body on its title page?

The claim that "man is the image of the world" belongs of course to an ancient tradition of microcosmic thinking, which persisted through the early modern period and connected the human body to the cosmos in material, spiritual, and metaphorical ways.[18] An essential part of the cosmographic literature, it is enshrined in the opening definitions of Ptolemy's *Geography*, which uses an anatomical analogy to differentiate between cosmography and chorography: "The goal of *chorographia* is an impression of a part as when one makes an image of just an ear or an eye; but [the goal] of [*geographia*] is a general view, analogous to making a portrait of the whole head."[19] Peter Apian's now much-discussed illustrations (fig. 5) of this contrast in his popular *Cosmographia* (1524) ensured its place in the canon of texts used by geographers of all stripes, but they also exploited a deeper disciplinary connection between anatomy and geography.[20] In the *Geography*, Ptolemy presents chorography as a kind of surgical operation, a process of cutting the body of the whole (cosmographic) world into pieces in order to better investigate the individual part. The verb he uses (*apotemnomai*) signifies an act of dissection, a cutting off, severing, or dividing up a body or an area.[21]

By the mid-sixteenth century, however, Ptolemy's language of geographic dissection would have found its closest analogue in another "new" discipline and genre—the anatomy, epitomized by Vesalius's *De humani corporis fabrica*, a book with its own claim to be the first "atlas." The chorographic cutting of the world into pieces for closer scrutiny paralleled the study of anatomy with its practice of dissection—the cutting open of the human body for closer scrutiny—which became widely practiced over the sixteenth century. Anatomy is frequently juxtaposed alongside cartography as a key harbinger of empiricism and modernity: in both cases, new techniques of gathering information produced new knowledge and necessitated new the-

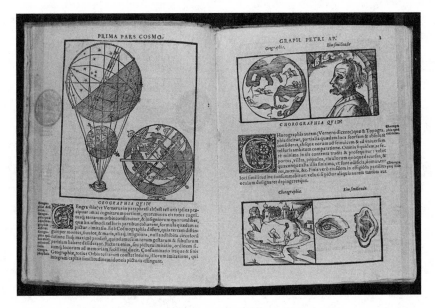

FIGURE 5. From Peter Apian, *Cosmographicus liber* (Landshut, 1574). Courtesy of the Beinecke Rare Book and Manuscript Library, Yale University.

oretical accounts of entire systems.[22] But as the frontispiece to Mercator's *Atlas* suggests, the two discourses did not proceed merely along parallel, analogous tracks. They were closely related in their epistemologies, their histories, their metaphorics, and their representational strategies, so much so that both disciplines produced similar textual genres that attempted to gather and systematize a comprehensive body of knowledge.

This conjunction signals how strongly interconnected was the development of the modern notions of self and world, as the mapping of the world and the "mapping" of the body seemed to take place through similar means. Rhetorically, the materialities of geographic and bodily space become powerfully fused in the early modern period, paving the way for subsequent analogies between the individual human body and the great body of the world. This was not only the consequence of a normative application of maps' geometric rationalization of space to the depiction of human bodies, as Valerie Traub has argued. The conjunctions between anatomy and cartography reveal a profound, almost subterranean set of images, metaphors, and vocabularies that helped join together the experience of simultaneously inhabiting both a particular body and the world at large.[23]

Medicine and geography overlapped on a practical level, since many cartographers had been trained as physicians—a partial list includes such notable figures as Gemma Frisius, Thomas Geminus, Wolfgang Lazius, Oronce Finé, Lorenz Fries, Charles de l'Escluse, Johannes Mellinger, and Johannes Sambucus. Resemblances between territorial and corporeal spaces were frequently made: the similarity between the circulation of blood and the earth's hydrography, for instance, was almost commonplace, while mathematical grids were used for anatomical drawing as well as for cartographic representation.[24] Theories about the relationship between the humoral body and the physical environment (*klimata*)—whether and why bodies were affected by place—would have been familiar from Ptolemy and Hippocrates and were hotly debated especially in the wake of travel to the torrid zones of Africa, Asia, and the New World.[25] The late seventeenth-century naming of the brain's hemispheres for the two halves of the earth may be a linguistic trace of these habits of analogy.[26] Other tropes, such as the figuration of uncharted lands as female, or analogies between land and ruler (or later, nation and national character) were conventional and exploited long-established medieval conceptions of the body politic. Classical theories of microcosm and macrocosm as well as the tradition of anthropomorphic geography in writers such as Strabo and Pliny all helped to construct powerful rhetorical links between body and world.

In this light, Mercator's implicit association of his book with Vesalius's anatomy deserves more attention since it reveals the *Atlas*'s desire to unite the human sphere of action with a new vision of the cosmos. It underscores the ways in which the redefinition of the world as a conceptual field was inextricable from the parallel reimagining of the human body and the self.[27] Anatomy was a powerful disciplinary analogue for geography since the anatomist displayed and described parts of the body that had hitherto never been observed, much as the mapmaker revealed parts of the world that had hitherto never been readily available for visual scrutiny.[28] Though still inspired by ancient analogies, these connections between body and world were increasingly based on empirical experience and observation: both Vesalius and Mercator claimed to surpass classical authorities (Galen, Ptolemy) and to forge new knowledge by resorting to autopsy and artisanal skill, typically neglected by learned schoolmen.[29] Similarly, both sought to transform their respective fields by aiming for a kind of universal synthesis. For Vesalius, *De . . . fabrica* represented an attempt to gather up and unite the scattered body of medical knowledge, much as Mercator would seek to unite all knowledge of the world in the *Atlas*.[30]

These parallels may even point to a specific link between the anatomist and mapmaker. Both were close associates of Gemma Frisius and may even have encountered each other in Louvain between 1535 and 1540. In *De . . . fabrica*, Vesalius tells the grisly tale of how Gemma helped him rob a corpse from the gallows outside Louvain, thus launching his career in dissection, an event that has been dated to 1535–37. At this time, the young Mercator, a part of Gemma's "familia" of students, was also in Louvain, working with him on a set of globes. It is through Gemma too that Mercator may have become exposed to the exciting contemporary developments in anatomy in the years right before the publication of Vesalius's groundbreaking book in 1543: in 1540, Gemma was appointed a professor of medicine at the University of Louvain, where, along with his friend Jeremy Thriverius, he was keenly involved in reforming medical education, most notably by enhancing the importance of anatomy; he may even have assisted Thriverius in public dissections.[31] To what extent Mercator himself may have been interested in these activities is unclear, but his second published map provides a clue.

Orbis imago: *The World Is a Heart*

Mercator's first world map, *Orbis imago* (fig. 6), was published in 1538 and features a bi-cordiform projection based on Oronce Finé's 1532 map, *Nova et integra universi orbis descriptio* (fig. 7). The heart-shaped double hemispheric map is the first trace of the analogy between body and world in Mercator's oeuvre, connecting him to a cosmographic-medical tradition, since both Johannes Stabius and Finé, the projection's originators, were trained as physicians. One of a handful of true cordiform maps, it remains something of an anomaly eluding interpretation, since the circumstances of its production and its intended audience remain obscure, though it was commercially successful.[32] However, when considered alongside the emerging interconnections between body and world, the map takes on a richer meaning.

The cordiform maps of the early sixteenth century might be seen as participating in a symbolic economy of the heart, which was associated with a range of medical, political, and spiritual discourses.[33] As the primary seat of perception and organ of knowledge, the main receptor of sensory images and the seat of memory, the heart was the governor of the microcosm of the human body. Cosmologically it represented the center of the world and had long been associated with the sun; as late as 1628, William Harvey would open his *Exercitatio anatomica de motu cordis* by invoking the heart as "fundamentum . . . vitae, princeps omnium, microcosmi sol" (the basis of life, its chief organ, the sun of its microcosm).[34] These terms were already astrological

FIGURE 6. Gerhard Mercator, *Orbis imago* (1538). Courtesy of the Rare Books Division, the New York Public Library, Astor, Lenox and Tilden Foundations.

Insulæ Iaconu

Insulæ Moluccæ

Terra alta

Iaua minor

Oceanus

Iaua maior

Indicus

Losroicos insula

meridio

nalis

Zanzibar

Madagascar vel S. Laureñj

C. Tomespro C. bene sper

interior

FRI

Ocea nus

Tropicus Capricorni

occiden

talis

mudan deducabat

Ne quæ ad nos tram orbis diuisi
onen pertinebant omitteremus studiose
lestor vbi locus deerat suffecimus i
Germania quidem F, S, W, H, T, lo
co Frisiæ, Saxoniæ, Westphaliæ, Has
siæ & Thuringæ, in Turcia verò
loco Ponti & Buhyniæj, Asiæ pro s
priæ z, Lyciæ 3, Pamphyliæ 4, Ciliciæ
5, Galatiæ 6, & Cappadociæ 7.

Insulæ infortunatæ

Circulus antarcticus

Terrates

Polus antart ticus

Menrata huc esse certum est
sed giñadas gruhilæ limitibus
finitas inuertam

Fretu acartheu
siue Magella
neu

Gi gantu regio

B. dc. mathia

Bareçaî Barças

B. Los contres

Presitia

AMERI

CAE pars me Tabes
ridionalis

D. todobi factu

R. de rees

Porto real

C. blāco

Aruaccas

FIGURE 7. Oronce Finé, *Nova et integra universi orbis descriptio* (Paris, 1532). Map reproduction courtesy of the Norman B. Leventhal Map Center at the Boston Public Library.

commonplaces a century before: in 1550, the physician Antoine Mizauld, who dedicated his dialogue between Urania, the muse of astronomy, and Aesculapius, the god of medicine, to Oronce Finé, devoted an entire chapter to "the analogy or correspondence between the sun and the heart."[35]

Recent scholarship on medical humanism in the Renaissance has turned the spotlight once again on the deep affinities between astrology and medicine that underlies such an iconography.[36] The medicine-astrology connection was easily extended to cosmography, since most cosmographers and mathematicians of the period, including Stabius, Finé, Gemma, and Mizauld, dabbled in astrology as a sister art. The recent discovery of Mercator's astrological interests, and Steven van den Broecke's studies on the importance of Louvain in astrological debates in the sixteenth century, suggests that Mercator too was involved in such inquiries.[37] He may even have known Mizauld, a friend of Gemma's, while his later correspondence with John Dee attests to his sustained interest in the cosmic dimensions of the human body.[38]

Against this background, *Orbis imago* arguably looks like an initial attempt to take on questions of cosmic scope through a multilayered approach that combined mathematics, geographic reportage, and cosmographic theory with an astrological-medical angle. From a cartographic perspective, the map stands out for its mathematical skill, its precision engraving, and its bold insertion into a contemporary conversation on the nature and disposition of the continents—and by extension, the materiality of the earthly sphere as a whole. It is one of the first maps to connect North and South America into a single continent and to separate it from Asia, claiming a place among early attempts to solve the cartographic puzzle of the New World's boundaries.[39] The projection, which divides the earth at the equator, also depicts a vast austral continent whose exact boundaries are "incertum."[40] Mercator's departure here from the spatial configurations he had engraved on Gemma Frisius's globes just a few years before reveals the practical challenges of world mapping. His choices reflect theoretical debates on the nature of continents and the balance between land and water masses on the planet rather than any new empirical observations, and they foreground his interest in cosmographic speculations that were increasingly the realm of physics rather than geography.[41]

These are early intimations of Mercator's self-fashioning as a worldmaker: he grapples with the problem of integrating Old and New Worlds into a single mathematical frame, experiments with and updates Ptolematic projections, and through the map's symbolic form connects geographic matters to human and spiritual realms. *Orbis imago* is a flashy beginning to a cosmographic career. Mercator asserts as much in the cartouche describing the map, which explains the scope of his project as a "division of the world along broad lines" that would be followed by individual maps of "particular regions."[42] *Orbis imago* may in fact be Mercator's earliest explicit statement about his plan for the great cosmographic project that would become the *Atlas*.

Like the *Atlas*, which explores connections between the body and the world through its iconographic frame, the form of *Orbis imago* signals cosmic interconnections between the image of a newly expanded globe and the most intimate core (*cor*, heart) of an individual.[43] Placing the heart at the emblematic center of such correspondences emphasized the harmony of the universe, for the Sacred Heart was a mystical symbol for Christ's compassion and love for humanity.[44] A corporeal organ endowed with divine meaning, the heart also alluded to the Incarnation, the manifestation of the divine in human form, and may even recall the superimposition of the body of Jesus upon the world in medieval *mappaemundi*.[45]

Even as the map glorifies the world as the supreme product of a divine maker, it also celebrates the cartographer's creative ability: Karrow notes that the map envisions the cartographer as "*homo faber;* he can split the world in unfamiliar and uncomfortable ways, but he also has the means, through mathematical reasoning, to bind it up so that it approaches once more the divine unity."[46] The implicit analogy to anatomical dissections, which were driven by the "claim that fragmentation is a means of getting at a unified truth," is once again present here.[47] Cutting open specimen bodies ultimately enabled the reconstruction of an idealized (universal) body by reconciling and integrating the range of human diversity.[48] Mercator's *Orbis imago* presents a kind of cartographic anatomy: the Americas seem flayed out, spread across three lobes of the map's two hearts; Asia and Africa are likewise spliced at their southern extremities. The logic of representation, driven by the power of geometric analysis, enables an opening up: it reveals a defamiliarized world, forcing us to look more closely.

This distortion, which makes it useless as a working tool, also makes *Orbis imago* a "mapmaker's map," a commentary on the craft of its maker.[49] Unlike Finé's map, which claims epistemic authority as a "new and complete description of the entire world," Mercator's map calls attention only to the task of representation—it is, simply, "an image of the world." The Latin *imago* aligns the map object with an aesthetic and poetic discourse of imitation usually associated with statues, busts, or portraits. It draws attention to the artist's hand, a particularly important matter to Mercator, since the map showed off not only his scholarly claims to cosmographic knowledge but also his artisanal skill: as a copperplate engraving in his own hand, *Orbis imago* surpassed Finé's woodcut map through the double mastery of its maker. The beautiful italic hand and fine shadings on display here would become the basis of Mercator's first technical publication, a treatise on calligraphy for maps, as well as the gold standard for subsequent cartographic lettering.[50] *Orbis imago* thus reflects a profound material connection between the body of the artisan and his physical reconstruction of the world. By replicating Finé's map, Mercator physically retraced, learned, and reinscribed a vision of the world that was, in a crucial sense, a product of his own body. As Pamela Smith has suggested, such an experiential artisanal epistemology, gleaned by copying a master, becomes an act of cognition in itself.[51] The process of making a world map—measuring, drawing, naming, and integrating partial pieces of knowledge into a global framework—produces a comprehensive understanding of the world through the very act of reconstruction.

Atlas: The World Is a Man

These synergies between body and world find their fullest expression in the framing of the *Atlas*, whose paratexts diverge from contemporary cosmographies. The title page, followed by a portrait of Mercator, introduces three distinct sections before getting to the maps: a long eulogizing section featuring Ghim's biography of Mercator and various dedicatory poems, epitaphs, and letters; Mercator's exegesis of the Atlas figure and Johannes Mercator's poem on his grandfather's admiration for Atlas; and finally, a hexameral treatise on the creation of the world (the first part of the book). A second title page entitled "Atlantis pars altera. Geographia nova totius mundi" (The next part of Atlas. A New Geography of the Entire World) precedes the maps themselves and once again prominently identifies Mercator as the "Authore." This arrangement accentuates the mapmaker's role and casts the figure of Atlas as both his prime product and his surrogate.

Despite this textual insistence, Mercator's association of his book with Atlas has frequently been dismissed as conventional classical allusion since the Titan who holds up the world is a fairly frequent emblem for the disciplines of astronomy and cosmography.[52] Several scholars have argued that there was no particular significance to Mercator's naming and it is largely due to chance that a book of maps is today called an "atlas" rather than a "theater."[53] Yet, Mercator seems to disavow this interpretation. The muscular man on the title page is utterly unlike conventional allegorical depictions of Atlas, which typically show an old man bent over, straining under the weight he carries on his back. Indeed, the book's context and paratexts suggest that this historical choice may actually reflect an important conceptual shift rather than mere accident.

Ortelius's *Theatrum orbis terrarum,* which set the standard for geographic atlases, exploits the favorite early modern metaphor of the world as a theater, a stage for human action, and envisions the map book—and by extension, the world—as a space of representation and performance.[54] But where for Ortelius the mapping of the world is a narrative enumeration or collection of worldly knowledge which invites contemplation of historical and political action, Mercator's book places the accent on the technical craft of making the world. The *Atlas* displays the world's underlying structure, much as Vesalius's *De . . . fabrica* sought to reveal the making and structure of the human body. For Mercator, the mapping of the world is a revelation that makes the invisible whole visible to human eyes at a glance; it is a meditation on form, order, and the act of making itself.

The *Atlas*'s Atlas is a maker and measurer—his fingers imitate the shape of a compass—and he towers over puny replicas of the great world. He sits on a large rock, which may represent the hilltop of contemplation, and controls the globe with effortless authority.[55] The finger-compass, a rich trope in sixteenth century paintings, prints, and illustrations, symbolizes intellectual exploration and technical expertise, political dominion, creative power, and moral salvation. Visually, the dynamic energy of his pose suggests depictions of the creator God in medieval and Renaissance painting; his gestures activate a rhetoric that recalls classical and Christian evocations of God as a *deus faber,* an *artifex maximus.*[56]

The composition of the title page suggests that the figure may also refer to the mapmaker himself, since the marble plinth proclaims the name of the author: Gerardus Mercator of Rupelmonde, Cosmographer to the Most Illustrious Duke of Jülich and Cleve. This identification is enhanced by the book's next image, a portrait of Mercator (fig. 8), also holding compasses and a globe, in a gesture that imitates both the Titan who challenged divine authority and the divine creator himself.[57] If the visual associations were not sufficiently explicit, Johannes Mercator's poem, "On *Atlas* / by his grandfather, Gerardus Mercator," which follows some pages later, claims that Mercator saw himself as an Atlas-figure: "And because the virtue of the loftiest men is to be imitated, / my grandfather took this man [Atlas] as an example for himself."[58]

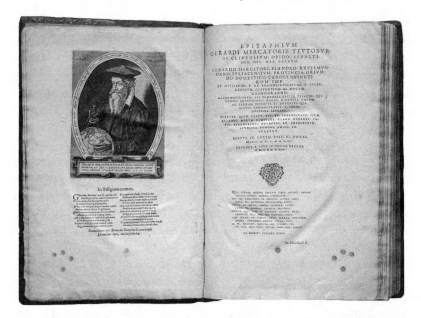

FIGURE 8. [Portrait] (after Frans Hogenberg), from Gerhard Mercator, *Atlas* (Duisburg, 1595). Courtesy of the Library of Congress; reproduced from an original in the Geography and Map Division.

This triple association of the body on the title page with classical mythology, creative (Christian) divinity, and the mapmaker's own craft suggests Mercator's quest for an emblem that would effectively represent the ambition of his book. Mercator's Atlas stands for all would-be worldmakers by emphasizing the centrality of the human body in attempts to comprehend the world—a world that can only be known through the partial vision and particular materiality of an individual. But that body also represents the knowing self who stands in the image of an all-knowing God, seeking to capture in a single cartographic gaze an omniscient consciousness of the whole. Here, the artisan-mapmaker is closest to the *deus artifex* in his physical task of recreating an image of the world since he too is a maker: the artisan's knowledge is elevated by analogy to the creative deity.

The history of scientific epistemology is also a history of the self, suggest Daston and Galison in their monumental account of the making of scientific objectivity.[59] The authority of the scientific atlas before 1800 and its claims to truth and epistemic legitimacy, they argue, depend on the mediation of a sagacious scientific practitioner who takes on an instrumental role in making the knowledge displayed in the book's representative images. Such a figure guaranteed the work's truth-value by virtue of his own skill and stature as a knower; the book and the body of the knowledge it put forth drew their power from the body of their maker. Though Daston and Galison focus on the eighteenth century as a distinctive moment in the development of this epistemic ideal, Mercator's shaping of a precise textual presence—his crafting of Atlas as a mythic doppelganger and "scientific self"—suggests how, at its very inception, the genre of the atlas relied on an epistemic bond between self and world. Far from eliminating the hand of the artist-scientist in the making of knowledge, Mercator flaunts it.[60] For the earliest atlases were anatomical and geographic, and, by the nature of their subjects, were forced into metaphoric and metaphysical territory: to open up a human body or to reveal the harmonious structure of the world demanded a god's-eye view, an imitation of the creator in the act of reconstruction.

Mercator's Atlas is therefore situated at the threshold between mortals and immortals, a "dieu homme" who literally embraces the cosmos even as he measures and contemplates it.[61] Following mythographers such as Conti and historians such as Diodorus Siculus, Mercator adopts a euhemeristic interpretation of classical myth, identifying Atlas as a Phoenician king, though the genealogy he deduces owes much to Hesiod, Homer, and Ovid.[62] Mercator's iconography rationalizes the task of holding up the world as a metaphor for the intellectual grasp of its structure: Diodorus explains the mythic depiction of Atlas as an expression of his command of astrology and knowledge of the spheres.[63] This naturalization of Atlas also opens the possibility for the deification of a heroic human—here, heroism associated with knowledge-making—and Mercator's adoption of the Titan as a model might suggest his own claim to such a transformative achievement.[64] Mythography and humanist letters here enhance and legitimate empirical science.

Significantly, while Mercator emphasizes the learning, skill, and nobility of Atlas in his explication, he strategically glosses over the best-known classical tale associated with him: his participation in the Titans' rebellion against Zeus, a tale of gigantomachy that in the Renaissance had become a charged allegory for challenges to divine and divinely sanctioned political authority. But the story lurks in his stated sources, Diodorus and Eusebius, and its conspicuous absence signals the dangers that a self-consciously *human* worldmaker posed to theological, political, and social orthodoxies. Mercator—an irenicist and an Erasmian—may have perceived, exploited, but shied away from an aggressive exploration of these potentials. To claim that humans, even modestly born artisans, could be worldmakers was to raise the dizzying possibility that access to cosmic form and global structure was not the privilege of an elite few but was attainable by the dint of reason and experience. More insidiously, it was to suggest a fluidity between — perhaps a substitution of — the divine by the human: the world's shape was perhaps ultimately knowable only through artificial human fabrications.

FIGURATIONS OF THE WORLD

If Mercator self-consciously envisions himself as a worldmaker, the arrangement and structure of the *Atlas* also point to his attempts to theorize the process of worldmaking and its products — that is, to theorize the world atlas as a genre. Among his cartographer-contemporaries, Mercator was uniquely positioned for this task. His authorial role as both mapmaker and atlas-compiler allowed for much greater uniformity and standardization of the maps in the *Atlas*. But this dual function also meant that the book reflects the combined skills of technical craftsmanship and strategic intellectual ordering. James Akerman notes that Mercator was a "master of critical map compilation from multiple sources," and declares that his atlas is "perhaps the purest sixteenth-century expression of the atlas idea." [65] Not only did he draw his own new maps to reflect the latest geographic knowledge, instead of collecting and reproducing others' work (as Ortelius had done in the *Theatrum*), but he also reimagined the arrangement of the maps within the atlas. His editorial and critical influence through the *Atlas* was also considerable: its structure became the norm for Dutch world atlases throughout the seventeenth century and affected the theory of atlas-making even beyond. [66]

Mercator's bibliographic innovation centers on the fusion of empirical observation and aesthetic form: the *Atlas* consolidates and names a genre that straddles the "two cultures" of art and science. As authoritative collections of images that identify a discipline's core objects of inquiry, atlases "set standards for how phenomena are to be seen and depicted." [67] They offer a pictorial taxonomy and calibrate the collective eye by establishing "how to describe, how to depict, how to see." [68] Atlases of the world in particular faced a unique challenge: not only were they charged with defining and authorizing a particular conception of the world, but in order to do so they had to transform an invisible abstraction into a visible scientific object that could sustain investigation. Unlike national, regional, or city maps, with their strong roots in land surveys and topographic verifications, world maps relied to a certain inevitable extent on speculative knowledge, subjective interpretation of data, and artful unification to help piece together a vision of totality. Achieving such a synthetic view of the whole rested as much on artistic execution and an aesthetic sensibility as it did on mathematics and empirical observation.[69]

Mercator alludes to this aesthetic substratum in his title through two important rhetorical markers: *mundus* and *figura*. In a notable departure from Ortelius's *Theatrum*, Mercator refers to the world by using the post-classical word "mundus" rather than the more commonly used "orbis terrarum" (circle of lands), with its spatial and political connotations. The Latin translation of the more wide-ranging Greek term *kosmos*, an aesthetic concept meaning both order and ornament, *mundus* described the beauty of neat structure, the perfect harmony of the whole.[70] Though subtle, the difference marks a profound theoretical divergence between the two projects and their respective attempts to envision and represent the world: the one circumscribed, the other all-encompassing.[71]

Ideals of wholeness, harmony, unity, and perfection were the province of artists, a point emphasized by Ptolemy's recourse to portraiture as a suitable analogy for geographic representation. His celebrated metaphor, which compares the human body to the body of the world, seems to literalize Aristotle's influential analogy of the organic whole.[72] If the world is like a body, every detail included must be integral to it. Thus geographic *mimesis* is also governed by aesthetic criteria of beauty, most clearly outlined in the *Poetics:* "a beautiful object, whether it be a living organism or any whole composed of parts, must not only have an orderly arrangement of parts, but must also be of a certain magnitude . . . which may be easily embraced in one view *(eusunopton)*." For world atlases, which by definition sought to provide a synoptic view of the whole, the notions of organic wholeness and *eusunopton* (that which can be easily taken in by the eye) were fundamental epistemological criteria.[73] Though aesthetic in origin, they correspond to what Daston and Galison identify as the prime epistemic value of early modernity—"truth-to-nature"—the desire to uncover and display an idealized form, "a reality accessible only with difficulty."[74]

For this reason, the paradigmatic image of the world that opens every atlas (or quasi-atlas) in the sixteenth century is the *typus orbis terrarum,* the synoptic view of the whole. The *typus* or reasoned image established the typical and the ideal exemplar, which organized a chaotic mass of information into an abstracted and coherent form. Following Ortelius's "Typus orbis terrarum" (fig. 9), which opens the *Theatrum,* almost all major

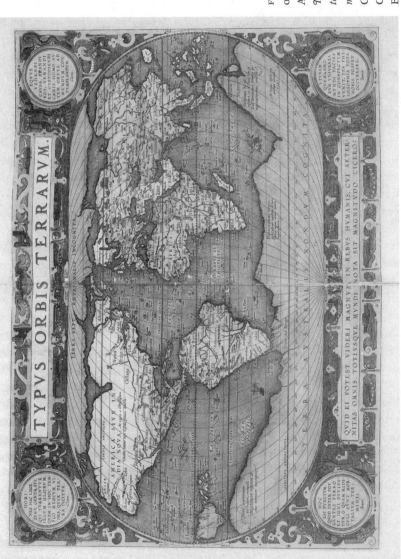

FIGURE 9. "Typus orbis terrarum" in Abraham Ortelius, *Theatre de l'univers, contenant les cartes de tout le monde* (Antwerp, 1587). Courtesy of the John Carter Brown Library at Brown University.

mapmakers—Hondius, Visscher, Plancius, Blaeu—produced significant world maps described as *typus*. Mercator is a notable exception. Not one of his world maps is described as a "typus," a term he seems to reject for other titles (*Orbis imago, Nova et aucta orbis terrae descriptio, Orbis terrae compendiosa descriptio*) that highlight the Ptolemaic-Aristotelian vision of aesthetic comprehension, wholeness, and totality.

In the *Atlas*, Mercator insists on this aspect of world-mapping by highlighting the book's attention to *fabricati figura*, the "image of the made world," a doubly aesthetic phrase that suggests a double mimesis, the representation of an artifact. *Figura*, itself a charged concept that migrates from aesthetic theory to biblical exegesis and finally to scientific atlases, offers a semantic trace of how the conception of the world also crosses a range of discourses before coming into focus in an atlas as a visual object for scrutiny. It begins as a term of art to describe a form or a shape usually associated with an act or object of imitation, then shifts to a historical register where it describes the hermeneutic and metaphysical connection between radically different cultures (Old and New Testaments, Jewish and Christian norms), and finally enters the vocabulary of the New Science, where the *figura*—figure or diagram—depicts the ideal, representative image of a subject. The world map, as an "image of the made world" follows this discursive trajectory: it is an artful imitation, a metaphysical reflection, a historical copula, and a working scientific object, bridging the spatial and temporal dimensions of the concept "world." Thus, Mercator's *fabricati figura*, which literally refers to the maps in the *Atlas*, attains a figurative fluidity that invites rumination on the manufacture (literally, making with the hand) of global knowledge, the nature of artful creation, and the historical link between artifact and fact. The ease with which world maps stand in for "the world" signals how concept and representation are inseparable.

If an atlas had to imitate the world's structure through the artful organization of its parts, the harmonious ordering of the book would ideally represent the harmonious order of nature. The mapmaker's compositional problem—how to arrange a collection of maps in an atlas—thus acquired a metaphysical dimension since it sought to reflect the "deep structure" of the world itself. The task was only more acute when the contours of the world seemed in flux. Mercator's signal contribution to the atlas as a genre might therefore be his influential reorganization of the book's form: he extricates it from the older substratum of the cosmography in which it was embedded and from which it would only gradually break free by the eighteenth century. For most of the early modern period, the atlas was indistinguish-

able from the cosmography, a compendious genre that included everything
from editions of Ptolemy's *Geography* to ever more unwieldy folios devoted
to enumerating various aspects of the world. Mercator initiates a generic
and formal break in two distinct ways: he decisively separates the ancient,
Ptolemaic world from the modern world of the Renaissance, and reorders
the maps in a world atlas, a structure that would be reproduced for almost
two hundred years thereafter.

Mercator's Ptolemy: A Visual History of "World"

A founding moment for the "modern" world as distinct from the classical or ancient world comes with the publication of Mercator's authoritative
edition of Ptolemy in 1578. This edition presents a radical departure from
the Ptolemaic genre. Simply entitled *Tabulae geographicae Cl. Ptolomei [The
Geographical Maps of Claudius Ptolemy]*, this slim volume strips away a century's worth of cartographic accretion and presents new, precisely engraved
versions of the twenty-seven original Ptolemaic maps alone. There is no at-
tached textual commentary, and Mercator does not even reprint the text of
the *Geography*. Only a four-page biographical preface and a list of Ptolemaic
tables accompany the images, which highlight the maps' strict technical adherence to the Ptolemaic grid, projection, and coordinates. The significance
of this volume lies in its dramatic separation of ancient and modern conceptions of the world. In Mercator's Ptolemy, a vision of the world—the one
encapsulated in the *Geography*—is rendered definitively historical; the scope
of Ptolemaic maps is now ancient geography, a field of study quite removed
from the task of representing the modern world.[75]

If the atlas trains the eye to pick out certain objects as exemplary and to
regard them in a certain way, Mercator's Ptolemy becomes a model both for
the historical and the scientific atlas. Its maps become the standard depiction of the ancient world and are repeatedly reproduced until 1730.[76] In
terms of the style, the spare, graphical focus of the work looks ahead to the
picture-oriented style of the great scientific atlases of the Enlightenment.
But Mercator also suggests the closure of the Ptolemaic corpus: as an atlas
of ancient geography, the book circumscribes and fixes one, now obsolete,
view of the world as a finite object.

The emergence of the modern world atlas is closely linked to the publishing history of Ptolemy's *Geography* in the sixteenth century, for the habit of
collecting a set of maps that claimed to describe the world coalesced around
the very text that theorized the task of mapping at both the global and re-

gional scales. While it remains a matter of debate as to whether the *Geography* was originally accompanied by maps, its textual tradition in the Renaissance, both in manuscript and print, included cartographic images.[77] From the 1477 Bologna edition, the first to include printed maps, editions of Ptolemy became a site for displays of state-of-the-art cartographic knowledge: the 1508 Rome edition of Ptolemy, for instance, featured Johannes Ruysch's *Universalior cogniti orbis tabula,* one of the first and most widely circulated printed maps to depict the New World, along with five "tabulae modernae" (modern maps) based on contemporary observations and data unavailable to Ptolemy. These hybrid mapbooks were thus also *de facto* world atlases. In their feverish attempt to keep up with new knowledge of the world through the continued addition of new maps, they offer a fascinating textual trace of the world's construction in the early modern period.

The modern world first comes into focus in the modestly titled "tabulae modernae" in editions of Ptolemy. These maps represent the slow aggregation of global knowledge around the central classical canon. Like steady encrustations, however, the ever-expanding corpus of new modern maps eventually overwhelmed and transformed the original core. By 1561, when Girolamo Ruscelli published the vernacular Venetian edition of the *Geography,* editions of Ptolemy had swelled from the original twenty-seven maps to sixty-four, as *tavole nuove* engulfed the *tavole antiche.*[78] The edition of "Ptolemy" was now a peculiar mix of ancient and modern, as maps based on classical geographic coordinates and Ptolemaic projections coexisted uncomfortably with maps based on updated mathematical grids and the latest information gleaned from explorers and voyagers to distant places. Ruscelli highlights the hiatus between the two systems of global representation by carefully separating out the two map sets—with the striking exception of the two world maps, which are bound consecutively as though to mark the temporal and spatial gap that divides them.[79] To further heighten the difference, he inserts a long commentary on the *Geography* between the graphical sections of the book as a kind of textual bridge between ancient and modern attempts to describe the world. Even the maps' cartographic styles are strategically different (fig. 10, fig. 11). Such editions materialized the collision between old and new knowledge of the world. In their juxtaposition of Ptolemaic and contemporary maps, they made visible contradictions and suppositions about the world that had otherwise survived undisturbed in their own corners. The seemingly immutable idea of "world" was now unexpectedly dynamic and mobile.[80]

When Ortelius's *Theatrum orbis terrarum* appeared in 1570, its indebted-

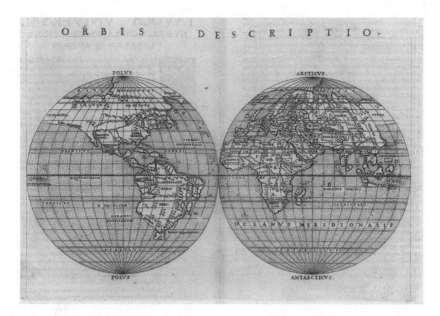

FIGURE 10. "Tavola universale nuova: Orbis descriptio" in Girolamo Ruscelli, *La geografia di Claudio Tolomeo* (Venice, 1561). Courtesy of the John Carter Brown Library at Brown University.

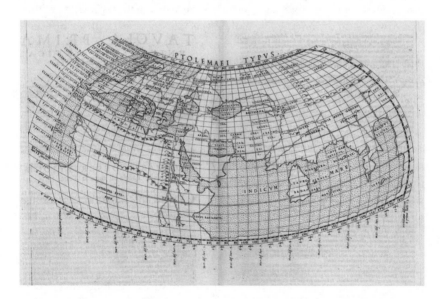

FIGURE 11. "Tavola prima universale antica: Ptolomei typus" in Girolamo Ruscelli, *La geografia di Claudio Tolomeo* (Venice, 1561). Courtesy of the John Carter Brown Library at Brown University.

ness to and emancipation from this Ptolemaic genre must have been striking. Not only did Ortelius compile and standardize the best maps of the world he could find, he freed them from their subordination to Ptolemaic maps even as he retained important aspects of the editions' architecture (such as the index of Ptolemaic place names, the cosmographic textual commentaries on the maps, and the arrangement of the maps themselves). Yet the conception of the world in the *Theatrum* remains enmeshed in the classical view. It is a humanist editor's dream and a philologist's delight—"the world" for Ortelius is firmly moored on a classical foundation, even if its external contours have been reshaped by contemporary observation. The text introducing the opening world map lists ancient and modern sources, while the map itself is framed with quotations from Cicero and Seneca. This is a world envisioned and theorized by Ptolemy, Pliny, Strabo, Solinus, Pomponius Mela, Dionysius Apher, Eustathius, Apuleius, and Diodorus Siculus (among a long list of others).[81]

Published just eight years later, Mercator's Ptolemy turns away from this classical framework and points instead toward a new conception of the "world" by the very act of historicizing another conception of it. By prying apart the awkward amalgamation of "tabulae modernae" from the Ptolemaic corpus, by enclosing the classical world and authorities within their own textual frame, he visually dissociates two ways of seeing—and of understanding—the scope of the world. Such a shift goes beyond a movement from a tripartite *oikoumene* to a theory of four (or five) continents, or from a simple conical projection to a conformal, cylindrical one. Rather, it articulates a more subversive epistemological shift that renders the totalizing category "world" itself historically contingent and subject to revision.

The allegorical frontispiece to Mercator's edition (fig. 12) anticipates many of the themes that find their full expression in the *Atlas*. An architectural façade supports a celestial globe on top and a terrestrial globe below, while the founding fathers of geography, Ptolemy and Marinus, positioned on either side of the frame, gesticulate toward the world. Though the image is fairly conventional, its focus on the world-as-representation illuminates the stakes of the work, which emphasize the spatial and temporal reimagining of the world. The same plate would be reused as a second title page in the *Atlas* (fig. 13), where it now introduces the modern maps of the world; but the framed terrestrial globe in these images—as both model and artifact—also gestures back to its maker. In the 1578 *Tabulae geographicae*, the original maker, Ptolemy, gets pride of place, but by the 1595 *Atlas*,

FIGURE 12. Frontispiece to Gerhard Mercator, *Tabulae geographicae Cl: Ptolomei* (1578). Courtesy of the Beinecke Rare Book and Manuscript Library, Yale University.

Mercator has supplanted his classical predecessor: he is the new Ptolemy of the age. After 1578, in fact, Mercator is increasingly hailed as a modern Ptolemy; the 1633 Mercator-Hondius world map depicts Mercator opposite Ptolemy as the modern progenitor of the global mapping enterprise. These slippages and substitutions of authors and images track a process of dis-

FIGURE 13. "Atlantis pars altera" (title page) in Gerhard Mercator, *Atlas* (Duisburg, 1595). Courtesy of the Library of Congress; reproduced from an original in the Geography and Map Division.

sociation and reinscription through which the modern world emerges into view. At the moment when Mercator separates out the ancient and modern worlds by placing them in different textual genres, he becomes a substitute for Ptolemy; in the same moment, the new, modern maps decisively replace the *tabulae geographicae* of Ptolemy as a picture of the world.

From Cosmography to Atlas

The gradual transition from the cosmography to the atlas marks the shift from an accumulative, compendious style of structuring global knowledge to an analytical, anatomizing style that sought to reveal the ideal form of the whole. Mercator's *Atlas* occupies a central place in a narrative of Enlightenment universalism and Weberian modernity because it is poised evenly between two genres whose linked literary histories trace a transition in ways of knowing the world and structuring knowledge about it. If cosmography is the defining mode of the sixteenth century with its centripetal energies of synthesis, the atlas becomes the standard of the eighteenth century, fueled by a centrifugal drive to analyze. In its adherence to *both* forms at once, Mercator's *Atlas* highlights the crux of the worldmaking enterprise: the difficulty of theorizing the relationship between part and whole, and of negotiating constant changes of scale in the movement from local to global.

Like Vesalius's famous anatomy, which progressively strips away layers from its "musclemen," Mercator too strips away layers of geographic distance from his depiction of the world, steadily zooming in from an external view of the whole in the opening planisphere to detailed individual maps of particular places from north to south. The body of the world is progressively opened up as the gaze of the mapmaker focuses on ever more minute locales. This parallel between the structural relations of anatomy and cosmography was already evident by the late fifteenth century when Leonardo da' Vinci invoked Ptolemy's *Geography* as a model for his own intended treatise on human anatomy: "There will be revealed to you in fifteen entire figures the cosmography of this *minor mundo* in the same order as was used by Ptolemy before me in his *Cosmographia*. And therefore I shall divide the members of this body as he divided the whole world into provinces, and then I shall define the function of the parts in every direction, placing before your eyes the perception of the whole figure."[82] For Leonardo, the analytic movement inward toward the specificity of the member-part leads back to the synthetic movement outward toward wholeness ("the whole world," "the whole figure"). In this prescient analogy, he anticipates the separation of the atlas's interest in formal order from the cosmography's accumulative drive.

Where cosmographies could be accused of speculative musing, leaps of imagination, or for glossing over detailed descriptions of parts in their quest to encompass everything, atlases reveled in their attention to detail. The most obvious difference between the two lies in their attitude to text and image: cosmographies remain primarily textual books, using maps as strategic illustrations, while atlases privilege the graphic image, relegating the text to an explanatory gloss. In this respect, Mercator's *Atlas* moves closer to the atlas ideal than Ortelius's *Theatrum,* in which detailed textual descriptions accompany maps. While the *Atlas* still retains the play of text and image, it seems to yearn for the graphic simplicity of the *Tabulae geographicae.* Mercator sheds the textual accretions that plague previous map compilations and privileges the table or diagram over the narrative, cosmographic description.[83] Later editions of the *Atlas* by Hondius, Jansson, and Blaeu adulterate this pictorial focus, reverting to the familiarity of cosmographic density, but the stylistic similarity between the 1595 edition and the great eighteenth-century French atlases such as Nicolas Sanson's *Atlas nouveau* or Gilles Robert de Vaugondy's *Atlas universel* is striking.[84]

More important, perhaps, Mercator's *Atlas* gives unprecedented importance to the arrangement and order of the maps themselves, recreating, in the turning of its pages, the tactile and visual experience of slowly scanning a globe. Early atlases drew on the original structure of the twenty-seven maps in the Ptolemaic corpus—Antonio Lafreri, the compiler of the Italian mapbooks which take his name (Lafreri atlases), even advertised his publications as being "after the style of Ptolemy." These typically moved through various scales of description from large to small, beginning with a map of the world, then the continents, and then individual regions and countries. But this seemingly intuitive structure was not common to all mapbooks in the period, which often privileged national, regional, or patrons' concerns.[85] Mercator's *Atlas* definitively establishes both a descending scalar movement from world to continent to region, and also stabilizes an order of maps based on the movement of the eye across the globe, from northwest (the British Isles and Scandinavia) to the northeast (Eastern Europe and Russia), and then from southwest (Spain, France) across to southeast (Greece, the Balkans).

As Mercator explains in the text accompanying the world map: "Since order always requires . . . [placing] things in general before particulars and the whole before the part . . . I, bound by the same law, ought to preface this first volume . . . with the image of the universal globe of the earth and its four parts . . . [thus] the reader will always have a complete work and description of the universal whole, even if he has purchased for himself only the delineation of his own country, and will not be deprived of this useful speculation. . . . The contemplation of things in general is pleasant and above all necessary for anyone wishing to have even a minimal knowledge of the world."[86] The atlas seeks to bridge opposing scales by describing parts in all their careful detail, even as it meditates on their proper relation within a harmonious whole.

Paratextual material and framing considerations are thus crucial in atlases, since they reflect on the relationship between the individual segments and pages. Mercator's *Atlas* is the first to demarcate clear sections for specific regions, each introduced by a new title page and brief textual description and diagram explaining the order of the regional maps. The maps themselves are numbered for the first time as part of a set (for instance, "Hibernia III Tabula," which identifies it as the third map in a series), and standardized to the same scale in such a way that they could even be pieced together to form a larger wall map of the whole.[87] These small details are instrumental in establishing a narrative within the atlas and acted as the glue that binds together pieces into an organic whole.[88] By the publication of the 1662 Blaeu *Atlas maior,* these structural cues had been expanded into an internal map to the eleven volume set, depicting the order of the entire work as a conceptual diagram that also described the structure of the world. The atlas becomes, in effect, a *harmonia macrocosmica.*[89]

MEDITATIONS ON THE WORLD

The 1572 French edition of Ortelius's *Theatrum* begins with an epigraph that reflects on the purpose of the world atlas: "Le cheval est creé pour porter et tirer: le Boeuf pour arer et labourer la terre: le Chien pour chasser et garder la maison: Mais l'Homme pour considerer et contempler des yeux de l'entendement la disposition du Monde universel." (The horse was created to draw and carry; the ox to plough and work the earth; the dog to hunt and keep watch over the house; but Man [was created] to observe and contemplate with the eyes of his understanding the disposition of the universal world.)[90] Adapted from Cicero's *De natura deorum,* the passage implies that the atlas fulfills the essence of human nature by facilitating the contemplation the world. Offering visual and conceptual clarity, the atlas displays the "universal world" as a harmonious unity. The cosmographer is thus charged with a metaphysical task.

But Ortelius's allusion also undercuts the analogies between the human body and the world that I have been exploring so far. In the Ciceronian dialogue, man is made to contemplate and imitate the world (*mundum contemplandum et imitandum*) but is doomed to inadequacy, for he is no more than "a particle of perfection," while "the world, as it comprehends all, and as nothing exists that is not contained in it, is entirely perfect."[91] Imperfect, partial, and particular, man faces the impossible task of imitating perfection and comprehending totality. The world's perfection, which signals its divinity, thus also emphasizes man's mortality; but it also offers a means of its transcendence through contemplation. This sentiment, with its mix of existential promise and tragedy, captures the Stoic approach to matters of global knowledge that was shared by a number of cosmographers in the sixteenth century. Recent work on figures such as Ortelius, Finé, Plantin, and by implication Mercator has suggested that rather than only placing their maps within a narrative of progressive improvement in science, we must also see in them an invitation to contemplate God's world in an act of spiritual transcendence.[92]

Mercator's alternative description for his atlas, *cosmographicae meditationes*, alludes to this philosophical context and suggests a quite different generic identification, one removed from the epistemological concerns of the scientific atlas but deeply engaged with the ethical and moral dimensions of human existence.[93] The association of human and global bodies emblematized by Atlas suggests not only a way of knowing the world but a way of being in the world. We use maps not only to trace geographic paths but to chart psychological itineraries—to imagine other lives and other selves. To envision the entirety of the world is also to reflect on one's particular place within it or in relation to it. The mapbook thus becomes a meditation on a sense of place, a site for ruminating on that nebulous but fundamental question about our emotional and spiritual location in the cosmos.[94] In early modern Europe, this question was bound up with theological positions as well as with classical philosophy, particularly Neo-Stoicism.

Cosmography, with its quest to capture the invisible whole, involved an alternate epistemology. The displacement of one's point of view upward involved, in Frank Lestringant's terms, an imaginative elevation "to the point of grasping instantaneously the entire convexity of the globe. At that imaginary point, the eye of the cosmographer ideally coincided with that of the creator."[95] This all-encompassing view—the exercise of *kataskopos*—was central to Stoic philosophy and to what Mercator describes as the "cosmographic meditation."[96] Associated with Scipio's vision of the world from the Milky Way in Macrobius's cosmographic *Dream of Scipio,* the 360-degree "view from above" symbolized philosophic exercises through which man could transform himself from being a prisoner within the world to becoming a spectator from without. As *specula* (mirrors) of the world, maps facilitated *speculationes* (reflections) on the world: in fact, in a little slip of the pen that signals the importance of this play on words, Walter Ghim refers to Mercator's book as *Atlas sive cosmographicae speculationis*.[97] This distancing enabled the elusive view of the whole, facilitating a "spiritual gaze that disclose[d] the beauty and order of the world beyond the shimmering of appearance and the limitations of human knowledge" and led to "a relativizing of human values and achievements."[98] Such visionary experience generates what Jean-Marc Besse describes as a "cosmic conscience," and an ability to shape one's destiny through the entwined knowledge of self and world, in full recognition of both the dignity and the insignificance of human existence.[99]

By the sixteenth century, the extensive Christianization of Stoic philosophy had led to the complete absorption of this perspective into various spiritual exercises and meditational practices. Cosmographies and world-mapping projects, which made *kataskopos* accessible, were deeply embedded in this religio-philosophical matrix. Ortelius's well-known "Typus orbis terrarum" (fig. 9) is a case in point. An authoritative image of early modern European claims to dominion and scientific advancement, based as it is on the 1569 Mercator projection, but it also undercuts such claims. The Latin inscriptions within the map's frame are Neo-Stoic quotations from Seneca and Cicero that celebrate worldly contemplation while mocking the human obsession with territorial boundaries. "Is this the pin-point which is divided by sword and fire among so many nations? How ridiculous are the boundaries of mortals," declares the top right hand quotation from Seneca.[100] A claim to mastery, the map also registers a tentative response to a shifting world, asserting the stability of (a divine) order to counteract the uncertainty of partial knowledge.

The brief textual accompaniment to the world map in Mercator's *Atlas* similarly culminates with a prefatory apostrophe to the reader and a call to experience *kataskopos*:

> [R]eader, farewell, enjoy, and think diligently of the glory of your dwelling-place, which is only temporarily granted to you, along with the poet George Buchanan, who thus compares it to the celestial realm in order to draw forth your souls, immersed in terrestrial and transitory affairs, and show the way to higher and eternal things.[101]

This valediction echoes the Senecan quote used by Ortelius as Mercator too argues that envisioning the world leads to meditative distanciation. But the power of this argument rests on the interplay between the world map that follows and the interpretive clue provided by the extended quotation of Buchanan's cosmographic poem, *De sphaera*:

May you perceive how small a portion of the universe it is
That we carve out with magnificent words into proud realms:
We divide with the sword, and purchase with spilled blood,
And lead triumphs on account of a little clod of earth.
... but if you compare it with heaven's starry roof, it is as
A dot or the seed from which the old Gargettian [i.e., Epicurus]
created innumerable worlds.

· ·

How tiny the part of the universe is where glory raises its head,
Wrath rages, fear sickens, grief burns, want
Compels wealth with the sword, and ambushes with flame and with poison;
And human affairs boil with tremulous uproar![102]

These lines contrast sharply with the mathematical mastery of the world on display in the accompanying map, *Orbis terrae compendiosa descriptio* (fig. 14). Buchanan traverses the changes of scale encountered within an atlas—from the "little clod of earth" to the vastness of the universe and back

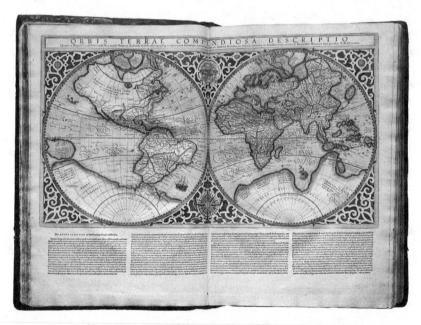

FIGURE 14. "Orbis terrae compendiosa descriptio" in Gerhard Mercator, *Atlas* (Duisburg, 1595). Image courtesy of the Library of Congress; reproduced from an original in the Geography and Map Division.

again—to articulate a Neo-Stoic vision of the relation between the human individual and the immensity of the cosmos. A miniature cosmopoesis, the poem refers to the (heretical) Epicurean theories of matter and the plurality of worlds, even as it deflates anthropocentric pretension by evoking the beauty of the natural world. Far from trumpeting geopolitical control, the map discloses the insignificant noise of human action when compared to cosmic amplitude. It thus generates a play of existential perspective that recurs repeatedly over the period, perhaps most famously in Milton's *Paradise Lost*. For imagining the world, even in an atlas, was ultimately inseparable from the dilemma of human ethical consciousness—a recognition that transforms the atlas from a geographic description into a cosmographic meditation. By framing his world map with exhortations to metaphysical contemplation and spiritual connection, Mercator binds the spatial and the ethical, the vast external spaces of the cosmos to the rooted sense of place within the self.[103]

In God's Image

In his own lifetime, Mercator became a symbol for the triumph of the moderns over the ancients and a scientific model for future generations. But he was most celebrated as a seer who held the universe in his mind's eye, a prophet who could reveal the true relations between the earth and the heavens. The prefatory matter of the *Atlas* emphasizes these aspects of Mercator's achievement. "With heaven as auspice I espied all the earth, / reconciling things below with those above. / Through me, the stars of heaven shine in maps," claims Mercator ventriloquized by Johann Metellus. More prosaically, Lambert Lithocomos insists on the totality of Mercator's reach, which "revealed the breadth of the universal orb, / and everything there was in the universal orb."[104] But it is the humanist Bernardus Furmerius who offers a striking poetic counterpart to the title image of the worldmaking Atlas:

> He joined the stars to the earth and added the sacred to the profane,
>> rectifying both at once,
> As a mathematician, he described the stars with his cunning rod,
>> and gave them to be observed in a little globe.
> He brought together the broad orb of the earth into maps . . .
> . . . .
> He uncovered the sacred mysteries of the prophets
>> and commanded Christ's four heralds to march in step.

> And he did these things so as to surpass all past
> artists, on his own, and by his own hand.[105]

Mythic Titan and mortal cosmographer merge in this invocation. The task of worldmaking, its unprecedented achievement and its all-encompassing scope, cut beyond traditional disciplinary boundaries, merging science, art, philosophy, and theology, as Mercator simultaneously analyzes details and brings them into synthesis and universal harmony.

The quest to describe the world as a whole and attain a complete understanding of its parts demands superhuman capacity and precipitates an implicit confrontation between human and divine makers. In Fumerius's epigraph, Mercator is no mere mapmaker. He joins the stars to the earth, combines the sacred and the profane, displays an entire vision of the world, demarcates the boundaries of its kingdoms, and establishes with certainty the progression of historical time. Mercator even uncovers the mysteries of the prophets and reconciles the four Gospels. The hymnic quality of these images, drawn in part from the Psalms, accentuates the implicit comparison to divine power and places the *Atlas* in a tradition that draws on Ptolemaic cosmography but transcends it. Where cosmographers such as Münster, Thevet, and Ananio carefully dissociate themselves from divine analogy, subordinating cosmography to theology, Furmerius's vision of Mercator suggests that the two disciplines may ultimately be inseparable in their scope.

The "Praefatio in Atlantis," presumably written by Mercator himself, further strengthens such a claim:

> I have set this man Atlas, so notable for his erudition, humaneness, and wisdom, as a model for my imitation, so far as my genius and strength suffice, as I begin to contemplate cosmography as though in the lofty mirror of the mind. . . . Beginning from the creation, I shall enumerate all its parts and contemplate them naturally so that the causes of things shall be evident. . . . Thus I shall lay out the whole world as though in a mirror, so that there shall be certain rudiments for finding the causes of things and attaining wisdom and prudence, sufficient to lead the reader to higher speculations.[106]

Mercator's language here echoes Paul to the Corinthians, particularly the famous image of seeing through a mirror darkly (1 Cor. 13:12): he imagines his own mind as a "lofty mirror," which finds a counterpart in the world whose contents will be laid out in his book, "as though in a mirror." Knowing self, world, and the book of cosmic knowledge become coextensive and inseparable; they are self-reflecting mirrors and images of each other. Mercator's vision is suffused with a rhetoric of revelation. He becomes a prophet-scientist of the cosmos enumerating "all its parts" and uncovering "the causes of things."

Significantly, his goal is not the accumulation of knowledge as an end in itself. Knowledge of the world is a means to moral knowledge ("good ends"); it begets "wisdom and prudence" and leads to "higher specula-tions." Mercator's emphasis on knowledge in the service of the good life reminds us that for much of the medieval and early modern periods, the world (*mundus*) was also associated with evils of worldliness that threatened to take humanity further away from salvation. By placing the world within a powerfully moral scheme and subordinating universal knowledge to indi-vidual ethical action, Mercator rewrites an ancient moralistic discourse that opposed knowledge of the world to a (much-desired) knowledge of God.

This recourse to divine omniscience also links world maps to the tradi-tion of *geographia sacra* (sacred geography). Cosmographers, like the natural philosophers, tended to follow Saint Paul's dictum that individuals reach toward divine intelligence through the contemplation of nature. Even when the world is subject to a mathematical graticule, the fact of measure, pro-portion, and calculation belongs to an invisible and higher order of things. European cosmography from the ninth to the seventeenth century thus overtly mobilizes Christian ideology, promoting readings of the "book of the world" that are intended to assure the viewer that nature reflects the genius of creation. Maps become aids to contemplation, used to apprehend divine intervention in the phenomenal world.[107] This is the explicit purpose of the cartographic program in the *Biblia Regia* (the Antwerp Royal Polyglot Bible), produced by Benito Arias Montanus at the Plantin Press between 1568 and 1573, the same period during which Ortelius published the *The-atrum* (also with Plantin) and Mercator produced his famous navigational world map projection, *Nova et aucta orbis terrae descriptio*. Studies of maps and religion in the early modern world have, however, been largely con-fined to explicitly biblical maps or those that overtly depict ecclesiastical power, thereby excluding connections between overtly religious and seem-ingly secular maps.[108] Though the sheer number of maps of the Holy Land produced in this period make it the most popular region to be depicted, almost as many world maps were made as maps of the Holy Land—a fact that highlights the theological dimensions of the cosmographic project.[109]

There is an important and repeated insistence throughout the sixteenth century on the relationship between cosmography and theology as an ex-pression of the analogy between human and divine worldmaking. By 1595, Mercator was drawing on an established set of tropes that would continue

FIGURE 15. Portrait of André Thevet (ca. 1550). Courtesy of the Bibliothèque nationale de France.

to be used well into the seventeenth century. An early portrait of the French royal cosmographer André Thevet (fig. 15), for instance, shows the author holding a compass poised over a globe. The image finds a counterpart in the preface to the *Cosmographie universelle* (1575), where Thevet describes God holding the world in his hands and turning it between two or three fingers.[110] Acutely aware of the parallel between himself and the Creator, he qualifies this connection at some length in the preface:

> This discipline of Cosmography, therefore, serves to reveal the vanity of that with which we stop ourselves; then, bending our pride, it directs our mind toward that which is great and no longer permits it [our mind] to stop itself over trivialities. And for this reason I think that there is no science, after Theology, that has a greater virtue in making us understand grandeur and divine power, and to hold these in admiration, than this one.[111]

This doubling of theology and cosmology, creator and cosmographer, emerges once again through the iconography of the finger-compass poised over a globe. Mercator's portrait and the figure of Atlas both evoke the image, which derives from a long medieval tradition that depicts God as a geometer (fig. 16).[112] The pose in fact is reproduced for a range of figures in the period, from Ptolemy to Hondius: a subsequent image in the *Atlas*, first used in the 1613 edition, shows a doubled version of the same *topos* as Mercator sits beside the printer Hondius and both hover over globes within an expanded cartouche (fig. 17).

Too often, these images of worldmaking have been reduced to expressions of conquest—whether intellectual dominion of the natural world or political dominion in the name of empire. They are also, however, images of contemplation, reflecting on the world as the product of divine handiwork. World maps and atlases were certainly implicated in shaping national identity, imperial desires, and international commercial enterprises—chapters 3

FIGURE 16. Codex Vindobonensis 2554 (ca. 1220s). Cod. 2554, fol. 1v. Courtesy of the Österreichische Nationalbibliothek, Vienna.

FIGURE 17. Gerhard Mercator and Jodocus Hondius, *Atlas* (Amsterdam, 1613). Courtesy of the Beinecke Rare Book and Manuscript Library, Yale University.

and 4 explore these aspects of cartography and worldmaking in greater detail. But, the philosophical contexts of such images reveal how cosmographers could also negotiate a position between two extremes. A celebration of scientific progress and the human intellect, mapmaking also sought to mimic God, the Original Maker, with appropriate deference. By displaying the symmetries underlying the world's creation, maps provided a metaphysical, and not merely geographic, orientation.

Spiritual Navigation

Mercator's *Atlas* was placed on the Index in 1603 for its intellectual and doctrinal over-reaching. Prime among the reasons for this decision was the inclusion of a short treatise on the creation of the world, "De mundi creatione ac fabrica" (On the Creation and Making of the World), which opens the volume, positioning the *Atlas* at the junction of cosmography, theology, and natural philosophy. Given Furmerius's quasi-divine presentation of Mercator and his own self-conscious worldmaking ambitions, it is not difficult to see why the book would run afoul of religious orthodoxies. Mercator's work, like that of several contemporaries, was shaped against the clamor of religious conflict in post-Reformation Europe and as a direct response to it. Cosmography was not, and perhaps could not be, morally disinterested, and its practice throws a refracted light onto the spiritual contexts that gave the discipline its most profound questions. Recent studies have sought to name the complex beliefs of Mercator, Ortelius, and other members of the Plantin circle as Lutheran, Erasmian, Melanchthonian, as "spirituals," or secret members of the Family of Love.[113] But such identifications are less important than the practical impact of the religious wars on the early modern reimagining of the world. The use of maps as propaganda in national and confessional battles has been well documented, but what of their spiritual and meditative power?

In a world whose religious center seemed to have fallen apart—the fall of Constantinople on 1453, the sack of Rome in 1527, the fragmentation of Christendom into small warring centers—geography, particularly the synoptic, unifying practice of world-mapping, was imbued with particular symbolic power. To bind up the fragmented world and give it coherent form provided a vision of harmony, a goal for spiritual contemplation that sought to counteract the chaos of ordinary experience. Mercator's biography registers the impact of such tensions: in 1544, he was accused of "lutherye" and narrowly escaped a death sentence. This early brush with the intolerance of religious authorities would profoundly mark his life and career. His first two maps—*Terra sancta* (a map of the Holy Land showing the Exodus) and *Orbis imago* (the cordiform world map)—both evoke Reformation symbolism and locate the mapping enterprise within a religious matrix.[114] But his near execution for heresy seems to have muted this audacious beginning.

This relation between mapping and religion in Mercator's work surfaces almost unexpectedly in 1569, which saw the publication of both the *Chronologia* and the *Nova et aucta orbis terrae descriptio*. Jerry Brotton has argued that this wall map, which launches the so-called Mercator projection, must also be seen as a call for religious toleration. A spatial counterpart to the temporal project of establishing a universal chronology, which required reconciling seemingly incompatible historical data, *Nova et aucta orbis* demanded a similar reconciliation of the spherical earth onto a rectangular, two-dimensional plane. Brotton suggests that the mathematics behind Mercator's straightened loxodromes celebrate a spiritual *cursus obliquus*, an oblique course through a worldly existence: "instead of exhibiting a confident Eurocentrism, Mercator's world map would provide an oblique rejection of such values and a search for a larger picture of harmony across universal space and time."[115] Whether or not one accepts his allegorical reading of the rhumb line, Brotton's intuition that the "navigational use" for which Mercator intended his great wall map was spiritual rather than imperial is utterly convincing.

For Mercator also dabbled in theological commentary, though he kept his work out of public view, given his brush with religious persecution. Walter Ghim speaks of a compilation of the Gospels to produce a "continuous evangelical history," commentaries on Paul's Letter to the Romans that attempt "to solve . . . several controversies of our time concerning divine providence, predestination and free will" and also "several chapters on Ezekiel, on the Apocalypse and others."[116] These reflections culminate in the posthumously published "De mundi creatione," an extended commentary on Genesis that opens the *Atlas* by returning to origins and imagining the material fabrication of man and world. In a striking departure from all previous mapbooks and would-be atlases, the first part of the *Atlas* is thus not a world map but rather an extended meditation on the Creation.

Mercator completed this treatise in the final years of his life, considering it his greatest achievement. In a letter from 1583, he writes, "Although this is the last part of my work, it will nonetheless be the most important, indeed the very base and summit of the whole. . . . This will be the goal of my labor, this will mark the end of my work."[117] "De mundi" begins with a rumination on "the scope of all cosmography," a claim that recalls the opening lines of Ptolemy's *Geographia* with its attempt to establish the scope of the discipline. But rather than technical definition, the reader is invited into a celebration of divinity: "For any mind considering . . . the construction of this earthly machine, it is axiomatic that God, its author, is of immeasurable power, wisdom, and goodness. . . . For this is our goal while we treat of cosmography: that from the marvelous harmony of all things toward God's sole end, and the unfathomable providence in their composition, God's wisdom will be seen to be infinite, and his goodness inexhaustible."[118] Neo-Stoic rhetoric and the worldmaking iconography of the book's prefatory material are crystallized here, as Mercator places the cosmographer's art in the shadow of divine creation. To witness the "frame of the world" is to witness the power of the deity. The atlas thus becomes a revelatory "book of nature."

The historical significance of "De mundi creatione" lies in its generic fluidity and unique position. Despite the theological pretensions of other geographers, no one had presumed to write a treatise on Genesis and incorporate it into a cosmography. To do so was to argue that mapping derived from a worldview based on the Bible and that it had an instrumental role in enhancing human understanding of the cosmos. Cosmographic works such as the *Atlas* were no mere imitations of the world; they were explications of it. If the figure of Atlas points to the humanistic learning of the Renaissance, "De mundi" indicates a close allegiance with post-Reformation commentaries and annotations on the Bible. The work rightly takes its place within the tradition of hexameral writing—texts in prose or poetry that celebrated God's creation of the world in six days. Though often considered a medieval genre, hexamera enjoyed a revival in the late sixteenth century; Mercator's treatise belongs alongside such works as Guillaume Du Bartas's *La sepmaine* and Torquato Tasso's *Sette giorni del mondo creato*.[119] It is also the intellectual and bibliographic heir to Hartmann Schedel's *Liber chronicarum* (Nuremberg, 1493), which offered a historic and geographic account of the world from the creation to the present. Ostensibly affirmations of the marvels of creation, these hexamera are poignant responses to a crisis of world order: they record apprehensions over the nature of the world and seek to allay intellectual fears with new, often ingenious interpretations of scriptural texts.

Mercator's prose hexameron is no exception. It reconciles new geographic discoveries with the account in Genesis and effectively reinterprets Biblical texts to explain the observations of explorers, mathematicians, and natural philosophers. If new knowledge of the world had fractured the intellectual consensus of the ancients, it also affected post-Reformation debates profoundly. Mercator's purpose is to reestablish the order and harmony of all things within a providential scheme: "For this is our goal . . . that from the marvelous harmony of all things . . . God's wisdom will be seen to be infinite, and his goodness inexhaustible." As the belief in marvelous harmony was being assaulted from all sides, Mercator's impossible worldmaking mission contains the seeds of the Enlightenment craving to catalogue the whole as well as the old mystical desire for unity and transcendence.

The philosophic interest of "De mundi creatione," like that of Du Bartas's *La sepmaine*, lies in its attempt to grapple with the nature of matter, specifically its origin and properties, without succumbing to Lucretian atheism. Mercator seeks to reconcile the classical idea of chaos and the doctrine of the creation ex nihilo; rejecting the Platonists' division between form and matter, he seems to suggest—quite heretically in some quarters—that the origin of all things is a divinely created "first matter" that suspiciously resembles chaos. This original, divine matter then becomes the basis for a theory of unity despite the immense diversity observed in all earthly forms.

Though the treatise has been dismissed as a Neoplatonizing account of creation, it participates in a complex pan-European meditation on the nature of the cosmos, its materiality and structure, as well as the place of human beings within that universal frame. Filled with catalogues of places, plants, animals, and even a miniature theory of the origin of human culture, the treatise deserves greater attention for the ways in which its philosophical underpinnings frame the compendious "description of the world" in the maps that follow.

Mercator's ambitions demand that the *Atlas* be read alongside such late sixteenth-century writers as Postel, Du Bartas, Lefèvre de la Boderie, Louis le Roy, Dee, and Spenser. Like these figures who fought to cut through masses of new information, see beyond diversity, and uncover the harmony of the cosmos, "De mundi" aims to transform how we see the world, penetrating into its originary mysteries even as the *Atlas* reveals the structure of the world to the mortal human gaze. Mercator delicately balances the desire for omniscience with the limits of human knowing as he calibrates the mapmaker's representational ability with creative speculation. The informal title he used for the treatise that would become "De mundi" captures this doubleness—"cosmopoeia," the creative making of the world.[120]

CHAPTER TWO

On Cosmographic Autobiography: Montaigne's *Essais*

"Our world has just discovered another world," writes Michel de Montaigne in the late 1580s, "and who will guarantee us that it is the last of its brothers, since the daemons, the Sibyls, and we ourselves have up to now been ignorant of this one?"[1] In "Des coches," the *Essais*'s long meditation on the Spanish discovery and conquest of the New World, the "very image of the world which glides along as we live in it" is like a mirage beyond the grasp of human knowledge, perpetually multiple in its many forms.[2] New knowledge had opened a wound, both spatial and temporal, from which the very conception of the world had yet to recover. As the euphoria of discovery gave way to conquest, commerce, and the scrutiny of long-cherished intellectual certainties, early modern Europe was faced with a teeming multiplicity that threatened to destroy traditional structures of authority and understanding. Montaigne captures this *esprit du temps* with an elegiac quotation from Lucretius: "This age is broken down, and broken down is the earth."[3]

SELF AND WORLD

Recording and reacting to this upheaval of historical and geographic proportions, Montaigne makes "the world" one of his primary subjects alongside the relentless self-exploration that is his professed theme. *Monde* [world] is a keyword for the *Essais*— an all-encompassing category that barely contains the plethora of customs, cultures, peoples, beliefs, opinions, attitudes, and anecdotes that fill the work's pages. The proliferating fullness of the world is the great bulwark against which Montaigne interrogates the self.[4] *Monde*

describes an expanding cosmos, a whole that is always just out of the essayist's reach. It is also the most frequently repeated and powerfully suggestive
analogue for the self, offering an external parallel for the endless variety
and vast spaces within. The wonder with which Montaigne encounters *le
moi* is the same wonder with which he encounters *le monde*.

Despite the *Essais*'s interest in the wider world, much scholarship on the
work extends from Montaigne's famous claim, "I myself am the matter of
my book."[5] Drawing a dichotomy between the world without and the self
within, he proclaims in the essay on presumption, "The world always looks
straight ahead; as for me, I turn my gaze inward."[6] The *Essais* have thus
come to be celebrated as one of the earliest forms of autobiography and
paradigmatic of the emergence of modern subjectivity. But a different picture emerges if we attend to the manner of Montaigne's telling, the mode of
that inward gaze. The texture of the *Essais* is interwoven with descriptions
and anecdotes drawn from far-flung parts of the globe. The essayist was
fascinated with cosmographic writing and owned a large number of works
related to cosmography, travel, and cosmology—contemporary travel anecdotes, geographic speculations, and ethnographic curiosities are almost
all drawn from cosmographic compendiae.[7] Indeed, the *Essais*'s intellectual
and moral style rests on the habit of blurring, contrasting, and meditating
on seeming binaries that emerge from a sense of spatial expanse: the essayist's interior spatiality in fact depends on and follows from the evocation of
global vastness. For Montaigne, self-making hinges on worldmaking and
vice versa.

This porous interpenetration of self and world in the *Essais* emerges in
the book's thematic architecture and the prominence of the New World.
America is famously the subject of "Des cannibales" (1.31) and "Des coches" (3.6), but it is also a crucial nexus in the second half of the "Apologie
de Raimond Sebond" (2.12), providing a powerful contemporary test case
of the limitations of human knowledge. These invocations of America unfold against reflections on universal history and geography, on the making and destruction of worlds. In the process, they become meditations on
the perspectival, partial nature of human knowing and on the centrality of
the singular self's desire for universalizing categories—a self defined paradoxically through its engagement with the world, just as the world itself
emerges into focus only through the lens of the self.

The New World essays thus dominate the structure of the work; they
stand as two pillars in the first and third volumes supporting the skeptical
assaults of the "Apologie."[8] Reflections on an emerging, new idea of "world"

precipitated by the encounter with a new continent become occasions in the *Essais* for reflections on epistemological inquiry, self-knowledge, and ethical action. The very existence of America tests the possibility of comprehensive knowledge of the world and puts renewed pressure on the vexed relationships between the ancients and the moderns, the Old and the New. Its invocation at strategic points in the *Essais* signals crucial interventions in debates over ideas of world order and creation—and it is finally through the New World that Montaigne meditates on the ties that bind intimations of the divine to the mortal self and world.

The Body of the World

Montaigne situates his authorial self in relation to the world in his 1580 preface to the reader ("Au lecteur"): "Que si j'eusse esté entre ces nations qu'on dict vivre encore sous la douce liberté des premieres loix de nature, je t'asseure que je m'y fusse tres-volontiers peint tout entier, et tout nud." (Had I been placed among those nations which are said to live still in the sweet freedom of nature's first laws, I assure you I should very gladly have portrayed myself here entire and wholly naked.)[9] Montaigne can only explain the book's mode of self-description, its intervention in the ancient battle between elaborate rhetorical excess and honest, stripped-down self-confession, through analogy with the American Indians' "douce liberté." Although he signals his readiness for complete self-exposure by imagining himself as a naked Indian, the comparison rests on the author's awareness that such images of a nude America were themselves sophisticated representations crafted in counterpoint to the portrait of a corrupt Old Europe. Readers of the *Essais* would have been familiar with the images in popular cosmographies such as André Thevet's *Singularitez de la France Antarctique* (1558) or the allegorical figures that adorned the frontispiece to Ortelius's *Theatrum orbis terrarum* (1570). This topical visual juxtaposition highlights the novelty of Montaigne's autobiographical reflection as a complex and compromising encounter, a return to an untainted point of origin that is both a retreat *into* the self as well as an imaginative journey beyond the boundaries of individual consciousness into the "new" world of America.

This desire for representative transparency itself becomes a rhetorical trope within the *Essais,* which repeatedly chart the struggle to reach outwards into the world. Throughout, the drama of oscillation between the observing self and the world is figured in the struggle to reach past images, representations, and *topoi,* the mirages of *vraisemblance,* toward the truth, *le vrai,* the genuineness of the real thing. For Montaigne, representation and object, like the map's mimesis of the world, must be recognized as distinct, even if they are seen as part of an epistemic continuum. Because the world can only be known *through* its images, as we have seen with the world atlas, knowledge of the world can only be attained through the body of an observer. The inward search for an authentic self parallels the movement outward into the world at large.

It is no accident that Montaigne situates this double quest for world and self in relation to the Americas. The emergence of the new continent opened immense possibilities for increasing knowledge even as it marked a sharp disruption of world order. The analogies it invited between geographic exploration and self-discovery became the logical extension of the parallels between anatomy and geography that animated the mapmaking enterprise in the sixteenth century. The spatialization of the self now became the counterpart of the corporealization of the world.

Thus, in his early essay on education, "De l'institution des enfans," written for Diane de Foix, comtesse de Gurson, Montaigne invokes the image of the world as a giant mirror for the self: "Ce grand monde que les uns multiplient encore comme especes soubs un genre, c'est le miroüer où il nous faut regarder pour nous connaistre de bon biais. Somme, je veux que ce soit le livre de mon escholier." (This great world, which some multiply further as being only a species under one genus, is the mirror in which we must look at ourselves and recognize ourselves from the proper angle. In short, I want it to be the book of my student.)[10] Montaigne explicitly invokes the medieval *speculum mundi* (mirror of the world), which, like its sixteenth-century cosmographic descendent, sought to encompass the world in a book.[11] But while the *speculum* represented the divine order of the cosmos with its hierarchical classifications and symmetries, Montaigne alters this original function. By linking it to two other distinct philosophical positions—the Epicurean belief in a plurality of worlds and the mystical correspondence between microcosm and macrocosm, in which the self is an image of the world—he reimagines the world's order. The problem of worldly plurality is elegantly enclosed within the classificatory metaphor that describes "the world" as a *summum genus,* an all-inclusive class that assimilates various subjects. "The world" can now logically consist of many worlds and still retain an orderly structure.

The image of the world as a mirror for the self, with the subtle play on the word *espece* (literally the profusion of worlds, "species," but etymologically derived from the Latin *specio,* to look) already suggests a li ⸗ between world order and subjective vision. As the metaphors of classification and reflection merge into each other, we glimpse the transition from one epistemological system to another, from an older method of organizing knowledge to a new one: the divinely established order of biblical creation is being replaced by a humanly invented system of apprehending the world.

But *espece* also evokes *espace*, the spatial and intellectual horizon, which expanded with the discovery of the American continent. If the world in the *Essais* constantly extends into the self and the text, it does so because Montaigne mimics the sense of spatial extension made possible by the New World. The shape-shifting world acts as both subject and metaphor for the dialectic between the known and the unknown, the pursuit of knowledge and its inevitable limitations. In a delicate modification on the metaphor of the book of nature, Montaigne now describes *the world* as a book—and there is the inevitable suggestion that the world is one with *his* book, this book of *Essais*, which is simultaneously a portrayal of himself "tout entier, et tout nud."[12]

This presentation of "the great world" as body, mirror, and book highlights Montaigne's thematization of the reciprocal relationship between self and world. Unlike the secure consciousness of an inquiring mind (the *cogito*) that gives the Cartesian self a rootedness and stability in relation to the external world, Montaigne resists the self-possessed separation of mind and world that would be Descartes's legacy. Maurice Merleau-Ponty notes Montaigne's distinctly interrelated consciousness of self and world: "The world is not for him a system of objects the idea of which he has in his possession; the self is not for him the purity of an intellectual consciousness. For Montaigne—as for Pascal later on—we are interested in a world we do not have the key to. We are equally incapable of dwelling in ourselves and in things, and are referred from them to ourselves and from ourselves to them."[13] Montaigne often turns to metaphors that apply interchangeably to self and world—he speaks often of the world's body and of the horizons of the self, mixing geographic, metaphysical, and moral registers.

In the 1580 (A-text) version of "De l'institution des enfans," Montaigne already articulates this problem of representation in his own literary practice as one of epistemology and autobiography: "Quant aux facultez naturelles qui sont en moy, dequoy c'est icy l'essay, je les sens flechir sous la charge. Mes conceptions et mon jugement ne marche qu'à tastons, chancelant, bronchant et chopant; et quand je suis allé le plus avant que je puis, si ne me suis-je aucunement satisfaict: je voy encore du païs au delà, mais d'une veuë trouble et en nuage, que je ne puis desmeler." (As for the natural faculties that are in me, of which this book is the essay, I feel them bending under the load. My conceptions and my judgment move only by groping, staggering, stumbling, and blundering; and when I have gone ahead as far as I can, still I am not at all satisfied: I can still see country beyond, but with a dim and clouded vision, so that I cannot clearly distinguish it.)[14] Once again, Montaigne returns to a cosmographic, spatial metaphor. Though he claims the essay concerns his own natural faculties, the metaphor he deploys is that of a journey to a distant country. *Essai, essayer* (trial, essay, experiment), that characteristic pun, here signals exploration in and of the world, as the quest for intellectual clarity and self-understanding is compared to the exploration of new lands. The vague outlines of land beyond the horizon are always dimly visible but just beyond reach.[15]

The halting, staggering style of thought that Montaigne describes here as he gropes toward a horizon of certain knowledge will become, in "De la vanité," the defining style of his text. In "Du repentir" (3.2) the same style and rhetoric famously describes the movement of the world:

> Le monde n'est qu'une branloire perenne. Toutes choses y branlent sans cesse: la terre, les rochers du Caucase, les pyramides d'Ægypte, et du branle public et du leur. La constance mesme n'est autre chose qu'un branle plus languissant. Je ne puis asseurer mon object. Il va trouble et chancelant, d'une yvresse naturelle. . . . Je ne peins pas l'estre. Je peints le passage. . . . Je pourray tantost changer, non de fortune seulement, mais aussi d'intention. C'est un contrerolle de divers et muables accidens et d'imaginations irresoluës et, quand il y eschet, contraires. . . . Si mon ame pouvoit prendre pied, je ne m'essaierois pas, je me resoudrois: elle est toujours en apprentissage et en espreuve.

> [The world is but a perennial movement. All things in it are in constant motion—the earth, the rocks of the Caucasus, the pyramids of Egypt—both with the common motion and with their own. Stability itself is nothing but a more languid motion. I cannot keep my subject still. It goes along befuddled and staggering, with a natural drunkenness. . . . I do not portray being: I portray passing. . . . I may presently change, not only by chance, but also by intention. This is a record of various and changeable occurrences, and of irresolute and, when it so befalls, contradictory ideas. . . . If my mind could gain a firm footing, I would not make essays, I would make decisions; but it is always in apprenticeship and on trial.][16]

This passage refers back explicitly to "De l'institution des enfans" (repeating the verbs "chancelant" and "essaierois") as the world merges into the self and vice versa. What had been a passing simile of land in the distance is here transformed into a fully developed comparison between the perennially changing world and the unstable self. The shifting spaces of the world may also look back to the preamble of "Des cannibales" (1.31) with its meditation on tectonic movements, and in particular to Montaigne's comparison of the world to a body: "Il semble qu'il y aye des mouvemens, naturels les uns, les autres fievreux, en ces grands corps comme aux nostres." (It seems there are some movements, some natural, others feverish, in these great bodies, just as in our own.)[17] A similar sentiment surfaces in the "Apologie de Raymond Sebond," where Montaigne describes the world as "ce grand corps que nous appellons le monde" (this great body that we call the world).[18] If "ce grand monde" is a mirror of the self in "De l'institution des enfans," here the world metaphorically becomes an image of the body—an important shift of emphasis that highlights Montaigne's increasing intuition that the world can only be experienced and known through the physicality of his own body.[19]

Order and Disorder

The analogy between Montaigne's own body and the materiality of the world precipitates a new approach toward shaping a clear concept of "the world" in the wake of the New World encounter. Faced with the proliferating multiplicity of the visible world, the essayist must yield to the partial, fragmented lens of his mind, giving up the dream of attaining an ordered, synthetic description of it. But in his cognitive and emotional fragmentation, Montaigne finds the perfect mirror for the equally unstable external world. The distempered, unknowable, and uncontrollable body of the world in essays such as "Des cannibals," the "Apologie," "Du repentir," and "Des coches" looks back to the essayist's own unruly body. In "De la force de l'imagination" (and later, "De l'experience") that vulnerable body is so easily susceptible to the power of the imagination that it sickens and heals, produces and destroys sexual arousal with no more than a thought.[20] Through such strategic overlapping of self and world, Montaigne evokes the ancient theory of microcosm and macrocosm only to change its scope radically. No longer are the body and world symbolic images of each other, asserting a symmetry between political and cosmic order, as they are for instance in the thirteenth-century *Ebstorf* map where the *oikoumene* is mapped onto the body of Christ.[21] Nor can the world be brought into coherence through powerful creative analogies, such as Mercator's vision of Atlas. Rather, in the *Essais*, self and world mirror each other in their *disorder*, in the contingency of the events that produce them and the events that occur in them—and finally, for the imaginative fashioning required to give them meaning.

This reorientation epitomizes the two primary dynamics of worldmaking thought in the late sixteenth-century—the one centripetal in its energies, seeking to gather up a burgeoning whole into a harmoniously ordered unity, the other centrifugal in its drive to chronicle the breakdown of an old world order and skeptical of claims to have a discovered a new synthesis. Montaigne often juxtaposes these contrasting energies through a counterpoint between Plato and Epicurus (or Lucretius), the two classical philosophers who respectively champion principles of order and contingency as underlying the pattern of the world. Neither alternative is entirely satisfactory, and Montaigne returns restlessly to their differences, seeking the key that will unlock the mysteries of world order and thus enable comprehensive knowledge of the whole. But, as the essayist discovers through the skeptical argument of the "Apologie de Raymond Sebond," there are no certain answers, since we are subject to the limitations of human reason.

The *Essais* therefore chronicle the failure of all overarching philosophic systems that claim to describe world order and locate instead the key to knowledge in the self's often contradictory experiences of the world. Montaigne strategically deconstructs the intellectual history of worldmaking efforts—and also shows why such projects are essential. In the process, he argues passionately that knowledge of the world resides in the experiencing self—not in folk wisdom, or in particular philosophers, and perhaps not even in the turn to divine revelation, for God is the ultimate unknowable Other. Where Mercator finds a stable center in his own technical mastery, his mirroring of the creative deity who makes the world, Montaigne asserts the insurmountable alterity of the divine. He argues for a moral center within, a consolidation of the self that alone can provide an ethical compass to make sense of the world. Worldly knowledge is thus not an external entity waiting to be discovered but rather one that is imaginatively constructed and always subject to error. We cannot know the world with certainty, and yet we are compelled to keep trying.

The problems of worldmaking occur in encounters with radical difference and become emblems for the always incomplete nature of human mastery over the world. Meditations on the crisis of knowledge and world order, whether in the New or Old World, magnify questions of diversity, difference, and cross-cultural engagement at both the local and global scales. If Montaigne's most famous essay, "Des cannibales," introduces the matter of cultural relativism and otherness into the texture of the book, the

"Apologie" and "Des coches" expand its socio-political ramifications into the cosmological and philosophical spheres, showing how the dilemma of world order leads back to the self.

Montaigne's desire to bring the world into coherence by clarifying the foundations of the self—by discovering the "forme maistresse" that governs his own being—presents an intriguing counterpart to Mercator's ambition to explicate the structure of the world. But in the essayist's hands, the underlying epistemic challenge of worldmaking acquires a new twist. No longer is the part (self) a hierarchically placed subset of a distinct, overarching whole (the world); rather, it is held in a horizontal, reciprocal relation that suggests that the world can never be grasped in its entirety and can only be only be known in partial fragments through the mediating self. Thus, the *essai*, simultaneously encyclopedic and egocentric in scope, evolves to capture the mutually constitutive interplay between self and world. Montaigne finds in this new genre a supple instrument to capture a new world-picture. He abandons systematic philosophical exposition in the scholastic tradition for the highly suggestive, if logically ambiguous, power of the imagination to fashion new knowledge.

The world atlas and the book of essays thus become generic counterparts, new forms that share surprisingly similar strategies of representation. If Ptolemy drew on portraiture as an analogy for geography, the *Essais* now use geographic metaphors for autobiographical portraiture. Both genres play with the mimetic barrier between word and image: maps mirror, speculate, and threaten to substitute the world with an image of it; the essay claims to witness, observe, and narrate a living self whose textual presence remains thoroughly mediated. Conceptually, too, the *Atlas* and the *Essais* find parallel methods for grappling with the key challenge of integrating a plethora of parts into a unified whole (though the *Essais* resist the very notion of wholeness). The essay and the atlas invest in the primacy of local detail while using the book's overarching structure to draw the parts into a unifying framework; if Mercator's *Atlas* uses a human body to figure the world, Montaigne's *Essais* are a global, all-encompassing atlas of the self.

Charles Taylor sees this investment as Montaigne's unique contribution to the history of modern identity: his aim "is not to find an intellectual order by which things in general can be surveyed, but rather to find the modes of expression which allow the particular not to be overlooked. . . . The Cartesian [enterprise] calls for a radical disengagement from ordinary experience; Montaigne requires a deeper engagement in our particularity."[22] This is why cosmographic metaphors are so productive for the *Essais*. The world

atlas circumscribes the whole by highlighting a collection of individual ex-
emplars and reveling in their minute particularity. As would a mapmaker,
Montaigne dwells on singularities as a means to reach out, with moral clarity, toward a vision of the whole.

FROM COSMOGRAPHY TO AUTOBIOGRAPHY: "DES CANNIBALES," ONCE MORE

The *Essais* wrestle with the desire for knowledge of the entire world and
seek ways to contain and regulate its profusion. If variety is the only universal law, as Montaigne argues in "De l'experience," to comprehend the
world, we must make sense of the parts. In essay after essay, Montaigne
seizes upon radical, surprising, often shocking examples that challenge European norms of custom and civility, using them as test cases to find the cosmic glue—the human, cultural, or natural denominator—that might hold
them together. In earlier essays, Montaigne sees self and world in terms of
productive binaries—the contrast of old and new worlds, of ancient and
modern, of barbarism and civilization, of nature and art. Gathering and
reconciling oppositional variety is the first step to grasping the contours
of an often self-contradictory image of the world. By the end of the book,
however, it is the essayist's own experiencing self which encounters the
world and becomes one with it: "Je m'estudie plus qu'autre subject. C'est
ma metaphisique, c'est ma phisique," he writes, infusing the unique singularity of his own body with macrocosmic resonance.[23]

Cosmography and autobiography in the *Essais* thus become parallel, analogous pursuits whose techniques of investigation, modes of understanding,
and rhetoric of description overlap and coalesce. Though the singularity
of autobiography typically stands against the collective whole that is the
world, Montaigne suggests that the painstaking processes of collection, artful organization, critical juxtaposition, and finally synthesis that characterize cosmographic thinking also define the autobiographic enterprise of his
book. In "Des cannibales," Montaigne explores these connections through
the confrontation between Old and New Worlds.

As a seeming instance of radical difference, cannibalism is both a trope
and a phenomenon in early modern Europe. It both divides and ties the
Old World to the New, and "Des cannibales" has thus become a touchstone
for early modern expressions of cultural relativism and politicized literary
practice. Commentaries on the essay continue to be dominated by Montaigne's own binary method, pursuing analyses of the innocent Amerinid-

ian savage and corrupt European sophisticate, *l'autre* and *l'être*.[24] However, Montaigne's cosmographical emphasis and his insistence on the literary metaphors embedded in descriptions of the Amerindians suggest another, more expansive agenda.

The confrontation of European and Amerindian early in the book presents in a miniature, personalized tableau the cosmic collision of old and new worlds. In "Des coches," Montaigne will expand this method to unfold how we know the world through the self, and how that process of knowing is closely linked to acts of imaginative making. "Des cannibales" demonstrates this process in its fabrications, unreliable testimonies, and allusions to famous philosophical thought experiments, all of which problematize the boundary between certain knowledge and speculative reasoning, between fact and fiction. The essay's preoccupation with the blurry distinctions between truth and verisimilitude (*vraisemblance*), reality and representation, will become in the "Apologie" the basis for Montaigne's skeptical attack on all systems with claims to comprehensive knowledge of the world. By emphasizing the human fashioning of fundamental intellectual categories, Montaigne implicitly questions the belief in an unshakeable, universal basis for world order.

Atlantic Speculations

The lengthy preamble to "Des cannibales" strategically situates the New World within cosmogonic philosophic texts, familiar from Plato and Aristotle, and cosmographic texts and travel accounts such as those of André Thevet and Jean de Léry. These were the two key genres used to imagine, explore, and construct the world in the sixteenth century, and their conjunction in "Des cannibales," and indeed all three New World essays, thus demands further examination.[25] Recalling Mercator's hybrid *Atlas*, Montaigne gestures here toward the double function of worldmaking as both description and re-conceptualization. The essay's often neglected but elaborate opening reflection highlights this effort and frames the subject of "Des cannibales" in global terms.

The essay opens with a crucial allusion to Plato's tale of Atlantis as a plausible explanation of the New World's existence—a reference that will return in later essays.[26] The story of Atlantis, gleaned by Solon from ancient Egyptian priests, suggests a kind of infinite historical regress, a constant unfolding of stories that subverts any attempt to establish a stable point of origin or preestablished order.[27] By invoking it, Montaigne signals a philosophical kinship with the best known dialogue of the Platonic corpus—the *Timaeus*. A text of crucial importance to the Latin Middle Ages and to early modern philosophy, the *Timaeus* is devoted to the project of worldmaking, to questions of cosmic origin, and to the vexed problem of distinguishing between the "real" world and its representations. In subsequent essays, Atlantis will become a prime marker for worldmaking claims. Montaigne highlights its importance in his revisions: the Atlantis tale is an integral part of the preamble to "Des cannibales," while it is a C-text addition in both the *Apologie* and "Des coches." He thus connects the three essays' themes, emphasizing the philosophical task of experiencing and expressing the world.

In "Des cannibales," Montaigne establishes a suggestive series of parallels between Brazil and Atlantis ("païs infini"), and between Montaigne and Solon as seekers of knowledge who journey to an unknown place that might hold the key to universal understanding. Cosmographic description thus merges into philosophical inquiry and vice versa: the geographic description of Atlantis, a textual map of the tripartite pre-Columbian world, derives almost verbatim from the *Timaeus* (24e–25a), but it is cloaked in cosmographic rhetoric derived from post-Columbian writers such as Apian, Münster, and Thevet. Such fusions of past and present knowledges identify the Atlantean allusions as metaphoric nodes, meeting points in the text where philosophic speculation and cosmographic description coalesce and disrupt each other.

By juxtaposing speculative accounts of a mythical land with the materiality of the American continent, Montaigne blends fact and fiction, a move familiar to the mapmakers as central to worldmaking. In contemporary writing about the New World, America, like Atlantis, hovered between history and legend, emerging into European consciousness through a mixture of actual eyewitness accounts, romance rhetoric, and imagined monstrosities recorded in classical and medieval sources. Like the story of Atlantis, which is orally transmitted to Solon by the Egyptian priest, the tales of the New World were garnered primarily by hearsay and transmitted, often orally, to cosmographers in Europe.[28] In metropolitan centers, miscellaneous information and anecdotes were collated into a coherent narrative, which inevitably required imaginative addition and edition to produce a coherent whole, permanently distorting the boundaries between fabricated stories and truthful accounts. Many such syntheses eventually acquired the status of authoritative reports and became the basis for the conquest, conversion, and extermination of native populations. By accenting the role of imaginative construction in these accounts, Montaigne reflects on the permeable boundaries between fact and fiction. More importantly, he discloses the epistemic effects as well as the political and ethical consequences of fiction-making on a global scale.

"Des cannibales" thematizes this hybridity and its slippery relationship to truth as the tricky problem of *vraisemblance*, and versions of the word are peppered throughout the essay. "Il est bien vray-semblable," he writes in the second paragraph, that a flood could have changed the face of the earth. Though it passes quickly, the evocation of likeness despite essential difference, of plausibility and imaginative resemblance, enters the verbal texture of the essay. Later, Montaigne will demand "un homme tres-fidelle, ou si simple qu'il n'ait pas dequoy *bastir et donner de la vray-semblance*, à des inventions fauces"; and most famously, "chacun appelle barbarie ce qui n'est pas de son usage; *comme de vray il semble que nous n'avons autre mire de la verité* et de la raison que l'exemple et idée des opinions et usances du païs où nous sommes" (my emphases).[29] *Vraisemblance*, which lies between historical veracity and fictional falsehood, is in fact what holds together a cohesive idea of "world" as a collective whole. Mirror-like, it names the convergence of similitudes, of representations, which construct a conceptual totality. But this quality of truth-likeness also discloses the contingency and human fashioning inherent in conceptions of wholeness.

The allusions to Plato's *Timaeus* enable Montaigne to harness the philosophical implications of *vraisemblance*, for they are at the very core of the dialogue. Its cosmological narrative (29b-c) hinges on the credibility of Timaeus's *eikos logos*, a "likely story" or truth-like account that acknowledges the need to rely on useful fictions of origin and order. The tale of Atlantis, a parable about political and moral order, acts as a prelude to this central discussion of cosmic ordering and marks it as an epistemological set piece, a thought experiment not unlike the famous the myth of the cave.[30] Here, unlike the *Republic*'s fierce expulsion of poets from the polis, Socrates embraces the imaginative capacity of fictions to make intelligible complex philosophical frameworks—and Montaigne implicitly draws on this authority for his own essay. At the same time, Timaeus's account of creation by a demiurge, often seen as the classical counterpart to Genesis, gestures to the role of the divine in worldmaking—a matter that Montaigne will address in the "Apologie."

If the Atlantis-frame of the *Timaeus* facilitates reflection on the making of political and cosmic order for Plato's classical audience, the New World, for Montaigne, acts as its modern epistemological counterpart.[31] Like Timaeus's *eikos logos*, the likely stories of America make it integral to the making of a modern world order, which can only emerge through the mixture of new evidence and imaginative reconstruction. The centrality of the Atlantic— the vast expanse beyond the Ptolemaic *oikoumene*—in both the classical and contemporary imaginations signals the essay's anxiety about worldmaking as an exercise in *poiesis*, a powerful and dangerous instance of the human ability to construct narratives, establish them as authoritative, and then use them as a legitimate basis for action, as though they were infallibly true and not carefully made up.

World-Description and Self-Fashioning

Montaigne emphasizes the borders between fictional tales and factual knowledge in his idealizing description of the cannibal culture as transcending even the wildest dreams of philosophers and poets.[32] This influential passage, cited in Shakespeare's *The Tempest* and often identified as the origin of the Enlightenment's noble savage myth, centers on an epistemological concern with experience and imagination. Montaigne claims that the nations of the New World have been seen "par experience," but the elusive, mirage-like quality of his description suggests that the paradisal state of the Tupinamba Indians owes more to the poets' visions than to eyewitness observation. There is a parallel here to the opening digression on the territorial expanse of the New World, which Montaigne locates at the vanishing point between Platonic myth-making and cosmographic world-description, between Atlantean politics and the erosion of the Bordeaux shoreline. Montaigne floods this passage with key epistemological terms to drive home the point: "voyons par experience," "ses inventions à feindre," "qu'il a imaginée" all suggest that the evocation of the cannibals is anything but factual.

Facticity and fabrication in cosmographic reportage is in fact the rhetorical bridge between Montaigne's initial geographic foray and his subsequent focus on the Tupinamba. Through the figure of his servant, who functions as the voice of the world traveler, Montaigne reflects on the workings of potentially unreliable though purportedly eyewitness accounts: [33]

Cet homme que j'avoy, estoit homme simple et grossier, qui est une condition propre à rendre veritable tesmoignage: car les fines gens remarquent bien plus curieusement et plus de choses, mais ils les glosent; et, pour faire valoir leur interpretation et la persuader, ils ne se peuvent garder d'alterer un peu l'Histoire. . . . Ou il faut un homme tres-fidelle, ou si simple qu'il n'ait pas de-quoy bastir et donner de la vray-semblance, à des inventions fauces; et qui n'ait rien espousé. Le mien estoit tel. . . . Ainsi je me contente de cette information, sans m'enquerir de ce que les cosmographes en disent.

Il nous faudroit des topographes qui nous fissent narration particuliere des endroits où ils ont esté. Mais pour avoir cet avantage sur nous d'avoir veu la Palestine, ils veulent jouir de ce privilege de nous conter nouvelles de tout le demeurant du monde. . . . Il entreprendra toutes-fois, pour faire courir ce petit lopin, d'escrire toute la physique.

[This man I had was a simple, crude fellow—a character capable of providing a true account; for clever people observe more things and more curiously, but they interpret them; and to lend weight and conviction to their interpretation, they cannot help altering history a little. . . . We need a man either very honest, or so simple that he has not the stuff to build up false inventions and give them plausibility; and wedded to no theory. Such was my man. . . . So I content myself with his information, without inquiring what the cosmographers say about it.

We ought to have topographers who would give us an exact account of the places where they have been. But because they have over us the advantage of having seen Palestine, they want to enjoy the privilege of telling us news about all the rest of the world. . . . However, to circulate this little scrap of knowledge, he will undertake to write the whole of physics.][34]

Montaigne's fulminations against lying cosmographers are often interpreted as a genuine call for a more careful, ethnographic method of reportage.[35] However, the return of *vraisemblance* and its associated keywords ("veritable tesmoignage," "glosent," "alterer un peu l'Histoire," "inventions fauces," "conter nouvelles") betrays his *own* method as guilty of the very cosmographic poetics that he claims to deplore.[36] The tongue-in-cheek ironies of this passage are unmistakable: Montaigne's insistence that people write only about things of which they have direct personal knowledge vitiates the authority of his own essay, which is based entirely on hearsay and describes a people he has never seen and a territory he has never visited.[37]

This strategic undermining of his own claims illustrates the practical effects of understanding the world as mediated through the self and vice versa. Here, words themselves manifest such mirroring as Montaigne exploits the auditory and visual senses of "histoire." If "histoire" plays on the etymological meaning of *histor* (one who has seen), then the pun on "histoire," picked up by the phrase "conter nouvelles" (telling tales), reminds us that claims to historical truth or fact cannot be extricated from the subjectivity of the reporter.[38] This play on words, which mimics a deeper conjunction between object and representation, foregrounds the essay's central theme: the epistemological problem of knowing and narrating the world, a process that depends on hearsay, mediation, embroidered narratives, and the imaginative conjuring of faraway places. Consequently, acts of reportage—both the servant's account of the New World at the essay's opening and Montaigne's own claim at the essay's end to have spoken to some Amerindians—become framing devices that emphasize the relationship between world-description and individual self-consciousness.[39] An eyewitness description of the whole world is an impossible task, as Montaigne suggests in the discussion of cosmographers and topographers. In order to achieve a narration of the whole, the would-be reporter must transform partial scraps of knowledge into "the whole of physics." We have already seen this dilemma in the process of mapping: the organization of scattered data into a single framework *requires* imaginative reconfigurations, so that the final account is more "false invention" than precise observation. Yet Montaigne introduces a radical twist to this familiar problem. By linking the narration of individual experience (in this case, travel) to the grand project of world description, Montaigne reveals how cosmography and autobiography might be fundamentally related practices.

From this perspective, "Des cannibales" offers a more nuanced view of early modern cannibals, one of the most deeply contested *topoi* in early modern literature.[40] The essay's long opening section, with its insistent emphasis on the fictive, imaginative underpinnings of New World reports, its allusions to Plato's likely stories, and its association of cosmographical and autobiographical narratives, effectively rules out certain interpretations of the Tupinamba. They are, for instance, clearly not to be seen as an idealized, utopian society, nor are they Rousseau's noble savages—Montaigne's irony is clear when he claims that they surpass the golden world of the poets and the philosopher's dreams, because his own cannibals derive, in part, from precisely such visions. More importantly, Montaigne's cosmographical frame suggests that Europe has *created* the cannibals and various New World narratives as insurmountably other. But this seemingly triumphant conquest is, in fact, a deeply deluded autobiographical act.

Montaigne is intensely self-consciousness about the complicity of Europeans in the rhetorical and political fashioning of their cultural others. His anatomization of how representations of the New World were integrated into a Eurocentric world-picture though the weaving together of varied and not always true accounts into a coherent whole effectively unravels the authority of the Eurocentric world. By revealing the teeming multiplicity of other worlds (*autres mondes*) contained beneath the appearance of a single world order, Montaigne questions and displaces authoritative (in this case, Eurocentric) world-theories as he reaches toward global coherence. While this critique remains implicit in "Des cannibales," it will become the subject of "Des coches."

But Montaigne goes even further. He argues that by constructing a quasi-fictional New World, Europe has created a world in order to fashion itself. This claim cuts beyond the familiar observation that Europeans used the New World as a mirror: to recognize in New World narratives an autobiography of Europe is to ask visceral questions about Europe's treatment not only of the Amerindians but of itself. "Des cannibales" may also be a powerful condemnation of the French wars of religion in which religious factions fought to the death for ideological reasons and in which cannibalism may have been practiced.[41] In this view, Tupinamba culture is an inversion of French aristocratic culture, a de-familiarizing device used to comment on European affairs rather than any real engagement with America.

And yet, Montaigne's analysis is more profound than mere nationalistic political critique. The essay does not reduce the New World narrative to a rhetorical ploy in the service of European self-scrutiny. Instead, it suggests that Europe is destroying itself (both in the wars of religion, but also in New World conquests) because it is unable to understand the link between self and world, internal fragmentation and external disorder. Montaigne thus plays a double-edged game: his essay dwells on the essential otherness of the New World, as European rhetoric typically presented it, even as it works to undermine those assumptions of distance and difference. Because Europeans do not understand the futility of drawing absolute ontological distinctions between self and other, New and Old worlds, Catholic and Huguenot, they are locked in violent battles that stem from the desire to assert radical difference where there is none.

This is the thrust of the essay's famous last words: "Tout cela ne va pas trop mal: mais quoy, ils ne portent point de haut de chausses." (All this is not too bad—but what's the use? They don't wear breeches.)[42] The most glaring difference between cannibal and European at first glance—nakedness and clothing—is the most tragic illusion of separation. Violent, unending cycles of conflict arise out of a mistaken belief in the possibility of separating out absolute truth and certain knowledge, of separating self from world. This desire for separation itself follows from an inability to grasp the mutually reinforcing circle formed by the outside world and the self inside. To overcome the barrier of breeches and recognize the self in the world, Montaigne suggests, would be to find a new kind of wholeness that withstands tectonic instability and embraces partial knowledge.

AN APOLOGY FOR WORLDMAKERS

"Des cannibales" sets out Montaigne's double concern with philosophical accounts of the world's origin and the porous boundaries between self and world in practice. The "Apologie de Raymond Sebond" consolidates these themes and tackles their consequences. It unveils a vision of Pyrrhonian worldmaking—an attempt to contain the world's profusion that depends not on system-making and hierarchies of cosmic order but on the expansive—if chaotic—embrace of singularities and experiences through the skeptic's suspension of judgment. For skepticism, observes Merleau-Ponty, has two sides: "It means nothing is true, but also that nothing is false. It rejects *all* opinions and *all* behavior as absurd, but it thereby deprives us of the means of rejecting any one as false."[43] The skeptic's probing toward truth annihilates all dogmatic, partial claims, but the very path cleared by this annihilation opens up an arena of intellectual possibility. The ruthless unmaking of the world in the "Apologie" thus facilitates a tentative remaking.

Montaigne's "Apologie" is often seen as a key text in the transmission of ancient skepticism into modern philosophic thought (and therefore as a crucial predecessor to Descartes). But while its skeptical method has been much admired, its conclusions are frequently dismissed as thoroughly conventional. This is because the essay seems to affirm a (medieval) conviction that knowledge of the world is only possible through divine revelation and the grace of God. The *Essais*, however, are contemporaries of Ortelius's *Theatrum* and Mercator's *Atlas*, books whose confident desire to show forth the world in maps have come to symbolize a scientific modernity. In this context, where ever-popular cosmographic and natural philosophical texts promised to disclose new knowledge as never before, Montaigne's seemingly reactionary position demands greater scrutiny. Placing the "Apologie" within a genealogy of early modern efforts to rethink the world restores its contemporary charge and reveals its surprising innovations. For the central question of the "Apologie" concerns the order of the world: can human beings read the "book of nature" and understand the underlying structure? And if so, how? If we begin with this problem of world order, which is also Montaigne's point of departure, we can begin to uncover his complex engagements with contemporary desires to comprehend the world's entirety.

The "Apologie" is positioned as a defense of the *Theologia naturalis* of Raymond of Sabunde (Raimond Sebond), a Catalan scholar, doctor in medicine and theology, who taught at the University of Toulouse and was its president from 1428 to 1435.[44] One of the central interpretive challenges the essay poses is in fact its relation to Sebond's book, a fifteenth-century encyclopedic compendium that aims to organize every aspect of the natural world according to its rightful order—in other words, a worldmaking text in its own right. Montaigne does not seek to defend the particular world-picture put forth by Sebond; rather, he seeks to defend Sebond against the charge that he is too glib, or too confident in his assertion that human beings can and should read the book of nature and thus acquire knowledge of the world. Montaigne's task therefore is to examine whether human reason is sufficient to know the world, or whether it requires divine aid in the form of faith, revelation, or grace. In the process, he raises fundamental questions about the conjunction between God and world and about the parallel connection between the human self and the world. He interrogates the relationship between the making of the world and knowledge of it, asking if a creature (man) can know the totality of creation (the world).

Montaigne troubles the long-accepted belief in a divinely established world order by wondering if we can ever access knowledge of it. And if we cannot know it, as he argues, is there such a preexisting order waiting to be discovered—or do we simply invent fictions of order for moral and emotional sustenance? The radical nature of skeptical doubting in this essay thus achieves the opposite of its ostensible claim: we cannot even know the world through deity, Montaigne suggests, because God is the ultimate insurmountable Other. Nor can we ever have absolute, determinate knowledge of the world. But what we can have is precious, partial, and hard-won: knowledge of the world through the flawed lens of the self and its astonishing capacity for speculation, a knowing that is grounded on doubt, and allows internal and external spaces to merge and mutually constitute each other.

Once again, cosmographic description, particularly of the New World, is central to the "Apologie." It foregrounds the problem of partial knowledge, the shock of recognizing similitude despite radical difference, and the seductive (if hubristic) desire to grasp a fixed vision of totality. The discovery of America (and Ptolemy's failure to know it) represents the climax in Montaigne's crescendo of examples that demonstrate the shortcomings of human reason. It becomes a lightning rod by which claims to absolute knowledge are judged and found wanting. But, at the same time, the discovery is also a powerful symbol for the promise of new ways of knowing and being in the world. To engage with the New World in theory or practice demanded a reimagining of old structures holding together a coherent vision of "world," as the cosmographers and mapmakers had demonstrated with visual clarity.

As the "Apologie" pounds the reader with examples of human limitation and presumption, deploying all its deconstructive force, it paradoxically draws its energies from a broader cultural desire for cosmic coherence and natural order. This desire is Montaigne's target as well as his inspiration. And it finds an apt emblem in Sebond's *Theologia*, the text he claims to defend. Sebond's book in fact occasioned Montaigne's first published work, a translation that appeared as the *Theologie naturelle* in 1569, more than ten years before the first edition of the *Essais*.[45] Little is known about the actual circumstances surrounding Montaigne's interest in Sebond, and speculation continues about the biographical circumstances that led to the translation. But a consideration of Sebond in historical context suggests other reasons for Montaigne's ongoing engagement with his predecessor's work.[46]

Sebond's encyclopedic catalogue of the divinely ordered natural world was immensely popular in the sixteenth century: there are twenty-five extant manuscripts, thirteen editions in Latin and twelve in various vernaculars (including eight editions of Montaigne's translation) between 1437 and 1648.[47] Among its distinguished readers are Lefèvre d'Etaples, Charles de Bovelles, Marguerite de Valois, even possibly Nicholas of Cusa, Pico della Mirandola, and Pascal.[48] Such wide dissemination signals a pervasive interest in the genre and concerns of Sebond's text: Montaigne may be downplaying the appeal and intellectual thrust of the work in the prefatory letter to his father with which his translation begins. Recent studies have highlighted important connections between Sebond's natural theology and early humanism that show how the *Theologia naturalis* is a characteristic *quattrocento* mix of Christian humanist apologetics. As in Pico della Mirandola's *Oratione* (written almost fifty years later), man is at the center of creation by virtue of his reason, and that reason buttresses the Catholic faith. [49]

The long-cherished characterization of Sebond's *Theologia* as an obsolete scholastic compendium written in turgid Latin thus begins to give way to a view of Sebond as a hesitant early humanist. The *Theologia* may even be a predecessor to the new natural philosophy with its intense focus on the natural world as a better path to an understanding of God's creation than Scripture itself. For this (and other) reasons, Sebond's work is no assertion of theological orthodoxy: it was placed on the Index in 1559, and even a revised decision in 1564 kept the Prologue on the list, requiring the book to be published with suitable emendations.[50] When Montaigne began his translation in the 1560s, this was no neutral, obscure book. Sebond's *Theologia* already offered a powerful response to a shifting world: in a period of intellectual turmoil, as the boundaries of knowledge, faith, and the known world were cast into flux, Sebond's text epitomized an impulse toward reflection and analysis, not unlike Montaigne's own. Situated at the beginning and end of the Renaissance, Sebond and Montaigne may have had much to say to each other. Despite the skeptical gulf which separates their deepest beliefs, Montaigne may have recognized an intellectual kinship.[51] When Sebond claims to free himself from scholastic jargon and write "ex intimo corde" (from the heart), we hear the voice of Montaigne.

Sebond and the Problem of World Order

Montaigne's dedicatory letter to the *Theologie naturelle* reveals a surprising link between his early and later works: it too returns once again to the New World. Montaigne mocks the scholastic Latin of the original and claims to have "taillé et dressé de ma main à Raymond Sebond . . . un accoustrement à la Françoise, et l'ay devestu, autant qu'il a esté en moy, de ce port farouche, et maintien Barbaresque . . . de manière qu'à mon opinion, il a meshuy assez de façon et d'entre-gent pour se présenter en toute bonne compaignie" (tailored and dressed Raymond Sebond by my hand . . . in a French costume, and have stripped him, to the best of my abilities, of his ferocious demeanor and barbarous bearing . . . such that, in my opinion, he has enough fashion and etiquette to be presented into any polite company).[52] Montaigne here describes dressing the savage, Latinate Sebond for polite, civilized French company. But this image, along with the strategic adjective "Barbaresque," metaphorically links Sebond to another *translatio*—that of the barbarous Tupinamba Indians in "Des cannibales."[53] Like the cannibals at Rouen, who don't wear breeches and who can only communicate through an unreliable interpreter, Sebond too needs his translator, Montaigne, who must cover and explain his unintelligible alterity to his modern audience. This metaphoric link between two translations—one across geographic space, the other across historical time—also reveals why Montaigne's engagement with Sebond might be central to the "Apologie," despite a critical tradition that has long considered the two texts to be locked in metaphysical conflict.

The precise appeal of Sebond's work for Montaigne may lie in its hybrid generic form and ambition. The work's original title, following the Toulouse manuscript, is *Scientia libri creaturarum seu naturae et de homine,* while the addition, *Theologia naturalis,* was only added after the second edition (Deventer, 1485). Sebond's work is therefore not only a theological intervention but must also be understood as a *liber creaturarum* or a *liber naturae*—a book of nature. Generically, the *liber creaturarum* belongs to that class of immense encyclopediae and compilations that sought to contain the visible world in language within the covers of a single book.[54] The burgeoning of new knowledge and hitherto unknown phenomena by the sixteenth century, however, rendered such projects quixotic.[55] Yet, such mammoth books offered a necessary reassurance and consolation: the very act of attempting to describe the indescribable world was an act of intellectual agency, a groping toward intelligibility and order. A cross between compendium, cosmography, and theology, the *Theologia* was a work that sought to establish the place of human beings within the cosmic hierarchy at a time when that place and that hierarchy were both in question. Though Sebond wrote sixty odd years before Columbus's explorations, the significance of his moral, physical, and cosmic ordering would resonate even more strongly in the aftermath.

Sebond's *Theologia* also goes beyond these generic comforts. Not only does it provide a clear account of world order according to late scholastic Catholic theology, but as one of the earliest statements of a theosophist natural religion, it consciously breaks down the traditional barriers between philosophy and theology, between the natural and supernatural orders. Like the *Ars magna* of Ramon Llul, the *Theologia naturalis* intends to prove the greatness of God (and the Catholic faith) through the rational demonstration of divine order in the natural world, since the book of nature is considered the incarnate counterpart of the book of sacred scripture. The *liber naturae* is therefore both a *speculum mundi* as well as a *speculum dei*. It offered a powerful exposition of the place of man in the created cosmos, defining a unified system of order within which everything had a determined place. Montaigne would use it as a practical instrument as well as a productive metaphor in "De l'institution des enfans," where he seems to echo Sebond's call to use the world as a textbook. [56]

The *Essais*, however, are situated in counterpoint to these monumental works that seek to mirror cosmic harmony: Montaigne self-consciously chronicles disorder, fragmentation, and multiplicity as the essays themselves seem to resist traditional structures or systems of order. Sebond's *Theologia* thus occupies a fraught position within Montaigne's essay. Its emphatic certainties provide an easy target for the essayist's rage against intellectual absolutism, but its unshakeable faith in the world order it describes simultaneously holds out the consolation of intellectual coherence in a disarrayed world. Montaigne cannot be anything but ambivalent about the cosmic architecture of Sebond's work since his *own* essay is based on, but also runs against, "a schematic tradition that had been holding knowledge together for the greater part of the century."[57] His early *Theologie naturelle,* and its traces in the "Apologie," become both a translation and a displacement of Sebond's *liber naturae* into a different historical moment, and this seemingly anachronistic insertion marks his engagement with a shifting epistemological paradigm.

The "Apologie" opens by situating its writer in a post-Reformation world, one in which ancient beliefs are shaken by "nouvelletez" on all sides. The precarious nature of belief in an age of religious turmoil provides a striking counterpoint to the world of metaphysical certainties that Sebond evokes. But the juxtaposition also throws into stark relief the danger of tying knowledge of the world to matters of theological dispute. This is the unspoken, ironic observation of the essay, which questions religious practice at various points, even as it claims to defend a theologically grounded text: if religious belief is itself unsettled, how can it be the foundation for our knowledge of the world—and vice versa?

One of the prime targets of Montaigne's rage against absolutes is the very notion that "men . . . should seek to bring God down to their measure"—an accusation that inverts in one stroke the anthropocentric Renaissance celebration of man as made in the image of God, destroying the symmetry between *homo faber* and *deus artifex.* Man no longer mirrors God the creator, but is arrogant to even imagine himself as a maker: "L'homme est bien insensé. Il ne sçauroit forger un ciron, et forge des Dieux à douzaines." (Man is certainly crazy. He could not make a mite, and he makes gods by the dozen.)[58] The analogy between human and divine making, so central to early modern representations of the world, is a hubristic deflation of the divine rather than an elevation of the mortal creature.[59] Through a series of quotations from a range of philosophers, Montaigne shows how analyses of the nature of god and the world—the ancient traditions of microcosmic and macrocosmic harmony, the hierarchies of being, the nature of the divine— are all finally driven and mediated by human desires: "Somme le bastiment et le desbastiment, les conditions de la divinité se forgent par l'homme, selon la relation à soy . . . Or donc, par ce mesme trein, pour nous sont les destinées, pour nous le monde; il luit, il tonne pour nous; et le createur et les creatures, tout est pour nous." [The construction and destruction of the deity, and its conditions, are wrought by man on the basis of his relationship to himself. . . . Now then, by this same reasoning, for us are the destinies, for us the world; it shines, it thunders for us; both the creator and the creatures, all are for us.][60] The excruciating irony here is unmistakable. The world and God emanate from the human self and fold back into it.

There is no separating ourselves from our so-called knowledge of the world. Montaigne's logic here seems to condemn Sebond's endeavor rather than to defend it, leaving the worldmaking enterprise in tatters.

This destructive nadir in Montaigne's skeptical attack also hides a promise, one that will hesitantly gain ground toward the end of the "Apologie" and blossom in the final essay, "De l'experience." The very logic that reveals human reasoning about the world as solipsistic and circular demonstrates the power of the human imagination and its capacity for fiction-making that enables us to make sense of the world. Montaigne's critique of vain philosophers thus culminates with an unexpected denunciation of philosophy as "sophisticated poetry": "Tout ainsi que les femmes employent des dents d'yvoire où les leurs naturelles leur manquent . . . ainsi faict la science (et nostre droict mesme a, dict-on, des fictions legitimes sur lesquelles il fonde la verité de sa justice)." [Just as women wear ivory teeth where their natural ones are lacking . . . so does science (and even our law has, they say, legitimate fictions on which it founds the truth of its justice).][61] Underlying this scathing put-down is an etymological play on words: cosmetic and cosmos share a common root that refers to ornament, symmetry, and order, signaling the aesthetic dimensions of epistemic inquiry. It is important to note that Montaigne's rage here is directed not at the act of fiction-making but rather at the easy slippage between fact and fiction; the "legitimate fictions" of the "Apologie" are reminiscent of the problem of *vraisemblance* in "Des cannibales."

Philosophical worldmaking is an exercise of the imagination no different from fashioning immensely persuasive fictions, and all philosophical systems are finally no more than heuristic devices. This epistemic power, even if it is imaginatively constructed, may also be their greatest strength. Toward the final sections of the "Apologie," Montaigne returns to the skeptical paradox by puzzling over whether philosophers such as Epicurus, Plato, and Pythagoras intended us to believe completely in their atoms, ideas, or numbers:

> Je ne me persuade pas aysement qu'Epicurus, Platon et Pythagoras nous ayent donné pour argent contant leurs Atomes, leurs Idées et leurs Nombres. Ils estoient trop sages pour establir leurs articles de foi de chose si incertaine et si debatable. Mais, en cette obscurité et ignorance du monde, chacun de ces grands personages s'est travaillé d'apporter une telle quelle image de lumiere, et ont promené leur ame à des inventions qui eussent au moins une plaisante et subtile apparence: pourveu que, toute fausse, elle se peust maintenir contre les oppositions contraires: *"unicuique ista pro ingenio finguntur, non ex scientiae vi."*

[I cannot easily persuade myself that Epicurus, Plato, and Pythagoras gave us their Atoms, their Ideas, and their Numbers as good coin of the realm. They were too wise to establish their articles of faith on anything so uncertain and so debatable. But into the obscurity and ignorance of this world, each one of those great men labored to bring some semblance of light, such as it was; and they exercised their minds on such conceptions as had at least a pleasant and subtle appearance, provided that, false as they might be, they could hold their own against opposing ideas. *"These are created by each man's imaginative genius, not by the power of his knowledge."*][62]

All Montaigne's reflections on truth and fiction, *tesmoignages vrais* and *tesmoignages fabuleux*, *articles de foi* and *fables*—familiar from "Des cannibales"—coalesce in this passage. We can never know the truth about the world or have absolute knowledge of it, but the philosophers can provide useful and persuasive fictions, which are truth-like (*vraisemblable*), probable, and serve important cultural functions. Knowledge here is in counterpoint to the imagination, which, though false in a metaphysically absolute sense, provides that necessary semblance of light against the darkness of ignorance. Though Montaigne attacks philosophical visions for their inherent fictive falsity, the philosophers' *own* consciousness of their theories as contingent constructs allows him to recuperate their therapeutic aim.[63]

This denunciation and recuperation of philosophy's theories of world order as epistemologically useful but always uncertain, partial, and individually imagined also provides a way through the skeptical bind. By separating out the question of world order from any particular philosophy, and casting doubt on the certainty of any knowledge about it, Montaigne liberates the idea of "world" from its attachment to particular philosophic and theological systems. His analysis suggests that *all* worldmaking attempts are legitimizing fictions that derive from the self and lead back to it—and should be recognized as such. This only becomes suspect if the fiction is confused for the thing itself.

The defense of Sebond is thus a paradoxical one. It depends on revealing the limitations of the *Theologia* as but one among many possible world-pictures. Sebond's *liber naturae* thus becomes a privileged locus for the intersection of skeptical and constructivist currents in Montaigne's thought: it sets out a rational, divinely sanctioned idea of world order whose premises the "Apologie" ruthlessly deconstructs; and yet, Sebond's faith in that order reveals how and why world-systems, even when recognized as contingent, humanly fashioned constructs, can still have meaning.[64] Like Plato's Ideas and Epicurus's Atoms, Sebond's *Theologia* too provides a therapeutic, culturally useful fiction of order. Montaigne's "Apologie" thus fulfills the claims of its title. It is a powerful defense (an *apologia*) of the need for world systems such as Sebond's even as it remains an apology (a regretful excuse) for such works.

Pyrrhonian Worldmaking and the New World

By the 1580s, the well-ordered world of Sebond's *liber creaturarum* may have seemed an *autre monde*, as removed in historical time as the New World was in geographic space. In the "Apologie," this movement from historical to geographic disruption is dramatized as a series of skeptical crises: following a "humanist crisis," where the classical philosophers are at odds with each other, Montaigne moves through the various sciences testing the author- ity of the ancients against the claims of the moderns. This survey moves through a strategic arc that traces the reasons for the breakdown of the world. Beginning with the Copernican revolution in astronomy, Montaigne touches on the Paracelsian challenge in medicine, the "novelties and . . . reforms in physics" (a reference that clearly alludes back to Reformation upheavals), the counterintuitive notions of Euclidean geometry, and finally, the apparent triumph of Ptolemaic geography. As the seeming pinnacle of modern achievement in the post-Columbian era, geography becomes the test of our limited knowledge about the world since it claims to have achieved the elusive goal of comprehensive global knowledge. But the mod- erns' accomplishment, suggests Montaigne, is only an illusion:

> Ptolemeus, qui a esté un grand personnage, avoit establi les bornes de nostre monde. . . . c'eust esté Pyrrhoniser, il y a mille ans, que de mettre en doute la science de la Cosmographie, et les opinions qui en estoient receuës d'un cha- cun; c'estoit heresie d'avouer des Antipodes: voilà de nostre siecle une gran- deur infinie de terre ferme, non pas une isle ou une contrée particuliere, mais une partie esgale à peu pres en grandeur à celle que nous cognoissons, qui vient d'estre descouverte. Les Geographes de ce temps ne faillent pas d'asseurer que meshuy tout est trouvé et que tout est veu. . . . Sçavoir mon, si Ptolomée s'y est trompé autrefois sur les fondemens de sa raison, si ce ne seroit pas sottise de me fier maintenant à ce que ceux cy en disent; et s'il n'est pas plus vray semblable que ce grand corps que nous appellons le monde, est chose bien autre que nous ne jugeons.

> [Ptolemy, who was a great man, had established the limits of our world. . . . It would have been Pyrrhonizing, a thousand years ago, to cast in doubt the sci- ence of cosmography, and the opinions that were accepted about it by one and

all; it was heresy to admit the existence of the Antipodes. Behold in our century an infinite extent of terra firma, not an island or one particular country, but a portion nearly equal in size to the one we know, which has just been discovered. The geographers of the present time do not fail to assure us that now all is discovered and all is seen. . . . The question is, if Ptolemy was once mistaken on the grounds of his reason, whether it would not be stupid for me now to trust to what these people say about it; and whether it is not more likely that this great body that we call the world is something quite different from what we judge.][65]

Cosmographic ambition and the incontrovertible fact of the New World crystallize here in an opposition that demonstrates the skeptical method and culminates in the period's epistemic crisis about the world. What would have been mere "Pyrrhonizing" before Columbus has now seemingly become fact, but that too could prove to be something other than what it appears.

The significance of this passage for the *Essais* as a whole emerges when we notice the carefully constructed parallel to the opening of "Des cannibales," whose main themes it replays. It also looks ahead to the similar moment at the center of "Des coches," establishing the intellectual architecture of the book. Despite its anxieties about representation, the essay on the Tupinamba still uses the imaginative potential of the New World to shape a dialectical vision of the world, one that emerges through the epistemic interactions between Europe and America, topography and cosmography. However, the "Apologie" attacks the possibility of a stable, single notion of "world," mercilessly dismantling the claims of cosmographers, geographers, and world-theorists that "tout est trouvé et que tout est veu." The very discovery of the New World should induce skeptical humility in modern geographers, but this is far from the case. Geographers and cosmographers still cling to a single concept of the whole ("tout"), but Montaigne argues that this certainty is only *one* possible position. Experience teaches us that we do not know enough or know with absolute certainty to make bolder claims.[66]

The experience of the New World as a true counterfactual leads the essayist back to a reflection on philosophical systems, producing a narrative arc that inverts the movement of "Des cannibales," just as the "grandeur infinie de terre ferme" in the "Apologie" alludes back to the "païs infini" of the early essay. Once again, Montaigne returns to classical *topoi* familiar from "Des cannibales"—Plato's *Timaeus*, Solon's journey to Saïs, the tale of Atlantis—but modifies them to emphasize the thematic continuities that bind the trope of the New World to the book's epistemological concerns.[67]

Seeking to explain the world's diversity—emblematized by the novelties of the Americas—Montaigne discusses various theories about the world's origin, eventually juxtaposing the contradictory views of Plato and Epicurus:

> [C] Platon tient qu'il [le monde] change de visage à tout sens. . . . Platon dict que ceux de la ville de Saïs ont des memoires par escrit de huit mille ans, et que la ville d'Athenes fut bastie mille ans avant ladicte ville de Saïs; [B] Epicurus, qu'en mesme temps que les choses sont icy comme nous les voyons, elles sont toutes pareilles, et en mesme façon, en plusieurs autres mondes. *Ce qu'il eust dit plus assuréement, s'il eust veu les similitudes et convenances de ce nouveau monde des Indes occidentales avec le nostre, presant et passé, en si estranges exemples.*

> [Plato holds that it (the world) changes its aspect in all regards. . . . Plato says that the people of the city of Saïs have written memories of eight thousand years, and that city of Athens was built a thousand years before the aforesaid city of Saïs; Epicurus, that while things here are as we have seen them, at the same time they are exactly like and of the same fashion in many other worlds. *Which he would have said with more assurance if he had seen the resemblances and parallels between this new world of the West Indies and our own, present and past, in such strange examples.*] (my emphases)[68]

The intricate analogies of this passage are astonishing—Platonic metaphysics compared to Epicurean physics, both of which are linked to the uncertain ontology and geography of the world in the sixteenth century. Once again, the moment must be measured against "Des cannibales." There, the allusion to the *Timaeus* highlighted the effectiveness of inventive philosophical frameworks as well as the symmetries between political and cosmic order. In the "Apologie," however, Plato's thought experiment becomes but one more example of the random and disorderly "plusieurs autres mondes."[69] This distinction and the strategic merging of Plato and Epicurus, Old and New Worlds, dramatizes a profound clash between ideas of world order evident across various disciplines in early modern Europe. Plato, the classical exemplar of rational order, divinely created cosmic harmony, and the symmetries of microcosm and macrocosm, stands in stark contrast to Epicurus, the standard-bearer for a universe perceived as random, contingent, and ever-changing. [70] By the late sixteenth century, the destabilization of traditional world-pictures and the rediscovery of the Presocratics frequently activated the question posed by the confrontation of Plato and Epicurus: is the fundamental nature of the world rational order or random chance?

The singular authority of classical philosophical systems thus crumbles in the "Apologie." "The world" emerges as an unstable and plural concept, skeptically suspended between the imaginary worlds of elegant Platonic fictions and contingent Epicurean particles. Old and New Worlds collide like the chance collisions of atoms, and their apparent similarity seems to prove the theory of the plurality of worlds. This passage leads to a long inventory of customs and beliefs in the New World that seem to anticipate Christianity and Old World societies, erasing (again) the appearance of radical alterity. Even though "there is nothing in which the world is so varied as in customs and laws," there are strange coincidences between the customs of the barbarous Amerindian natives and the elaborate rituals of post-Reformation Europe.[71] This cosmographic accumulation of anecdotes comparing Old and New World practices of religion finally brings the "Apologie" full circle, returning to the underlying theological context and questions against which the essay began. The reduction of religion to the status of variable "custom" brings Montaigne's skeptical method to bear on the most thorny question of all, raised by the juxtaposition of Plato and Epicurus: what is the relationship between God and world?

Worldly diversity itself provides a clue. Multiplicity, Montaigne argues in his closing peroration, does not conceal an underlying order—it *is* the only order we can know. We can only intuit the existence of God in complete opposition to the fluctuations and ambiguities of the world. The "Apologie" finally establishes God and the world as the only discernible ontological binary:

> Finalement, il n'y a aucune constante existence, ny de nostre estre, ny de celuy des objects. Et nous, et nostre jugement, et toutes choses mortelles, vont coulant et roulant sans cesse . . . toutes choses sont en fluxion, muance et variation perpetuelle . . . qu'il ne se pouvoit trouver une substance mortelle deux fois en mesme estat, car, par soudaineté et legereté de changement, tantost elle dissipe, tantost elle rassemble; elle vient et puis s'en va. . . . Mais qu'est-ce donc qui est veritablement? Ce qui est eternel, c'est à dire qui n'a jamais eu de naissance, ny n'aura jamais fin; à qui le temps n'apporte jamais aucune mutation. . . . Parquoy il faut conclurre que Dieu seul est, non poinct selon aucune mesure du temps, mais selon une eternité immuable et immobile, non mesurée par temps, ny subjecte à aucune declinaison.

> [Finally, there is no existence that is constant, either of our being or of that of objects. And we, and our judgment, and all mortal things go on flowering and rolling unceasingly. . . . All things are in perpetual flux, change, and variation. . . . And that no mortal substance can be found twice in the same state; for by the suddenness and nimbleness of its change, it is now dissipated, now reassembled; it comes and then goes. . . . But then, what really is? That which is

eternal: that is to say, what never had birth, nor will ever have an end; to which time never brings any change. . . . Wherefore we must conclude that God alone *is*—not at all according to any measure of time, but according to an eternity immutable and immobile, not measured by time or subject to any decline.][72]

Montaigne seeks to reconcile the experience of relentless change with the desire for a stable order of things, human agency with divine control. Suspended between Platonic and Epicurean visions of the world, the essayist seeks a way out of the skeptical bind. This conclusion, with its famous fideistic turn, confronts the problem of relentless change with a sudden, climactic appeal to time and eternity where skepticism appears to be countered by a turn to faith. And yet, readers of Montaigne have long been troubled by the unexpectedness and tragic passion of this finale. Does it represent a true expression of faith in God? Is it a duplicitous cover for the philosopher's skeptical debunking of all systems of thought, including religion? Or is there some third way out of the skeptical bind?

Montaigne makes a crucial choice here between two versions of world order, firmly choosing an Epicurean vision of random contingency over any sense of Platonic harmony. Drawing on Plutarch and Lucretius, Montaigne's moving evocation of unceasing cosmic flux underpins an argument for a skeptical approach to human knowledge of the world. Self and world fuse one final time as they mirror each other in variation and relentless change; like all "mortal things," they "go on flowering and rolling unceasingly," both equally subject to mutability and time. What certain knowledge can there be when self and world, observer and object, are so enmeshed and labile?

This fundamental skepticism also underlies Montaigne's "proof" for the existence of God as the final metaphysical other. This idea emerges in response to the Epicurean vision of worldly existence, but it is no consolation. To know God is only to imagine God as a stable, single center outside the moving frame of the world. The mutable world, however, can be known through the ever changing self. To do so is to confront the problems of instability and incertitude with integrity—not to flee them by seeking the fixity of the divine.

THE DEATH OF THE WORLD

As he invokes the Epicurean vision of cosmic change in the final pages of the "Apologie," Montaigne slips in one final quotation. It is taken from the fifth book of Lucretius's *De rerum natura* and was soon to become one of the Renaissance's favorite commonplaces from the epic:

> Mutat enim mundi naturam totius aetas
> Ex alioque alius excipere omnia debet,
> Nec manet ulla sui similis res: omnia migrant
> Omnia commutat natura et vertere cogit.
>
> [For time changes the nature of the whole world,
> and one state of things must pass into another,
> and nothing remains as it was: all things move,
> all are changed by nature and compelled to alter.][73]

Lucretius articulates one of the most contested Epicurean theories — the belief in the decline of this world and the inevitable emergence of new ones — which directly contradicted Platonic and Christian belief in the eternity of God's cosmos.[74] The citation initially amplifies Montaigne's reflection on the fear of mortality and the inevitable passing of human generations. But the cosmological subtext once again connects the individual human life to the life of the world: both are mortal and subject to the forces of time, mutability, and decay. There is little true consolation in the knowledge that new generations and new worlds will simply replace the old.

For Montaigne and other early modern thinkers, Lucretius seemed to articulate their own contemporary experience. The fragmentation of a singular world order by the discovery of the New World and its peoples gave the Lucretian account of cosmic and civilizational cycles new credibility. While Montaigne does not pursue this line of inquiry in the "Apologie," he returns to it at some length in "Des coches." In this final New World essay, the Spanish conquest of the Americas reverses and reenacts the Atlantean invasion of the European mainland. Ironically, this ongoing westward *translatio imperii* confronts its epistemic double as the destruction of the ancient Aztec and Inca empires recalls the burning towers of Ilium.[75] As the rise and fall of worlds in "Des coches" recalls the changing shorelines and sunken cities of "Des cannibales," the apocalypse of Atlantis and Troy mirrors the apocalypse of Cuzco and Tenochtitlán.

The essay's pattern of doubling pairs classical examples of magnificence with New World ones and pits Europe against the alterity of America once more. At its center, many worlds collide, fuse, are created and erased. The effect is one of apocalyptic simultaneity, destruction as well as creation: "Our world has just discovered another world . . . no less great, full, and well-limbed than itself, yet so new and so infantile that it is still being taught

its A B C; not fifty years ago it knew neither letters, nor weights and measures, nor clothes, nor wheat, nor vines. . . . If we are right to infer the end
of our world, and that poet [Lucretius] is right about the youth of his own
age, this other world will only be coming to light when ours is leaving it.
The universe will fall into paralysis; one member will be crippled, the other
full of vigor."[76] These images build on the preamble to "Des cannibales" and
take inspiration from the end of the "Apologie," mimicking the rise and fall
of philosophic world-systems. Here, however, envisioning the death of the
world becomes an indictment of empire and a meditation on the violence of
worldmaking in practice.

In "Des coches," America is framed by the now familiar double allusion
to Plato's Atlantis and Lucretius's Epicurean plural worlds. Verbal echoes of
earlier essays resonate through this passage, from the crescendo of negative
description ("il ne sçavoit ny lettres, ny pois, ny mesure, ny vestements, ny
bleds, ny vignes") reminiscent of the cannibals, to the geographic discovery
that acts as a moral warning for overweening intellectuals and would-be
conquerors in the "Apologie." But now, the descent from wonder into cata-
clysm is swift and unstoppable. And it is expressed through a paraphrase of
De rerum natura. Montaigne returns to the Lucretian citation in the "Apologie" and now completes it:

> namque aliud putrescit et aevo debile languet,
> porro aliud [suc]crescit et [e] contemptibus exit.
> sic igitur mundi naturam totius aetas
> mutat, et ex alio terram status excipit alter . . .

> [For one thing decays and grows faint and weak with age,
> another grows up and comes forth from contempt.
> So therefore time changes the nature of the whole world,
> and one state of the earth gives place to another . . .][77]

The eruption of the New World into European awareness signifies an end-
ing. But in an ironic twist characteristic of the *Essais*, the Lucretian subtext
does not describe the certain death of the Old World of Europe as we might
expect. It foretells the annihilation of the infant world of America. Lucretius's impersonal cosmic cycle is suddenly personified in the form of the
conquistadores who invert the natural progression of things, seize control
with brutal violence, and refuse to cede their place.

Montaigne now offers his own myth of the world's mortality presented
as an account of Aztec cosmology: "Thus they judged as we do, that the
universe was near its end, and they took as a sign of this the desolation that

we brought upon them."[78] What follows is a syncretic account of the Meso-american Legend of the Suns, the story of the creation and destruction of the world through a succession of periods or ages.[79] In Montaigne's version, there are five cosmic epochs, "four [of which] had already run their time," such that Amerindian cosmological history resonates eerily with European cosmogonic narratives:

> The first perished with all other creatures by a universal flood of water. The second by the heavens falling on us, which suffocated every living thing; to which age they assign the giants. . . . The third, by fire, which burned and consumed everything. The fourth, by a turbulence of the air and wind which beat down even many mountains. . . . After the death of the fourth sun, the world was twenty-five years in perpetual darkness, in the fifteenth of which a man and a woman were created who remade the human race; ten years later, on a certain day of their calendar the sun appeared newly created, and since then they reckon their years from that day. The third day after its creation the old gods died; the new ones have been born since little by little. . . .[80]

In this miniature narrative of cosmogony and apocalypse, based on López de Gómara's *Historia general,* Montaigne grafts the Lucretian allusion of worldly decay onto Mexican mythology and defamiliarizes the world known to Europeans by presenting it through the lens of yet another world-theory. There are strong parallels to Plato, Lucretius, and perhaps most importantly, to Genesis, in this passage, as the Aztec myth of the five suns includes visions of elemental destruction, human creation, the death of old gods, and the birth of new ones.[81]

As in "Des cannibales," the myth of the Suns appears just as Montaigne meditates on the radical difference of the Amerindians, but here the allusion erases that alterity. As the Aztec myth develops the Lucretian theme of the age of the world, "Des coches" becomes an elegy not only for a fallen civilization that rivaled classical antiquity but also for a concept of the world that must inevitably die violently with the discovery of a new one. The theological inflections of this passage, its millennarian flavor and intimations of "plusieurs grandes alterations et nouvelletez au monde," remind us that America was not only a geographic location conquered by imperial design, disease, and the modern technologies of guns, maps, and intercontinental trade. Rather, it embodied a symbolic space within which several cosmographic visions competed for explanatory success: Montaigne's digression into Aztec cosmology represents a showdown between Mexican and European *imagines mundi* that compete to gain conceptual control over the world as a whole.[82]

Montaigne's evocation of the Aztec perspective shows how the dream of empire and its apocalypse remained tied to a culture's cosmic myths about the world, a connection also crucial to Spanish colonial ideology. The notorious Requerimiento of 1510, for instance, set out the Genesis narrative from the world's origins to the present day in a form similar to Montaigne's account of the Aztecs' cosmic beliefs.[83] These stories, which acquired the authority of world-theories, served both a consoling and constructive function: they legitimated conquest and rationalized defeat. As the Spanish were well aware, the capture of territory had to be followed by an ideological conquest that reestablished the conceptual foundations of key terms such as "the world." Symbolic spatialities had to be conquered by other symbolic spatialities. The Aztec *cinq soleils* represent the passing of an entire world— the political entity that was the Aztec empire, the concept "world" that was constructed and maintained by its cosmology, but also the pre-Columbian world of the European *mappaemundi*.

Edwin Duval argues that Montaigne's New World essays function as intellectual exercises in the Book of the World, teaching the imagination to recognize its limitations as well as the power of its constructions. By resisting totalizing conceptions of "the world," whether philosophical or political, Montaigne forces us to examine ceaselessly the premises upon which ethical and philosophical authority rests. The oblique manner of narration, with its patchwork of details deftly sewn into a coherent whole, simulates this active process of discovery, so that we *"experience* them through the very process of reading."[84] Reading the *Essais* simulates the experience of imaginative identification, suggesting that comprehension of the world finally depends upon individual empathy, on the ability to identify with another despite seeming diversity and difference. In "Des coches," this takes the form of a final appeal to the reader to absorb the worldview of the Amerindian and to recount it.[85] "Contez, dis-je, aux conquerans cette disparité" (Eliminate this disparity, I say), Montaigne urges as he chronicles the disadvantages of the Indians when faced with the Spanish invaders. The exhortation is both for the *Essais*'s readers in Europe, and through them, for the defeated Amerindians. For it is only through the telling of tales that the imbalance of the encounter may be redressed ("contez" signifies the narration of tales, *contes*, as well as counteraction). As Timothy Hampton points out, the epistemic and ethical trajectory of this command moves from the imaginary to the actual: "It enjoins the reader to move from an imaginary reconsideration of the past, through the viewpoint of the American, toward a speech act . . . that will debunk the self-assurance of the Spaniards."[86]

Narrating the Amerindian perspective becomes a political act of resistance as well as a philosophical act of renewal, undermining the hegemony of the conqueror's worldview through an imaginative identification with the other. This telling and retelling is also a powerful act of worldmaking since the encounter with alterity (on both sides) transforms the self and its conception of "the world" through narration.

Imagining New Worlds

"Nous embrassons tout, mais nous n'étreignons que du vent" (We embrace everything, but we clasp only wind), writes Montaigne in "Des cannibales," speaking of the discovery of the New World. "Des coches" amplifies the echo of that early essay as Montaigne reflects on our doomed desire for total understanding: "Nous n'allons point, nous rodons plustost, et tournoions çà et là. Nous nous promenons sur nos pas. Je crains que nostre cognoissance soit foible en tous sens, nous ne voyons ny gueres loin, ny guerre arriere; elle embrasse peu et vit peu, courte et en estandue de temps et en estandue de matiere." (We do not go in a straight line; we rather ramble, and turn this way and that. We retrace our steps. I fear that our knowledge is weak in every direction; we do not see very far ahead or very far behind. It embraces little and has a short life, short in both extent of time and extent of matter.)[87] In the quest to understand the entire world, human knowledge reaches out across vast temporal and spatial expanses, but its scope is finally limited and small. But this emphasis on limitation, on contingency, on historical recursiveness and random encounter, is countered by the insistent patterns of the *Essais,* which return to the very philosophical frameworks that Montaigne both undermines and exploits. Even this passage on the impossibility of total knowledge is followed by a return to the familiar image of Solon at Saïs and Plato's tale of Atlantis.[88]

Once again, the Platonic *topos* is juxtaposed with an Epicurean image of a plural world in which mind and matter interpenetrate: "And Solon's story of what he had heard from the priests of Egypt . . . does not seem to me to be a testimony to be rejected in this consideration. *If we could view that expanse of countries and ages, boundless in every direction, into which the mind, plunging and spreading itself, travels so far and wide that it can find no limit where it can stop, there would appear in that immensity an infinite capacity to produce innumerable forms.*"[89] Montaigne returns allusively to cosmic concerns, invoking the *Timaeus* and transforming it by juxtaposition with a citation from Cicero's *De natura deorum.* If Solon's visit to the Egyptians suggests a never-ending excavation of knowledge about the world, the Ciceronian allusion suggests instead that an individual mind may acquire and create new knowledge of the world on its own. Faced with new knowledge, the individual mind suddenly finds that conventional understandings of "the world" have been undermined. But in the boundless spaces that have emerged, the mind discerns the pathway to new intellectual coherence through an individual reimagining of the world, and thus, of itself. Montaigne's citation suggests that the very existence of the New World has unlocked the door to an infinite universe.

Even as "Des coches" sings the requiem for a particular world-picture, it announces the birth of a new epistemological consciousness about worldmaking. The spatial, almost cartographic metaphor of the mind's travels reveals an inextricable link between the imagining self and the world which it inhabits but also seeks to define from *within*: the mind only discovers its *own* capacity for "innumerable forms" in the act of imagining the scope of the world. Montaigne here takes a step beyond the epistemological leap associated with sixteenth-century cosmographers. He now unequivocally associates the task of imagining and describing the world with the humanist and moral philosopher's parallel task of imagining and defining the self.

Yet this triumphant assertion of worldmaking powers also brings troubling implications. The image of the mind hurtling through the cosmos is drawn from a famous moment in Cicero's dialogue in which Velleius articulates the atheistic foundations of Epicurean philosophy, rejecting the need for a divine creator, and instead provides a naturalistic explanation for the world's creation: "You would certainly feel no need for [a god's] agency if you had before your eyes the expanse of [a] region, unmeasured and on every side unbounded, upon which the mind may fasten and concentrate itself, and where it may wander far and wide without seeing any farthermost limit upon which to be able to rest."[90] A counterpart to the citations of Lucretius throughout the essay, this allusion underlines the new philosophical climate of the late sixteenth century: for Montaigne and his contemporaries, the final Epicurean vision of a cosmos stripped of divine intelligence, where all coherence is produced by human intellect alone, captured the psychological impact of a changing world-picture.[91] Montaigne, however, strategically modifies key elements in this Epicurean picture of a dynamic, random cosmos: he insists on the cognitive dimension of worldmaking as opposed to the naturalistic basis of the Ciceronian world-picture. If nature produces innumerable worlds in the cosmos, the mind too can produce innumerable worlds in the imagination. To understand worldmaking as an act of imaginative fashioning is thus to recognize a novel means of revising a torn world-picture: a clear concept or structure of "the world" cannot exist independently of human agency.

As cherished systems of world order disintegrated and seemed only to be replaced by a burgeoning collection of strange places, Montaigne's call for establishing intellectual coherence through individual imaginative reconstruction marks a new direction in early modern thought. His replacement of Epicurean atoms with Platonic "forms" in this passage marks it as an intervention in the ancient philosophical debate on the relative importance of chance and order in the universe. But he offers a compromise. By emphasizing the creative, constructive capacity of the human intellect to fashion worlds—a position that balances a desire for world order with a skeptical acknowledgement of the inevitable perspectivism of individual imaginations—Montaigne navigates a third way between two extremes. The double reference to Platonic and Epicurean traditions—one that celebrates philosophic order, the other natural contingency—and Montaigne's simultaneous use and dismissal of them, suggests a dialectical relationship between the two positions as possible responses to the epistemological crisis of world-description. It is a dialectic to which the early moderns would return again and again.

CHAPTER THREE

Cosmic Politics: The Worldly Epics of Camões and Spenser

The year 1580 marked a turning point in the game of international politics for Portugal and England. On August 25, the Habsburg king Philip II of Spain won the decisive Battle of Alcântara against Dom António, the pretender to the Portuguese throne. The rout and sack of Lisbon that followed paved the way for Philip's coronation as King of Portugal the following year, establishing the most extensive empire Europe had ever seen. The Iberian kingdoms and their vast overseas territories in Asia and the Americas were now consolidated under Habsburg dominion. With the Iberian Union, the Portuguese saw a moment of national glory extinguished by a Spanish monarch whose motto declared *non sufficit orbis* (the world is not enough). Entwining his own afterlife with the imperial nation he had celebrated, the epic poet Luís vaz de Camões would lament that he "was content to die not only in it [Portugal] but with it."[1]

Almost exactly one month after the victory at Alcântara, on September 26, a triumphant Francis Drake returned to London having successfully circumnavigated the globe. Laden with gold from the Spanish New World and flushed with the success of superseding Magellan's 1519–22 expedition, Drake's glory seemed to herald the beginning of England's competition with Spain over territories in the Atlantic. In another gesture toward imperial expansion, Elizabeth I had dispatched Lord Grey to Ireland earlier that year on a covert mission against the Spanish—this one would end in the infamous massacre at Smerwick on November 8, 1580. Among the members of Grey's entourage was his new secretary, Edmund Spenser, Anglo-Irish settler and the future epic poet of Elizabethan England.

MAKING EMPIRES

With Portugal defeated and England newly assertive against Spain's imperial fringe, with one epic poet dead and another on the rise—early modern Europe's literary and cultural fortunes paralleled inter-imperial ones.
Desire for global dominion underlay international competition as nascent
nation-states sought to define themselves within and against that newly
emerging entity "the world." Chief among the contenders was Habsburg
Spain, the first global empire, against which all other European states measured their might, and where the emergence of the metropolitan nation
was shaped by colonial experience.[2] In Portugal and England, both ruled
by houses with dynastic ties to Philip II, a sense of vulnerability to Spain's
territorial claims intensified the struggle to delineate national identities and
imperial desires.[3] All three, however, sought to lay claim to a common entity, the world. They also turned to a shared rhetoric to name this desire,
drawn from the ancient Roman *imperium* and its symbolic heirs in Christendom—a language in which nation and world frequently merge, united
by the vision of an empire without end.

This imagined apotheosis finds expression in the Renaissance epic, the
quintessential genre of empire and nationalism. Tracing its own foundations to Virgil's celebration of imperial Rome under Augustus in the *Aeneid*,
the form was fueled anew by the expansionist impulses of early modern European polities. The unifying force of the imperial epic, with its centripetal
and centralizing energies, its teleological drive and its desire to consolidate
a singular, unified nation against the multiplicity of the world beyond, offered a powerful template to Rome's would-be successors.[4] But the imperial epic of the sixteenth century had to go beyond these well-established
generic and political paradigms. If the world was acquiring a new form in
the period, the epic too was acquiring a new subject.

Epitomized in quite different ways by the poems of Camões and Spenser, the post-Columbian sixteenth-century epic confronts a double challenge: not only must the poem articulate a nation's will to empire and
function as a cultural tool to shape national and imperial identity, but it
must also grapple with a shifting world-picture and the destabilization of
the grounds of cosmic authority. The *oikoumene* could no longer be mapped
onto the *imperium* as seamlessly as in the *Aeneid*, where the Roman Empire
seemed to swallow up the known world.[5] In early modern Europe, the
world's horizon was always in view, but was always just out of reach. It
is therefore in the hybrid space of epic—at once fictional and historical—

that we find the sharpest investigations of the imagined relations among nation, empire, and world.

Epic Expansions

In his *Discorsi del poema eroico* (1594), which crystallizes the Renaissance theory and practice of epic, Torquato Tasso reflects on the encyclopedic nature of the genre, which seems uniquely suited to the task of representing the world. Early in his treatise, Tasso describes the sweeping perspective of the poet as it moves from the local to the global, from the national-political to the cosmic-philosophical in scope:

> Poetic matter then seems vast beyond all others, since it embraces things lofty and lowly . . . unfamiliar and familiar, new and old, national and foreign, sacred and secular, civilized and natural, human and divine, so that its boundaries seem to be not the mountains and seas that divide Italy from Spain, nor Taurus, Atlas, Bactra, Thule, nor south, north, east, or west, but heaven and earth. . . . Thus Dante, rising from the center, ascends above all the fixed stars and all the heavenly spheres; and Virgil and Homer describe not only what is below the earth but also what the intellect can barely contemplate. . . . The variety of things they and others who poetized before or afterward created is therefore immense; immense too is the diversity of opinions, or rather the contradiction in judgments, the transformation of languages, customs, laws, rites, republics, kingdoms, emperors, and almost of the world itself, which seems to have changed its face and to present itself to us in another form and another guise.[6]

The catalogue of poetic subjects here has the flavor of a list taken from the pages of Münster or Thevet's *compendiae*, but its effortless merging into a reflection on the changing face of "the world itself" establishes a kinship between epic poetry and cosmography.

By the late sixteenth century, the epic was effectively competing against other genres that aimed to describe the world and stake imperial claims on it. In the wake of the voyages of exploration, cosmographies, collections of travel accounts, atlases, and chronicles all began to occupy the thematic and ideological space long reserved for epic narrative. The epic trope of the heroic voyage, inherited from the *Odyssey*, for instance, became commonplace in travel accounts and shipwreck narratives, while conquests of indigenous peoples were cloaked in the rhetoric of martial valor and chivalric romance.[7] Tasso's own rhetoric echoes some of this generic *contaminatio* and gestures toward the new challenge facing the epic poet: to encompass an expanding world, while realizing the epic's traditional task of celebrating the imperial nation.

Tasso's survey follows the intermingling of poetry and world, tracing a movement from the familiar, political realm of historical action to the cosmic, contemplative realm of speculation and cosmographic meditation. It thus captures a fundamental duality at the heart of the form: if the epic poem is a "little world," as Tasso asserts elsewhere, it must attend to both the (national) part and the (global) whole.

While Tasso powerfully articulates a cosmographic poetics for the late-Renaissance epic, his own epic practice remains carefully within the confines of the pre-Columbian, medieval world of the Crusades. It is Camões (and later Spenser) who first experiments with the epic poem's full cosmographic potential, joining together sustained meditations on national history, empire-building, and cosmological speculation with the descriptive modes of geography and travel narrative. In *Os Lusíadas*, the voyage narrative, a sign of errant romance, becomes the means to epic celebration. Though much criticized for its anti-epic lack of a martial plot or myth of national founding, Camões's poem is an encyclopedia of contemporary modes of worldmaking, which *generates* Portuguese nationhood as a consequence. These modes include the individual travel account, chronicles of national origins and imperial conquests, and, finally, a cosmographic meditation. Scholars have called attention to the seeming tension between the teleologies of epic and voyage in *Os Lusíadas*, pointing to conflicts between the aristocratic martial ethos and bourgeois commercial dealings of the poem's protagonists. But these fissures betray changes to the genre's form and subject matter under pressure from a changing world.[8] Similar tensions are at work in Spenser's epic-romance, *The Faerie Queene*, with its political allegories and cosmic myths, or his pastoral *Colin Clouts Come Home Againe*, where the landscape expands to include a quasi-epic voyage echoing New World themes, the politics of royal patronage, colonial settlement, and a story of cosmological origins that share affinities with the vaster epic. Here, too, an expanding world exerts pressure on the boundaries of genre.

Camões, Spenser, and Worldly Epic

As traditional epic themes of nationhood and empire collide with a new subject—the world—the genre itself evolves new worldmaking strategies. These, in turn, come to redefine the concerns of the imperial nation. The poems of Camões and Spenser offer two revealing case studies of how a national imaginary comes to be shaped in relation to a cosmic whole. Situated at opposite ends of the historical spectrum—Camões's *Os Lusíadas* sings of the zenith of Portuguese imperial power as it was already moving into decline, while Spenser's *Faerie Queene* envisages a glorious imperial future that is not yet in actual view—both poems are marked by a tension between internal, nationalistic concerns and panoramic visions of cosmic order. As they assert their cultural and political capital, both poems invoke the image and discourse of the world, conceived as a totality that supersedes and surpasses the imperial nation. But seen against the vastness of the world, empires, even ones with global reach like Spain's, are revealed to be only partial; at the same time, that enticing expanse fuels the nation's own internal consolidation through the desire for conquest and expansion.

Both poems introduce the world into the texture of their poems by exploiting an opening in the Virgilian epic: they unite the imperial legacy of the *Aeneid* with the didactic, cosmological epic tradition closely associated with Virgil's predecessor Lucretius, whose *De rerum natura* sought to explain the nature of the cosmos. But this combination, already present within the *Aeneid,* is a vexed one. Its intertextual play captures the conceptual tension between the scales of national, imperial, and cosmic authority. If the Virgilian epic unfolds a strongly political vision of worldmaking as the centralizing, annexing drive of empire, the Lucretian epic offers a compelling supra-national alternative. Through a cosmic, universalizing vision across space and time, Lucretius suggests how worldmaking can be subversive and demystifying—all-encompassing in philosophic scope but contingent in its historical expression. The political undertow of *De rerum natura* thus poses a challenge to all forms of institutional authority, whether of religion or the state. In *Os Lusíadas* and *The Faerie Queene,* these two strands emerge most obviously in the set pieces that juxtapose national chronicles of imperial achievement (both actual and imagined) against contemplative visions of the cosmos as a whole, thereby staging a collision between different scales of order. As the recursive, violent history of the nation is reinterpreted in the context of world history and cosmic order, it gains a coherent narrative form that shapes a national consciousness. At the same time, the nation's dependence on the metaphysical authority of cosmic structure highlights its own contingent emergence in time.[9]

The tension between these two modes of epic worldmaking—imperial and cosmic—also emerges in the poems' allegories of gigantomachy, which deconstruct the political and metaphysical authority of the pagan gods and the cultural values that they represent. In a now famous definition, Thomas

Greene described the heroic action of the epic as cosmic politics, invoking a vast landscape that encompassed human and divine matters.[10] By the sixteenth century, however, cosmic politics had taken on a new meaning. Human action no longer reflected a divine plan imposed from above; rather, the human political animal sought cosmic affirmation of his own desires through a reinterpretation of natural signs in the world. This is particularly evident in the problematic status of allegory in both *Os Lusíadas* and *The Faerie Queene*, which repeatedly chronicle the breakdown of a synergy between human and divine, the separation of the numinous from the natural. Camões's sea goddess Tethys reveals the mysteries of the Ptolemaic world to Vasco da Gama, but she also demystifies herself and the divine world she stands for; in Spenser's Mutabilitie Cantos, Nature passes judgment on and rules over the gods before vanishing from view. Both suggest a skeptical questioning of the very foundations of world order and cosmic authority on which, finally, the fate of nations and empires depends.

To understand these narratives of nation and world in the epics of Camões and Spenser as interrelated but fundamentally distinct is to revisit a longstanding truism about early modern epic and empire. This view privileges the perspective of the emergent nation-state and holds that the ideological scope of epic—and the cultural-historical matrix within which it develops—is thoroughly political, concerned with the historical bases of nationhood and with its imperial designs. Worldmaking, from this position, becomes synonymous with European imperial ventures across the globe; any gesture toward the universal is implicated in a hegemonic framework driven by Great Powers.[11] While such a political vision of worldmaking was very much a part of the cultural consciousness of early modern Europe, attention to the cosmographic, Lucretian strain in the post-Columbian epic suggests how focusing on the world, instead of the nation, yields a different set of concerns. It points to the evolution of a new, distinctly modern epic concern about the world, its structure, and its order, one that finds its fullest early modern expression in Milton's *Paradise Lost,* and which continues to fuel the tradition as late as A. R. Ammons's *Sphere: The Form of a Motion* (1974). On a wider cultural plane, this new focus on the world points toward universalizing desires that build on spiritual, philosophical, and naturalistic foundations that transcend local politics and particular national aims. To focus on the epic's concern with the world is, finally, also to understand how the imperial nation itself is called into being by the lure of worldmaking—and also subsumed within that greater whole.

This oppositional tension between nation and world begins to explain the strongly conflicted recent reception of both *Os Lusíadas* and *The Faerie Queene*, which is torn between denunciation and rehabilitation. The two poets' investments in and celebration of oppressive imperial regimes fit uncomfortably with the surprising "other voices" within their poems, because those voices seem to be constantly on the verge of undoing the very ideologies that the poems stridently assert.[12] Both poets were active participants in imperial projects—Camões spent fourteen years in Portuguese Asia as a minor administrator, sailor, and soldier, while Spenser served as a secretary, administrator, and planter in England's efforts to colonize Ireland. Their firsthand experience with the everyday practices of empire inevitably colors their exhortation to conquest, but it also creates a peculiar ambivalence in their representation of subject territories and peoples. If Camões's most eloquent poetry is reserved for the lush beauty of Asia, his empathy distilled into the thundering passion of the giant Adamastor, Spenser remains haunted by the Irish rivers and forests that would become his home, and his most powerful writing celebrates Mutabilitie, the dangerously lovely chthonic figure who seems to erupt out of the Irish landscape. At the same time, there is no equivocating about the violent, ethnocentric attitudes and actions of Camões's Portuguese heroes, or the brutal extermination of political opponents in the notorious fifth book of the *Faerie Queene* (to say nothing of Spenser's *View of the Present State of Ireland*).

Alongside their self-reflexive exposure of imperial aspirations and fervent nationalisms, the poems of Camões and Spenser also repeatedly invoke the Stoic practice of *kataskopos*, the cosmic view from above, which offered a vantage point outside of time and history. In these moments of cosmographic meditation, the expanse of the world provides a Stoic reflection on the pettiness of inter-imperial rivalry, like Ortelius's famous world map with its Ciceronian motto: "What in human affairs can appear great to him who is familiar with all eternity and the expanse of the whole world?"[13] But the practice of *kataskopos* is also double-edged, as the poets want to have it both ways. The very mastery of worldly knowledge that permits the *kataskopos* can become the basis of a claim to global, imperial control. As Camões and Spenser reflect on their own location as exiles writing from the colonial margins (Goa, Mozambique, Ireland), they observe and chronicle the emerging, twinned discourses of modern imperialism and globalism.

CAMONIAN VOYAGES OF DESIRE

Perhaps more than any other Renaissance text, Camões's *Os Lusíadas*, that "most triumphalist of Western epics," dramatizes the imperial dimension of early modern European worldmaking with its unapologetic celebration of the Portuguese empire in Asia.[14] Famously described as "the first epic poem, which in its grandeur and universality speaks for the modern world," the poem tells of Vasco da Gama's voyage to India by rounding the Cape of Good Hope, the inaugural expedition of what would become the *carreira da Índia* (the sea route from Lisbon to Goa).[15] The narrative is framed in accounts of national founding and prophecies of imperial greatness, which unfold against a vast cosmic backdrop. *Os Lusíadas* is in fact marked by a striking generic proximity to the cosmography: the poem's alternating cantos set forth detailed descriptions of Europe (3.6–20), Africa (5.4–15), and India (7.17–22) before culminating in a spectacular vision of the cosmos (10.79–141). Following the physical movement of the navigator-heroes as they leave familiar European waters, round the African coasts, and cross the Indian Ocean, Camões uses the trajectory of the voyage to trace a conceptual circuit across the globe. The crossing of the oceans from Atlantic to Indian is a national victory for Portugal, but more importantly it represents a conceptual breakthrough in global understanding.

This fascination with an emergent understanding of the world's structure becomes the basis for the poem's claim to national exceptionalism. Camões uses this new global knowledge as the foundation of Portuguese glory, defining Portuguese identity by its global understanding rather than any inherent characteristic or claim to an extensive land empire. This explains why the epic's more conventional focus on the history of Portugal (cantos 3–4; briefly reprised by Paulo da Gama in canto 7), and the prophecy of Portuguese conquests in the East (canto 10), are strategically interwoven with the poem's global preoccupations. Though the poem has become entwined with the idea and expression of Portuguese nationhood, *Os Lusíadas* celebrates the birth of the post-Columbian world and the vistas for imagination, travel, conquest, and contemplation that it invited.

Camões's poem illustrates the emergence of a hegemonic Eurocentrism in its expansionary themes, its imperial ideology, its strident tone, and its union of the ambitions of a small, marginal nation with a powerful, all-encompassing vision of the world. It is also an uncomfortable harbinger of the darker legacies of the Enlightenment. Set within this paradigm of an unabashed and predatory global imperialism, it has become something of an embarrassment to liberal readers in a "post-imperial" twenty-first century, who still struggle with its political and intellectual heritage.[16] But the contemporary unease with Camões's poem, generated by recent waves of decolonization and globalization, also suggests that the reception of *Os Lusíadas* is a bellwether for attitudes toward nationalism, imperialism, and globalism. This is why it remains crucial to address these questions in the poem once more.

For Camões's contradictory and self-conscious exposure of both the collusions and cleavages between imperial ideology and worldmaking illuminates the still-linked histories of nation, empire, and world. Its celebration of empire, an alliance between religion, science, and commerce, points to a profound anxiety about the destabilization of world order by the mid-sixteenth century. It thus reveals an important link between the epistemic anxiety of a culture and its ever more strident declarations of imperial dominion.

Cosmos and Imperium

In the final episode of the poem, the sea goddess Tethys sets before Vasco da Gama and his men "um globo . . . uniforme, perfeito" (a globe . . . uniform, perfect, 10.77, 79). She explains: "reduzido / Em pequeno volume, aqui te dou / Do Mundo aos olhos teus, pera que vejas / Por onde vas e irás e o que desejas" (condensed / in this small space, here I give you / the World so you may see / where you are going and will go, / and what you desire).[17] As she offers the Portuguese a miniature scale model (a "trasunto," which evokes the mimetic sense of a copy or reproduction), bestowing the world itself upon the nation, Os Lusíadas seems to have reached its ideological pinnacle. Tethys's gesture invokes the tradition of the *translatio imperii* on a cosmic plane. The conjuring of the globe and the almost ritualistic "aqui te dou / Do Mundo" dramatizes with little subtlety the yearning for a clear transfer of imperial authority that haunted the imagination of early modern European nation-states who all laid claim to the mantle of the Roman Empire. Camões amplifies the imperial dream to reflect worldmaking desire.

But this straightforwardly imperialist gesture is complicated by its setting and intertexts. Tethys's globe comes into view in a field strewn with emeralds and rubies on the notorious Ilha dos Amores, one of the most-discussed and puzzling locales in the epic, a place located outside the historical spaces of the world that the poem almost obsessively surveys. Having sailed to India from Lisbon by the Cape route, the Portuguese mariners are rewarded on their way home with a lush erotic idyll on a floating island somewhere in the Indian Ocean, which Venus has populated with delectable nymphs. It is here, after a hasty collective marriage of the sailors to the licentious maidens and a sumptuous banquet, that the poem reaches its climax in a double prophecy. The first, told by a "bela Ninfa" whose voice resembles an "angélica Sirena" (10.5-6), recounts the efforts of future Portuguese heroes in Asia. This tale of Portuguese expansion is finally followed by Tethys's revelation of the Ptolemaic cosmos, which turns out not to be a transfer of imperial power per se but a revelation of cosmic structure, universal order, and a cosmographic tour of the earth. Though this final canto juxtaposes a celebration of Portuguese imperial achievement against an unfolding of cosmic order, it does so in an exotic, sensual place associated with the excesses of Dido and Carthage—that is, at the very antithesis of the imperial center, a place mythically associated with threats to and a failure of the imperial mission.[18] The Nymph who sings of the Portuguese exploits is linked to Homer's Sirens in the *Odyssey* who lure sailors to their deaths with the promise of knowledge and glory; Odysseus must lash himself to the mast and have his sailors stop their ears with wax to guard against their menace.

Camões's choice to conclude his epic in this manner is thoroughly unorthodox in at least two ways. The episode of sexual abandon and fulfillment on the Ilha dos Amores transforms the traditional failures of romance into the culminating success of the epic: shockingly, unlike almost every other epic poem, there seems to be no moral problem with the heroes' sexual promiscuity. "Nothing bad happens," as it does for instance on the enchanted islands of Circe or Alcina where sexual desire results in a loss of humanity, or at Carthage where Dido's tragic suicide dooms Rome to the Punic Wars.[19] Nor does the Nymph's sirenic voice lure the Portuguese mariners to their doom, though her song is tragic and chronicles national failures as well as successes. Secondly, the poem's concluding scenes of prophecy typically appear in the middle space of epic as proleptic gestures toward a national-imperial history whose scope is beyond the actual events of the poem's action but which the epic legitimizes through its narrative. Instead, by crowning his epic poem with a visionary tour of the world set within a *locus amoenus*, Camões signals a revolutionary change not only in the scope of the imperial epic but also in the interrelated discourses of nation, empire, and world that the genre reflects and sustains. *Os Lusíadas*, these revisions suggest, is more concerned with the foundations of world order and its implications for a nascent imperial nation-state than with the ideology of imperial expansion itself.

The poem's focus on the world—its subordination of the Portuguese *imperium* to the Ptolemaic cosmos in the final canto—exposes and revises the relationship between *cosmos* and *imperium* that Philip Hardie has identified as central to the ideology of Virgil's *Aeneid* and the Augustan Empire.[20] By exploring the connections between these two poles—universal order and particular political dominion—*Os Lusíadas* points to the ways in which questions of empire, and the establishment of nation-states in the early modern period, were transformed by a new apprehension of global space and a parallel desire to uncover a new world order.[21] Historians and political theorists have begun to grapple with the legacy of this shift by charting the complex, reciprocal relationships between colonies and the metropole in the sixteenth century and beyond; Camões's epic offers a contemporary commentary on the transition by engaging with a rich imaginative tradition that sought to make sense of the relations between local and global concerns.

This tradition, which emphasized connections between the state and the natural world, encompassed myth and history, cosmography and scientific treatise, and had a venerable classical pedigree. Stretching from Hesiod's *Theogony* and Ennius's *Epimarchus* to Lucretius's *De rerum natura* and Ovid's *Metamorphoses,* its antecedents also included such philosophical works as Plato's *Republic* (which concludes with the cosmic Myth of Er) and the cosmological-political *Timaeus-Critias-Hermocrates* as well as Cicero's *De re publica* with its cosmographic *Somnium Scipionis,* well known through the commentary of Macrobius.[22] These poetic and philosophical texts frequently found their way into Renaissance cosmographies, taking their place alongside Ptolemy, Pliny, Pomponius Mela, and Strabo, who begins his *Geography* by identifying Homer as the founder of the geographic sciences.[23] When Camões situates his epic at the literary crossroads of these related traditions, he suggests why the epic poem was perhaps the genre best suited to address both the imaginative and ideological impact of the post-Columbian world on traditional notions of nation and empire: it was the genre within which these relations had initially been explored, most famously by Virgil. As Virgil harnessed the power of the cosmological epic derived primarily from Hesiod and Lucretius to the service of empire, Camões revisits this intertextual nexus and reinvents it for his own poem. Cosmological undercurrents enhance the Roman *imperium* in the *Aeneid,* reflecting the appropriation of scientific and sacral motifs by Augustan imperial ideology. But in *Os Lusíadas,* as in other post-Columbian epics, a new understanding of world order (*cosmos*) makes possible a new kind of *imperium.* This new, modern *imperium* is based not on territorial expansion but on knowledge of the natural world—not solely on the subjection of foreign peoples, but on a mastery of the elements. In Camões's hands, the Virgilian subordination of the cosmos to the *imperium* is reversed. Instead, it becomes the source for the Portuguese "sea-borne empire" that inherits and transcends the legacy of Rome.[24]

Canto 10 of *Os Lusíadas* crystallizes these new relationships between cosmos and *imperium*, and between nation and world, through its programmatic rewriting of Virgil. In the *Aeneid*, cosmic exposition provides a background for the central political narrative: in canto 1, the cosmic themes of Iopas's song in Carthage (1.742–46) precede and frame Aeneas's (political) narrative of the fall of Troy (book 2), while Anchises begins his narrative of the future Roman *imperium* in canto 6 with a cosmogonic account.[25] In Camões's poem, however, this narrative trajectory is strategically inverted, as historical, political action leads up to climactic cosmic understanding. The national chronicle of cantos 3 and 4 culminates in the mini-cosmography of canto 5, while in the final canto Tethys takes the role of Iopas as cosmic bard, leaving the Nymph to assume the role of Aeneas/Anchises and narrate the Portuguese imperial future.

Camões's explicit allusion to Virgil's Iopas thus demands attention, for the Carthaginian bard is certainly not the obvious classical precedent for the Nymph's song of Portuguese imperial glory:

> Matéria é de coturno, a não de soco,
> A que a Ninfa aprendeu no imenso lago;
> Qual Iopas não soube, ou Demodoco,
> Entre os Feaces um, outro em Cartago.

> [What she learned there in the ocean depths
> Was in the tragic not the comic mode,
> And not known to Iopas of Carthage
> Nor to Demodocus among the Phaeacians.]

> (10.8.1–4)

Camões's allusion here to Iopas and Demodocus, the Phaeacian singer in the *Odyssey*, takes a familiar form as the poet invokes his classical predecessors only to argue that he is superseding their accomplishment. But the passage also juxtaposes the Nymph's prophetic knowledge acquired from the Ocean itself ("imenso lago") with a distinct tradition of bardic song associated with foreign hospitality and divine revelation. Hardie notes that Virgil's own invocation of Demodocus in the figure of Iopas aligns Dido's bard with a cosmological tradition associated with Homer and Atlas, from whom Iopas has learned his craft (1.741).[26] Atlas, as we have seen, is a favorite Renaissance allegory for human knowledge of the world, and Iopas's song itself is a miniature *De rerum natura*: "Hic canit errantem lunam solisque labores; / unde hominum genus et pecudes; unde imber et ignes . . ." (He sang the straying moon and toiling sun, / The origin of mankind and the beasts, / Of rain and fire . . .).[27] Like Demodocus's song of Ares and Aphrodite, it reveals divine matters usually beyond the bounds of human understanding. Condensed, therefore, into this brief allusion is an indication of Camões's main focus in *Os Lusíadas*—not just the particular imperial conquests of Portugal but the vaster matter of cosmic revelation, the divine unfolding of the world's structure made intelligible to human intellect within the imaginative space of epic fiction. The Nymph's song is therefore just a prelude to the poem's climax, which celebrates Tethys's symbolic gift of knowledge to Gama and his mariners.

Wooing the World

This revision of the epic's emphases from political to cosmic themes highlights a related shift in the poem's gender relations. Instead of overcoming the feminine spaces of romance in the quest for the (masculine) epic mission, the Portuguese heroes' eventual triumph is symbolically marked by women. The llha dos Amores, a Venerean island seemingly populated only by women, is closely associated with Carthage and its female ruler, Dido. This link is established in canto 9, where the sailors' initial actions upon landing (hunting, exploring) allude to the Trojans' arrival at Carthage in book 1 of the *Aeneid,* and is amplified in canto 10 by the explicit allusions to Iopas and Demodocus (10.8).[28] These shifts produce dizzying temporal and spatial distortions. *Os Lusíadas* appears to conclude at the place where the *Aeneid* begins, as though reversing the trajectory and values of Virgil's poem by emphasizing cosmic knowledge over territorial ownership. More importantly, Camões seems to suggest that women (Venus, Tethys, the Nymph) hold the key to this cosmic understanding; to possess the nymphs sexually—to win Carthage—is therefore also to possess the world.

Scholars have reacted to this cluster of associations by suggesting that the episode on the island is no more than an illustration of "willing native girls in a thin mythological disguise," an early version of "Luso-Tropicalismo" or an overheated account of a semi-deliberate Portuguese policy of colonization through intermarriage due to their innate "amorous disposition."[29] But the systematic association of women and worldly knowledge may also signal a profound shift in thinking about the foundations of empire that goes beyond the familiar tropes of rape, miscegenation, and sexual control. For women in *Os Lusíadas* are not associated with indigenous peoples so much as with the power of the natural world itself, with the generativity of the earth, and with what Josiah Blackmore describes as "the fecundity of civilization."[30] Consequently, they also stand for the moral center of the nation (as the inset episode of Inês de Castro suggests) and its successful continuation, which is often under attack from inept or weak male rulers; the sexual idyll on the island, as Anna Klobucka convincingly argues, may also be a call for the young Sebastião to abandon hunting and procreate, that is, to turn from violent pastimes to the business of securing the succession.[31] National concerns, once again, reflect larger cosmic patterns.

Thus, Venus, the tutelary goddess of the Portuguese, recalls not only Virgil's Roman guardian but Lucretius's "alma Venus," "quae quoniam rerum naturam sola gubernas" (since therefore you alone govern the nature of things) (1.2, 21), the goddess who commands the natural world and secures generational transfer through procreation: "per te quoniam genus omne animantum / concipitur" (through you every kind of living thing is conceived, 1.4–5). Indeed, in his 1639 edition of Os Lusíadas, still considered the most learned commentary on the poem, Manuel Faria e Sousa explicitly cites the Lucretian intertexts for Camões's Venus, contrasting them to Virgil.[32] Venus may be the mother of Aeneas and the goddess who helps settle the Trojans in Latium, but in the sensual, elemental, and expansive world of Os Lusíadas she becomes the erotic principle that seems to govern the entire cosmos, from bending the will of Jupiter to taming the fury of the winds and from preventing gigantomachy by petrifying Adamastor to rewarding the heroes with sexual and intellectual rapture. In Os Lusíadas, references to Carthage and Virgil's Lucretian Dido, sensual, thwarted, and in thrall to the passion inflicted by Venus, inform both the fury of Adamastor and the extravagant pleasures of the Ilha dos Amores. The allusions to (Lucretian) female or feminized figures thus become a vehicle for both the transgressive nature and political potential of the Portuguese voyage to India.

Venus and the various nymphs associated with her powers of seduction and generation also shed light on the precise nature of the Portuguese quest and its challenges. To secure the sea route to India, the Portuguese must labor against the elements, as evinced by the poem's central focus on storms at sea—particularly the paired storms and the rounding of the "Cape of Storms" in canto 5. Their success is measured by their mastery of those elements and of the natural world itself—often symbolized by Venus's meteorological interventions in their favor. The Portuguese control of the world—the very basis of their seaborne empire—depends on the elemental knowledge figured by poem's divine women and the cosmic forces at their control rather than on territorial battles that showcase masculine martial valor.

This allegorical feminization of the natural world expands a Lucretian strand in the epic tradition that becomes a familiar trope in sixteenth-century natural philosophy.[33] In cosmographies and philosophical treatises, the Lucretian emphasis on matter and a feminized Nature/Venus invited sexualized descriptions of the scientific process where understanding of the natural world was frequently figured as an erotic unveiling or possession of a female body.[34] The imperial dimensions of these metaphors are certainly well established by the early seventeenth century, when, for instance, Francis Bacon's *Novum organum* (1620) overtly aligns a new empirical understanding of the physical world with the rhetoric of imperial dominance and control. Already by the mid-sixteenth century, however, *Os Lusíadas* draws together similar metaphors that unite natural knowledge, female bodies, and the will to empire.

Camões achieves this conceptual synthesis not only through his allegorical, Lucretian intertexts but also by engaging the well-established tradition of anthropomorphic geography, which we have seen at play in Mercator's *Atlas*.[35] Cantos 3 and 7, which stage recitals of Portuguese history, begin with descriptive surveys of Europe and Asia respectively, and in each case the regions are metaphorically compared to bodies. If Portugal is the head of Europe—"Eis aqui, quase cume de cabeça / De Europa toda, o Reino Lusitano" (And here as is crowning Europe's / Head, is the little Kingdom of Portugal, 3.20)—Calicut is, correspondingly, the head of India—"Calecu tem a ilustre dignidade / De cabeça de Império, rica e bela" (Here of all other towns, Calicut / Is undisputed head, beautiful / And prosperous, 7.22). Camões's deliberate symmetry here once again sets two feminized spaces/bodies against each other, as the female land (*terra*) is the implicit subject of both portraits.[36] The careful geographic surveys enact the movement of a measuring, scientific gaze that has come to be associated with cartographic imperialism, a naming that is itself an act of possession and control.

If the rhetoric of these images gestures toward a nexus between scientific knowledge and empire, particularly to the new cartographic technologies and their ideological freight, the identification of an imperial power as the head of a geographic body looks backward to a classical Roman commonplace: "mea Roma caput orbis terrarum sit" (my Rome is the head of the world), writes Livy in the *History of Rome*.[37] Camões modifies the absolutism of the *caput mundi* trope to mourn the absence of a universal (Christian) empire, lamenting the inter-Christian rivalry that has produced a patchwork of many "empires" across the early modern world.[38] Calicut is the head of the "Império," that is, of the Indian subcontinent, but the seventh canto opens with a stern admonishment to the Portuguese that compares their conquests to those of Germany, England, and France. Portugal itself is but a "pequena parte . . . no mundo" (a small part . . . of the world, 7. 2). How then can a small nation, dwarfed by larger competitors and the expanse of the world itself, achieve imperial renown and defend its claim to be the "cabeça / De Europa" and lords of all the world?

Os Lusíadas compares two different responses to this dilemma, one religious and the other scientific-commercial, each of which outlines a distinct claim on the world.[39] The epic is simultaneously a call to arms for a new global crusade as well as a celebration of a technologically driven trade monopoly that would characterize a range of modern empires, whether Portuguese, English, or Dutch. But Camões does not merely use religion to cloak the bourgeois, commercial core of the poem in the rhetoric of chivalric heroism.[40] The poem reflects two principal modes of envisioning global unity in sixteenth century Europe: a spiritual vision of a globe united under a single faith, and a new, geographic understanding of the terraqueous globe as a unified composite of land and water.[41] The former rested on claims to universal monarchy and empire, drawing on a rich classical and medieval tradition in which the memory of the Roman imperium was grafted onto a dream of global pan-Christian harmony.[42] The expulsion of the Moors, the "discovery" of the Americas and the sea route to Asia, along with related millennarian currents in late fifteenth-century Iberia, revived this dream on a truly global scale, even as the sharpening inter-imperial rivalries of the sixteenth century undercut any real possibility of its revival. But it was the latter cosmographic innovation that enabled the Portuguese empire, which rested on the idea of the world as a maritime entity rather than a topographic one. By the sixteenth century, controlling the world was measured no longer by the acquisition of land but by the control of the oceans and trade routes that were the foundation of a vast global commercial network. In a convenient historical coincidence, such control also dealt a blow to the monopolies held by primarily Muslim traders in the Indian Ocean. Thus, the voyage narrative at the center of the poem can celebrate the technological dominance of the Portuguese, look ahead to their innovative form of commercial empire based on the control of trading routes by sea, and also act symbolically as a crusade, a cosmic victory over the Saracen.

Epistemology and Empire

Os Lusíadas marks an initial and revolutionary moment in the development of a modern, scientifically grounded imperialism that diverged from the traditional Roman model of territorial acquisition. Camões's key innovation on the theme of the *imperium* in *Os Lusíadas* thus rests on a new understanding of the world. He translates the land-based, Christian-Virgillian epic (such as the kind Tasso would write in the *Gerusalemme liberata)*, which unfolded within the enclosed Ptolemaic *orbis terrarum,* into the open, uncertain spaces of the world's oceans, the post-Columbian *mundus.* If, for Virgil, the empire was bounded by the ocean (*Aeneid* 1.286: "imperium oceano"), for Camões, the ocean is the very basis of empire. The "imenso lago" where the Nymph learns of the Portuguese successes in Asia symbolically represents the source and foundation of their power. For Alexander Humboldt, therefore, Camões is first and foremost a "great sea painter"; in his magisterial two-volume *Kosmos* (1846–48), the geographer and naturalist singles out Camões as a scientifically grounded poet, commenting at length on his acute descriptions of physical phenomena, particularly "the never ceasing mutual relations between the air and sea," and his panoramic surveys of the world. "The whole poem," he notes, "contains absolutely no trace of graphical description of the vegetation of the tropics, and its peculiar physiognomy and forms," a characteristic of travel writing and description from Columbus onward.[43] This surprising absence signals the poet's innovative revisions to the genres of epic and cosmography by reorienting them toward the sea itself.

The reader's first glimpse of the Portuguese heroes is, fittingly, on the high seas: "Agora vedes bem que, cometendo / O duvidoso mar num lenho leve, / Por vias nunca usadas, não temendo / De Áfrico e Noto a força, a mais se atreve" (Now you can watch them, risking all / In frail timbers on treacherous seas, / By routes never charted, and only / Emboldened by opposing winds, 1.27). The emphasis on the unknowability and the uncontrollable power of the natural world is paramount here, and over the course of the poem it becomes the undifferentiated, impersonal force against which the Portuguese prove their mettle. Already Camões introduces the root verb that will characterize his protagonists—*atrever-se* (to dare, to venture)—which associates their brand of heroism with boldness, audacity, and insolence (later linked to the *atrevimento* of crossing from the Atlantic into the Indian Ocean). These epithets, associated since antiquity with the female quality of *mêtis* (cunning) rather than the male *bie* (strength), foreground intellectual capabilities rather than physical might and foreshadow the poem's feminization of the epic mission.[44]

But this tableau, where the small but daring national community confronts and withstands the amorphous world figured by the openness of the ocean, also hints at a greater metaphysical challenge. As Bernhard Klein has noted, the epistemology of Camões's poem is in fact closely allied with that of the sea chart used in deep-sea navigation, evoking a mobile, contingent, transitory space and constantly shifting responses to it.[45] The challenge of mastering the sea thus reflects a larger metaphysical problem of confronting the unknown and the ever changing, a perspective captured by the Spanish navigator Martín Cortes, who notes that "[sea voyages] differ from viages by lande in thre thynges. For the lande is fyrme and stedfast. But this [i.e. the sea] is fluxible, wavering, and moveable. That of lande, is knowen and termined by markes, signes and limittes. But this of the Sea, is uncerten and unknowen. And if in viages by lande, there are lylles, mountaynes, rockes and craggie places, the Sea payeth the same even fold with tormentes and tempestes."[46] Camões seems to have almost programmatically incorporated each of these contrasts into his poem, using them to enhance the valor of his protagonists; if the poem's opening stanza celebrates the Portuguese mastery of "mares nunca dantes navegados" (1.1.3), their central antagonist will take the form of "tempestes" at the Cape of Storms. The classical land-based epic is thus no match for the modern maritime one. While the former operates within a preestablished territorial and chivalric system, the latter struggles against elemental uncertainty and flux, seeking to establish dominion over a subject that is constantly changing.[47]

This epistemology of flux is, as we have seen in Montaigne's *Essais*, a characteristic early modern portrayal of and response to a shifting worldpicture. In Camões, as in Montaigne, it finds expression through a Lucretian rhetoric that focuses on contingency and change and that examines the appropriate human response to such mobility and partial knowledge. Camões's adoption of this logic as the basis of a new kind of empire—flexible in ideology, uncertain in its boundaries, overreaching in its desire to grasp the world—is what lends *Os Lusíadas* its peculiar, contradictory voice.

"Todo o mundo é composto de mudança, / Tomando sempre novas qualidades" (The world *is* change, which forever / takes on new qualities), writes Camões in a famous sonnet, evoking the restless movement and mutations that characterize the world in the sixteenth century.[48] Here, the world, like the ocean in *Os Lusíadas*, is unstable and incomprehensible. The sentiment is echoed in "Correm turvas as águas deste rio," for instance, where nothing is as it seems, and the confusion of the world even resists the order of time, seemingly forgotten by God himself: "O mundo, não; mas anda tão confuso, / Que parece que dele Deus se esquece" (not this world, whose chaotic strife / seems almost forgotten by God). These evocations of change and dissolution have long seemed to stand in contrast to the certainties of *Os Lusíadas*. Upon closer inspection, however, they articulate explicitly the epistemic uncertainty that underlies the epic and which accounts for the poem's skeptical undertow. This skeptical voice, which surfaces at unexpected moments (in the speech of Adamastor, the demystification of the pagan gods, the peculiar emphasis on Portuguese failures at home and abroad), has proved a challenge to the poem's readers, who struggle to reconcile it with the overweening self-confidence of the nationalistic bard. Helder Macedo, for instance, notes Camões's skepticism in his lyrics, locating it in the poet's "notion of love as the intelligible form of the unknown . . . the blind guide to somewhere not yet deciphered and from where there is no possible return"; it is this erotic instability, he argues, that also informs the epic.[49] But there may be a different source for both the poet's erotic and epic anxieties: his own prescient understanding of the impact of a new world-picture upon traditional notions of selfhood, nation, and empire.

The final canto, where nation, empire, and world all converge in an eroticized insular space, exposes this epistemic uncertainty and skepticism. For all the harmonious symmetries of cosmic order in Tethys's globe and the Nymph's prophetic assurances of future national greatness, the episode itself takes place in an unstable place—a floating island—associated with imaginative desire and suspect cartographic practices rather than certain, empirical knowledge.[50] And for all its triumphalism, the actual text of Camões's poem betrays a peculiar ambivalence about empire. The word "Império" appears only eight times in the entire poem, and it is not until the very end of the tenth canto that it is used to refer to the Portuguese control of territories in Asia and the Americas (10.62).[51] While the poem's invocation of the Christian "Império" in the second stanza celebrates medieval Christendom's dream of universal empire, Camões never quite argues that the Portuguese have achieved—or will achieve—anything close it. Instead, his use of "Império" to refer to a series of rival kingdoms (the Germanic states, Abyssinia, India) alongside the Portuguese attempt to expand their own sphere of influence evokes the complex political patchwork of the early modern world. The opening declaration of a desire to propagate a universal empire thus becomes both a lens for interpreting the cosmic vision of Tethys and also an ironic reflection on what she actually shows: her expansive vision of the cosmic whole dwarfs the stretches of Portuguese control, which are repeatedly shown to be tenuous and purchased at great human cost.

The Nymph's Anchises-like recital of future Portuguese exploits similarly has a peculiar narrative rhythm of rise and fall. Each hero, from Duarte Pacheco Pereira, "the Portuguese Achilles," is characterized by a tragic end that undercuts his success. Pereira is mistreated by an ungrateful sovereign (10.22–23); Albuquerque, the conqueror of Goa, is "condena" by his cruelty to a compatriot (10.45); Lopo vaz de Sampaio ousts Pedro de Mascarenhas in a competitive power struggle (10.59)—the tale is endless, recursive, violent, and certainly not an inevitable, teleological march toward imperial dominion. These historical details belie a tale of glorious political ascendency, mirroring a similar rhythm in the national historical chronicles of cantos 3 and 4. There too, the history of Portugal is marred by internecine conflict, deceit, and injustice, even as Camões claims to celebrate the glory of the House of Avis. Much to the poet's chagrin, the nation in *Os Lusíadas* struggles to constitute itself.

In contrast to these cycles of national glory and defeat, Camões juxtaposes the perfection of the world as revealed by Tethys. There is an important and conscious slippage in Tethys's narrative between a conception of "world" as the geographic expanse of the earth subject to time and history and "world" as a figure for the entire cosmos understood as existing beyond time and encompassing all space. This slippage emerges semantically as the graphic distinction of capitalization between *mundo* and *Mundo*—earth and universe—but Camões's strategic use of the Latinate word with its more ample reach suggests an important bond between temporal and cosmic scales of order. This conceptual blurring is essential for the epic's thematic unity. For as in other post-Columbian epics, one of the key conceptual tasks facing the poet of *Os Lusíadas*, is reimagining the relationship between the two realms, the earthly and the universal. In Tethys's revelation, Camões thus returns to the Stoic tradition of *kataskopos*, the view from above that formed the basis of the cosmographic meditation. Its grandeur both elevates and diminishes human endeavor, juxtaposing the immensity of the cosmos against its finitude.

Camões's invocation of the geocentric Ptolemaic world in canto 10 has rarely drawn comment, though given the poet's extensive cosmographic and natural philosophical knowledge, it is surprising that he should choose an ancient explanation of world order that was increasingly under attack.[52] A clue that Camões may not simply be reiterating Ptolemaic orthodoxy is his initial description of the cosmic globe: "Vês aqui a grande máquina do Mundo, / Etérea e elemental" (Look here at the great machine of the world, ethereal and elemental, 10.80). Sacrobosco's *De sphaera*, the favorite early modern textbook of cosmology, is usually cited as Camões's source here, marking the entire episode as a throwback to medieval views of the cosmos.[53] But Camões most likely knew Sacrobosco's text in Pedro Nunes's accomplished translation and commentary (1537), which discusses the cosmological problems of the model in careful detail, while making significant corrections to it.[54] And he would have certainly recognized the classical antecedents of the striking phrase "máquina do mundo" in Lucretius and Chalcidius's commentary on the *Timaeus*.[55] More tellingly, the initial view of Tethys's globe is followed by her radical demystification of the pagan gods and rationalization of the zodiac—a gesture associated with the Lucretian use of allegory in *De rerum natura*—to which I will return at the end of this chapter. The cosmic vision signals epistemic crisis even as it appears to resolve it.

Like the disjunctions created by the Ilha dos Amores itself, the Ptolemaic globe thus looks more fraught on closer inspection. Instead of a confidently Eurocentric evocation of cosmic order, it begins to look like the last coherent system available in a climate of intellectual disarray; Nunes's commentary was the last significant revision of the Ptolemaic hypothesis before the advent of Copernicus.[56] *Os Lusíadas* is therefore poised on the cusp of a momentous shift in the image of the world, a shift that it uneasily gestures toward but whose consequences it declines to investigate. From the perspective of the 1570s, the voyages of 1490s seem like a moment of certainty before the storm to come.

SPENSERIAN COSMOGRAPHIES

In the wake of Vesalius and Copernicus, over a half century after the initial New World encounters, the beautiful symmetries of the closed Ptolemaic world were under assault and on the verge of disintegration. Born in 1552, Edmund Spenser would come of age in a world of unprecedented openness but also of stifling political constraint and internal, national turmoil.[57] While his epic ambitions, like Camões's, begin with a desire to celebrate the imperial nation, *The Faerie Queene* must now explicitly confront the skepticism implicit in Camões's national and global juxapositions. Unlike the geographic specificities of *Os Lusíadas*, Spenser's epic takes shape in the fecund space of unknown possibility opened up by the European voyages of exploration. In the Proem to book 2 of his epic-romance, he invokes the discovery of the Americas in order to defend the existence of his fictional land of Faery. If "daily . . . through hardy enterprize, / Many great Regions are discouered, / Which to late age were neuer mentioned," he notes, "Why then should witlesse man so much misween / That nothing is but that which he hath seene?" (2.Proem.2–3).[58] The poet's claim here, which equates the once unknown and now discovered territories of Virginia and Peru with the literary space of Faeryland, has seemed like an elaborate rhetorical joke. But the epistemological insight behind his claim—that fiction and fact seem surprisingly interchangeable in the wake of the New World voyages—moves the epic itself into the arena of philosophic inquiry where slippages between real and imagined spaces can become opportunities for skeptical questioning and intellectual experimentation.

The Proem, then, signals Spenser's sustained engagement with the revitalized contemporary discourses of cartography, cosmography, and natural philosophy. The comparison between the possible discovery of Faeryland and the hitherto unknown New World allows Spenser to include his epic within a distinguished tradition of imaginative mapping which routinely asserted the existence of *terrae nondum cognitae* (lands not yet known), such as the fabled *Terra australis* (the great southern continent), or the mythical Thule to the north. The repeated rhyme, "show / know," foregrounds Spenser's recognition of that fundamental epistemological problem confronting sixteenth-century historians, poets, and mapmakers alike: the difficulty of achieving comprehensive knowledge of the world, most of which remained unknown and unseen. In contrast to Camões's insistence on experience and eyewitness observation, *The Faerie Queene* explores the historical and moral effects of the speculative imagination, which trumps the partial knowledge offered by particular, individual reportage. Where Camões seeks to eradicate doubt even as he acknowledges it, Spenser engages it and exploits it.

Despite the global reach of *The Faerie Queene*, much scholarship has focused on the chorographic instinct of Spenser's poetry and its nationalistic rhetoric. The poet's cosmographic desire to encompass the world—in fact, "worldes"—has received less attention; it remains commonplace to limit the geographic imagination of *The Faerie Queene* to Britain and Ireland, its first colonial conquest.[59] However, in its fluid movement from the territorial specificity of Peru and Virginia to the innovative (and heretical) theories of the cosmos, the Proem to book 2 establishes the vast scope of Spenser's poetic subject. Both the second and fifth books expand the poem's scope beyond the British Isles and onto a world-map extending from the Americas, through continental Europe, and into Persia and Asia. As the poem's grasp exceeds the frontiers of the nation to occupy the spaces where East and West meet, Faeryland—now associated symbolically with both India and America (2.10.72)—forms the metaphorical seam at the margins of the world that makes it a global whole. But Spenser's ambition also thrusts against this territorial specificity and stretches further to take on fundamental questions of cosmology and natural philosophy in the famously difficult Gardens of Adonis and the *Cantos of Mutabilitie*, episodes whose thought experiments look ahead to the cosmic centers of poems such as *Colin Clouts Come Home Againe* and the *Fowre Hymnes*.[60] The Proem to book 2 anticipates this expansiveness. *The Faerie Queene* is certainly concerned with the political worldmaking of monarchs and nations (the images of Amazonian conquest and "fruitfullest Virginia" refer to specific Spanish and English attempts to colonize the New World). But it also takes seriously the possible existence of "other worldes" and thereby considers questions of cosmic order and the nature of things.

The Proem opens two clear trajectories within the poem—what may be understood as two parallel "plot-lines"—that balance and reflect on each other: the relationship between nation and world, and the broader philosophic problem of world order itself. Like *Os Lusíadas*, which wrestles with the hierarchies of nation, empire, and world, searching for the appropriate relation between these scales of order, *The Faerie Queene* continually reflects on the links between political and metaphysical authority. But unlike Camões, who struggles to align local and global perspectives, Spenser decisively shifts his focus. If the 1590 installment of his epic makes conventional, nationalistic claims, by 1596 the still-incomplete *Faerie Queene* decisively transforms the scope of the imperial Roman epic to write the world—instead of merely "writing the nation." In doing so, it reframes narrow questions of political order as vast speculations on the nature and origins of world order. Spenser, like Camões before him, systematically blurs the generic boundaries of epic and romance; but he goes further than Camões, who crowns his poem with a cosmic vision but shies away from the full force of its destabilizing implications. The difference between the two installments of *The Faerie Queene*, in fact, marks a self-conscious "cosmological turn," one further emphasized by that pendant "parcell," the Mutabilitie Cantos, with its allegorical demystification of various scales and hierarchies of order. Taken together, Camões's epic and Spenser's poems capture a broad shift in political thought, from the particular political forms of nationhood to the imagined form of a global system.

On Chronicles: Reimagining Space and Time

The Faerie Queene has the peculiar habit of both collapsing and dilating time and space: the archaic time of Arthur both precedes and coexists with that of Gloriana-Elizabeth, Faeryland is both Britain and the world, the internal dramas of characters are both identical to—and distinct from—the events played out across the poetic landscape. The Allegory of Temperance, for instance, which moves the poem from the eschatological to the worldly realm, intentionally oscillates between allegorical tableaux of the conventional psychomachia (in figures such as Pyrocles, Furor, Occasion) and the historical-geographical events of the post-Columbian world. The Cave of Mammon and the Bower of Bliss are simultaneously allegories of greed and concupiscence, but are also reminiscent of the seductive promises of New World wealth and bliss.[61] Similarly, the Faery and Briton chronicles in the House of Alma set out a dual vision of local and world history, at once national *and* global. This fluidity is underlined by Spenser's strategic omission, during Guyon's Odyssean journey to the Bower of Bliss, of the verses prophesying Columbus's discovery of America in what is otherwise a fairly close imitation of canto 15 of the *Gerusalemme liberata*.[62] Where Tasso clearly demarcates the historic and geographic space of his poem, limiting it to the Ptolemaic *oikoumene* and gesturing prophetically to an as yet undiscovered New World, Spenser deliberately allows medievalizing moral allegory and contemporary travel accounts to merge by assuming as a given both his readers' and characters' existence in a post-Columbian world.

Contrary to the typically specific genealogies and geographies of epic, this historical hybridity in *The Faerie Queene* owes much to Spenser's close engagement with cosmography, but it is frustratingly difficult to set out any clear chronology or map for the events of the poem. In a manner quite unlike Camões's use of similar material, Spenser exploits the cosmographers' practice of mingling myth and history and tracing topical political events to large-scale cosmological processes (variously imagined in classical and/ or Christian terms).[63] Like the narratives of Münster or Apian, Spenser too mixes eyewitness narrative with imaginative speculation in order to bridge the epistemic gulf between the part and the whole. Perhaps the most quixotic endeavor of the Renaissance, the cosmography sought to encompass the world in a book; in its attempt to contain an infinite world in finite language, Spenser found there a resonant model for *The Faerie Queene*.

Spenser anatomizes the faculties involved in such acts of imaginative reconstruction, as well as their cultural, political, and intellectual-historical significance, in the House of Alma, where he describes the functioning of the mind specifically in terms of its relationship with the world.[64] In the three rooms of the Turret, commonly understood to be an allegory for prudence or the temperate mind, Spenser situates "three sages" typically identified as aspects of the sensitive soul—Phantastes (imagination), the unnamed figure "in the middest" (judgment), and Eumenestes (memory). Each is given functions specifically associated with ways of describing or thinking about the world as a whole. Phantastes's room is filled with "Infinite shapes of things dispersed thin; / Some such as in the world were neuer yit, / Ne can deuized be of mortall wit" (2.9.50). Eumenestes records, "As all things else, the which this world doth weld" (2.9.56). Housed between them, the unnamed central figure, usually associated with Spenser himself, suggests how the poet draws on imagination and memory to produce the orderly intellectual constructs that define and describe the world: "*all* artes, *all* science, *all* Philosophy / *all* that in the *world* was aye thought wittily" (2.9.53; my emphases); between the undiscriminating extremes of fictional excess (Phantastes) and factual accumulation (Eumenestes), the Spenserian poet selects, distills, and organizes details into wholes. His room is filled with "all"-encompassing images of intellectual discrimination and synthesis. Accommodating past, present, and future, the turret-brain provides one of the poem's most important self-reflexive figures, revealing the nature of Spenserian *poiesis* and its global subject.[65] Framing the worldmaking poet, Phantastes signifies the crucial (if fallible) openness to new knowledge and the possibility for future discovery, while Eumenestes's "immortall scrine" provides the foundation with which such novelty must be reconciled. It is the play between these two figures, mediated by the poet in the middle, that makes possible the Faery and Briton chronicles which Guyon and Arthur read in Eumenestes's library of "infinite remembrance."

Recent critical interest in theories of the nation and nationalism has redirected attention toward this passage, primarily to point to the chronicles' construction of Britain as an imperial nation and the uses of history in the service of an emergent nationalism.[66] But its immediate context, within an allegory of the mind, suggests a vaster purpose. These paired chronicles, like the twin narratives in the final canto of *Os Lusíadas*, reprise a now familiar *topos*, juxtaposing national history with cosmology. While the *Briton moniments* narrate a tale of national founding and political development, the *Antiquitie of Faerie Lond* presents a cosmogony and its historical consequences. Where the Briton chronicles depict violent patterns of historical repetition and regress—each of the four cycles of monarchy ends with the lack of a male heir—the Faery chronicles celebrate an unbroken line of monarchs, where "succession itself is success."[67] The interrupted and conflict-ridden history of Britain contrasts sharply with the smooth, indeed unmemorable, progression of similarly named kings—a distinction that underlines the narratives' respective emphases on national political struggle and harmonious cosmic process. Both present (imaginary) foundational myths, both connect these fabulous stories of origin to the historical record, and both mirror each other in surprising ways.[68] However, the epistemological scope of each chronicle is fundamentally different: while one focuses on the origin of the British nation, the other provides a uniquely Spenserian myth about the origin of the world.

These contrasts highlight Spenser's engagement with established epic patterns: even as he reflects back on the imperial Virgilian tradition, he points toward cosmic, Lucretian trajectories within his own poem. The chronicles of book 2 look forward to the Trojan and Roman genealogies of book 3 as well as to the poem's syncretic philosophical set-pieces such as the Gardens of Adonis with its allegory of cosmic, cyclical interchange. Spenser, like Camões, returns to the Virgilian drama of cosmos and empire to alter its classical emphases: here again, the order of chronicles is inverted from the *Aeneid*, as the cosmological narrative follows the national history rather than vice versa. It is, once again, a feminized space (the Castle of Alma) that is the site for the exposition of new knowledge, and here too we see the effects of an anthropomorphic geography as Arthur and Guyon find their respective histories within an allegorized human body. Spenser may never have read Camões's poem, but their concerns and allusions touch on remarkably similar themes. But Spenser moves beyond Camões's static juxtaposition of parallel prophecies, innovatively enlarging this ubiquitous epic *topos* into a set of double, allegorically interrelated chronicles that seek to expand the scope of epic itself. In the historically grounded *Briton moniments* and the entirely fictional *Antiquitie of Faerie Lond*, Spenser interrogates the effects of an increasingly sophisticated historiography that was responding to the new spatial challenge of world-description.[69]

At first glance, the contrast between the two chronicles, which exploits the perspectival differences between chorographic and cosmographic narrative, seems to develop a nationalistic poetics in clear opposition to any larger conception of the world.[70] The *Briton moniments* are captivated by a spirit of place and are constantly drawn to comment on specific landmarks that define the nation by linking its present geography to its historical memory.[71] In a narrative crescendo that will culminate with the catalogue of rivers attending the marriage of the Thames and Medway in book 4, Spenser emphasizes particularities of "the land" (the word, which begins the Briton chronicle, appears fourteen times in the tenth canto of the *Briton moniments* alone), clearly identifying the concept of nationhood with territorial specificity.[72] The word "world," however, appears only three times in the course of the entire Briton chronicle, and each time, Spenser uses it in the context of an attack on national sovereignty: it first refers to the invasion of the Huns (2.10.15), then to the internal attack by Lear's daughters (2.10.28), and finally to the marauding Picts (2.10.63). In each instance "the world" stands for a chaotic and unknown expanse beyond the borders of "this land." It is the great geographic other that threatens the autonomy of the nation. Even in the Lear story, "the world," in the rhetoric of the malicious daughter Regan, stands for all the material possessions of mortal existence contrasted with her immaterial filial love; in fact, that love corresponds to a divisive desire for a piece of the kingdom.

This historiographic model, which develops a vision of national integrity as a cultural response to hostile incursions from an uncontrollable external world, has often been identified as one of the oldest and most enduring forms of nationhood.[73] It is the basis not only for early national histories, but for wide-ranging "universal" histories, which despite their professed ambitions, remain concerned with the nation as the core unit of importance. Such a perspective remained bound by Ptolemaic geography and constrained by the availability of mostly local knowledge. The historian Sanjay Subrahmanyam describes the spatial imagination of such a historiography as the contrast between "two complementary zones, the inner core (namely the area to which the author usually belongs) and its outer counterpart."[74] Though Spenser does not explicitly engage with the "complementary zone" in the *Briton moniments*, the sustained land/world dichotomy suggests this older mode, which understands the nation and the external world as counterparts.

And yet, there is also a strong tendency within the Briton chronicle *against* this circumscribed vision of the nation and the world. As historiographical modes changed over the course of the sixteenth century in response to the new geography and the attendant shift in historical consciousness, Spenser and his contemporaries witnessed the writing of a new kind of history in texts such as Raleigh's *History of the World*. These world histories, though clearly drawing on older traditions of universal history and cosmography, differed in crucial ways. First, and most importantly, even *if* national history remained the author's primary interest, it was written from a perspective that recognized the subordination of the nation within that larger geographic entity, the world. Subrahmanyam explains that "world history is based on the recognition of the need for completeness, for full coverage—as it were—of the world." Moreover, world history, he argues, is also distinct in its adherence to a particular set of "aesthetic criteria," that is, it is "accumulative in character, often disordered, and certainly not symmetrical in nature," as the universal histories before it were.[75]

Spenser's cosmogony in the *Antiquitie* reflects the symmetrical ordering of an early univ ·salist history, even as it gropes toward the new model of comprehensive orld history, a model whose aesthetic imprint is also evident in the disorderly recursiveness of the *Briton moniments,* which seeks to be exhaustive rather than elegant (unlike, for instance, Merlin's more traditional chronicle of a providential national history in book 3). These aesthetic criteria signal a more profound engagement with the idea of world history, which transcends, and even undermines, the superficial dichotomy of British land and invading world. Instead, *The Faerie Queene* charts the absorption of nationalistic narratives into an unfolding tale of global scope, investigating causalities and searching for a comprehensive view of the whole.

New Origins for a New World

Spenser's use of the New World is a critical element in his expansion of the epic from a primarily national (and nationalistic) poem into a cosmic, worldmaking poem. As in Montaigne's *Essais,* the rhetorical turn to the New World in *The Faerie Queene* is both a test case for the limitations of human knowledge as well as a metaphor for epistemic possibility. It offers a new way of situating national history within a global context, even as it challenges conventional accounts of the world's creation and invites the poet to craft new tales of cosmic origins. Spenser's allusions to the New World thus exploit the metaphoric elasticity of the phrase, which refers both to the Americas and to a newly expanded understanding of the known world.

In the *Briton moniments,* the founding of Britain is described as a simultaneously ancient and modern event: it draws on the legendary tale of Roman settlement but encases it within a rhetoric of New World discovery. As the chronicle opens, Britain is an as yet undiscovered mass of land at the edge of the world. Far away from the Roman center "amid the Ocean waves," it "was saluage wildernesse, / Vnpeopled, vnmannurd, vnproud, vnpraysd" (2.10.5). The accumulative insistence of negative phrases and negating adjectives in the fifth stanza ("Ne was it Island then, ne was it paysd / . . . ne was it sought / Of merchants farre, for profits therein praysd") recalls Montaigne's description of the Tupinamba in "Des cannibales" and suggests a savage realm outside culture (emphasized by the play on the various senses of the Latin *cultus, colere*). There are obvious parallels here between the ancient Roman imagination of Britain and the contemporary European imagination of America—both Britain and America represent distant, uncivilized New Worlds that need to be conquered, cultivated, and transformed by agents from an established but decadent Old World.[76] Lest this parallel be unclear, "antique times" echoes the "antique history" and "antiquities" of the Proem (2.Proem.1), which has already been used to describe Faeryland *and* the New World. This peculiar fusion of Old and New World histories, which also occurs in *Colin Clouts Come Home Againe,* runs contrary to the medievalizing frame of *The Faerie Queene,* but through it the poem stages a rapprochement between a medieval vision of the nation and an avant-garde historiography for a post-Columbian world.

The following stanza proceeds to describe Brutus, the eponymous legendary founder, as "that venturous Mariner" who first successfully charts a course through "those white rocks . . . / Which all along the Southerne sea-coast lay, / Threatning vnheedy wrecke and rash decay" (2.10.6). The mariner-founder crosses a natural boundary and discovers a new sea route and a new land, which until then did not "deserue a name to haue." The subtle pun here on finding and founding, the quasi-divine act of naming and thus calling into existence a land across the ocean, and the final decision to invade and settle ("Finding in fit ports for fishers trade, / Gan more the same frequent, and further inuade," 2.10.6)—all this cannot but call to mind the voyages of two generations of explorers from Columbus to Drake. Further details in the early history of Britain, as Spenser recounts it, add to this picture. The tale of cannibalistic native giants, for instance, who must be exterminated in order to resettle the land, has recently been identified with the planting of colonies in Ireland. But it equally suggests the cannibalistic Amerindians who later appear in the sixth book of *The Faerie Queene* or Spanish methods of settlement in the New World (supposedly adapted by the English to quell Irish rebellion).[77] Brutus is thus *both* an Aeneas-figure *and* a Columbus-figure; the *translatio imperii* that he effects through the founding of Britain prefigures that other, contemporary westward movement of empire to America even as it reenacts the earlier *translatio* from Troy to Rome.[78]

In a rhetorical sleight-of-hand, Spenser establishes a story of national origins that mythically constructs Britain both as a New Rome and an Old America, once a colony and now an empire about to come into its own. The pattern is further expanded within *The Faerie Queene*: the future beyond the Briton chronicle (which extends from the Roman founding to the time of King Arthur) is completed by Merlin's chronicle (which narrates the monarchs from Arthur to Elizabeth), while its Virgilian past is told by Paridell and Britomart (Aeneas's journey from Troy to Rome). This tripartite narrative draws on Virgil but moves beyond the parochial frame of nationalistic histories. By reframing the myth of national origin, Spenser clearly situates Britain within a world-history and a post-Columbian geography. Britain is in effect the center of the world as Spenser sees it, situated "in the middest" both historically and geographically between classical Rome and modern America. The amplitude of the poet's global imagination as it emerges in these passages evokes what the historian Serge Gruzinski has described as "another modernity," "a state of mind, a sensibility, an experience of the world born from the confrontation between an empire with global ambitions and other societies, other civilizations."[79] Here, Spenser transforms an older concept of nationhood by relocating the peripheral island-nation on a powerfully imagined world scale.[80]

If the *Briton moniments* programmatically sets out a global context for British history, the *Antiquitie* strategically moves away from such a Virgilian strain, turning instead to the mutable forms of Ovidian cosmogony. In the Faery chronicle, the initial moment of founding is a tale not of geographic discovery but of genealogical worldmaking:

> It told, how first *Prometheus* did create
> A man, of many partes from beasts deriued,
> And then stole fire from heauen, to animate
> His worke, for which he was by *Ioue* depriued
> Of life him selfe, and hart-strings of an Ægle riued.
>
> That man so made, he called *Elfe*, to weet
> Quick, the first authour of all Elfin kynd:
> Who wandring through the world with wearie feet,
> Did in the gardins of *Adonis* fynd
> A goodly creature, whom he deemd in mynd
> To be no earthly wight, but either Spright,
> Or Angell, th'authour of all woman kynd;
> Therefore a *Fay* he her according hight,
> Of whom all *Faeryes* spring, and fetch their lignage right.
>
> (2.10.70-71)

The world of Faeryland resembles Ovid's cosmos of mutating forms and also an unfallen Eden where the archetypal male and female, Elfe and Fay, find each other and establish world dominion. This peculiar, radical synthesis, however, highlights how the Faery chronicle is more than an imperial dream of conquest, though it certainly imagines a renewed alignment between cosmos and *imperium*. Its untroubled, well-ordered narrative of Elfin origin and development seems to enact the meaning of *kosmos* (order, harmony), while the dynasty achieves that long dreamed-of mastery of the entire world extending from India in the East to America in the West: "Of these a mightie people shortly grew . . . / And to them selues all Nations did subdew" (2.10.71). But the origin tale, which yokes together a materialist, classical narrative of human creation with the biblical account, suggests that Elfe and his clan represent all humankind rather than any particular nation, race, or ethnicity. The harmonious quality of their unbroken rule becomes a dream of human unity on a global scale in the face of inter-imperial rivalry, political unrest, and fragmentation. Through this lens, Spenser's Faery myth represents a nostalgic desire for order in the face of an increasingly plural world where the whole is never in sight — a point poignantly reinforced by the abrupt truncation of a narrative never completed.

Spenser's tale of world origins is not a cosmological account or a cosmographic meditation, though it points toward both in its allusion to the Gardens of Adonis episode in book 3. A powerful, hybrid cosmological myth, book 3.6 seeks to combine the materialism of Epicurus — with its emphasis on matter, on the chance collision of atoms, and the inevitable flux that governs the universe — and the beautifully ordered symmetries of the Platonic cosmos with its ontological dualism and belief in the Forms.[81] With elegant ambiguity, the Garden presents the interaction of philosophic paradigms in terms of the sexualized union between Venus and Adonis: "she her selfe, when euer that she will, / Possesseth him, and of his sweetnesse takes her fill." Adonis, on the other hand, is made immortal: "All be he subiect to mortalitie, / Yet is eterne in mutabilitie, / And by succession made perpetuall, / Transformed oft, and chaunged diuerslie: / For him the Father of all formes they call; / Therefore needs mote he liue, that liuing giues to all" (3.6.46–47). The basic contours of this sharply gendered myth of cosmic order are already evident in two distinct stories from the Faery chronicle, both emphasizing the materiality of the world and the basis of world order: one recounts Prometheus's creation and its consequences, and the other Elfe's discovery of Fay, "th'authour of all woman kind," and perhaps the principle of generation itself. If Prometheus is an artisan who crafts Elfe from many parts and is himself associated with the ancient chthonic powers of the earth, the location of Fay in the Gardens of Adonis associates her with the Lucretian principles of matter, generation, and the allegorized female body that Spenser celebrates there.

The Lucretian Venus, invoked so resonantly in the *De rerum natura*, appears repeatedly in *The Faerie Queene* as an emblem for cosmic principles traditionally associated with female power. In the Gardens of Adonis, she stands for the dynamic energies of Lucretian materialism—the generative fecundity of matter, the creative energy of chance collision between atoms, the cyclical renewal of change. The Faery chronicle aligns Fay with this ancient figure, suggesting that she too emblematizes the principles of Faeryland as a whole. But the image of the Gardens of Adonis itself occurs in Plato's *Phaedrus*, while other details such as the transmigration of souls draw on the Myth of Er in the *Republic* and the *Timaeus*. Adonis, "the father of all forms," thus figures the Platonic paradigm of artisanal worldmaking that privileges phenomena traditionally coded as masculine: analytic reason, rational order, and the abstraction of Form. The interaction of Elfe and Fay, like that of Venus and Adonis, thus embodies longstanding philosophic conflicts about the nature of the world through the masculine epic's familiar generic encounter with the female space of romance. Like Camões's use of similar material, Spenser uses the older form of cosmological, Lucretian epic to reorient the direction of his own poem toward a cosmic scale, as the errant Elfe finds his dynastic, imperial, and genealogical success in the sexual fulfillment and cosmological knowledge figured by Fay in the Garden.

CLASH OF THE TITANS

"What is authority?" asks Hannah Arendt in a famous essay, proceeding to offer an answer that connects a cosmic sense of metaphysical disarray to the profound failure of political systems in the early twentieth century.[82] But the question itself, suggests Arendt, is a misnomer: we are tempted to raise it precisely because authority seems to have vanished from the modern world. "Most will agree," she argues, "that a constant, ever-widening and deepening crisis of authority has accompanied the development of the modern world in our century."[83]

Though it has its origin in the world she saw after the Second World War, Arendt's insight into the link between a perceived loss of authority and the emergence of a "modern world" is telling for the early modern period too. It suggests how modernity and the world order associated with it emerge with a double sense of novelty and loss—sentiments that permeate both Camões and Spenser's epic responses to the events of the late sixteenth century. Both *Os Lusíadas* and *The Faerie Queene* grapple with the relationship between political and metaphysical order and reflect on their own emergent modernity. But in their attempts to construct robust poetic fictions that assert national and cosmic foundations, they also register the effects of "a crisis of authority," a sense of a world in flux. Within the Renaissance epic, the most resonant trope for this crisis is the gigantomachy.

An ancient symbol of the clash between the forces of order and chaos for control of the universe, the battle of the sky-dwelling Olympian Gods and the earth-born Titans or Giants simultaneously signaled political conflict and philosophical dispute at a cosmic plane. Familiar to early modern writers from such sources as Hesiod's *Theogony* and Ovid's *Metamorphoses*, allusions to gigantomachy are integral to Virgil's imperial narrative in the *Aeneid*, which celebrates "the attempt by the forces of Rome, of order, of civilization, to defeat the forces of barbarism and chaos."[84] Gigantomachy is a staple of medieval and Renaissance epic as well, appearing with reliable frequency from Dante to Milton, usually to assert the triumph of institutional authority and order over plural, disorderly rebellion.

But the trope also has a distinguished literary history in the philosophical tradition, appearing in Plato's *Sophist* to symbolize the clash between materialist, Presocratic philosophies (giants) and the rational Platonic theory of forms (gods). This pattern is reproduced in Aristotle but finds a striking reversal in Lucretius's *De rerum natura*, where the myth of gigantomachy paradoxically stands for the Epicurean resistance (giants) to the fixed, obscurantist nexus between religion and philosophy that upholds the state (gods).[85] Thus, through a counterintuitive revision, Lucretius uses the myth to figure the demystification of pagan mythology itself, as the rational, materialist, and euhemerist explanations of the Epicureans threaten to displace belief in the supernatural.

These perspectives converge in the epics of Camões and Spenser, who amplify the cosmic dimensions of the gigantomachy to interrogate the relations between various scales of order, from the ethics of individual action to national policy and universal order. If Virgil uses the cosmic implications of gigantomachy to buttress the political, imperial claims of the Roman state, the subversive use of the trope in Lucretius suggests how gigantomachy could also celebrate resistance and transformation, ushering in a new vision of cosmic order contrary to the status quo.

Demystification and Modernity: Adamastor's Promise

The most celebrated episode in *Os Lusíadas* is its variation on the theme of gigantomachy. In the central confrontation between the giant Adamastor and Vasco da Gama, Camões stages an encounter between old and new forms of worldly knowledge and control.[86] An autochthonous son of the earth, Adamastor ("the untamed one") is allied with his fellow giants in the war against the gods, when his passion for the nymph Thetis proves to be his undoing: he turns into the landmass of the Cape of Good Hope, whose stormy squalls reflect his anger and unrequited passion. Camões's myth-making here opens into a rich range of signification. Adamastor stands for the African natives oppressed by the Portuguese, for the untamed African continent as it resists European onslaughts, and for the hubris and pride of the Portuguese mariners who transgress the boundaries of the known world.[87] A conflation of Homer's Polyphemos and Virgil's Dido, Adamastor gives voice in his curse to the desire for retributive justice against the victors of epic. But in his Ovidian metamorphosis from giant to cape, the monstrous figure also points to the naturalization and demystification of ancient knowledge, symbolized here by epic *topoi* themselves.

Adamastor thus elucidates both aspects of gigantomachy: the imperial Virgilian celebration of order and civilization against barbarism and disorder, and the Lucretian menace of ideological destabilization against an *ancien régime*. The former reading has become a commonplace understanding of the episode, which self-consciously mythologizes the confrontations between colonizer and colonized even as it acts to efface the voices of resistance from the imperial agenda. But the secondary function of gigantomachy here, long neglected, bears further analysis, for Camões's Adamastor has an unlikely Lucretian intertext as well.

As a mythologized version of the waterspout described so vividly at the beginning of the fifth canto, the earth-born giant is linked to the forces of the natural world—waterspouts, whirlwinds, tempests—which the Portuguese must master. While Camões's description of these forces, particularly his eyewitness account of the waterspout (5.19–23), has been celebrated as anticipating the demystifying gaze of the New Science, that description is also indebted to the sixth book of the *De rerum natura*, where Lucretius had already offered a naturalistic explanation of the phenomenon at some length.[88] Camões, in fact, follows Lucretius ("tamquam demissa columna") in describing the waterspout as "Tal a grande coluna, enchendo, aumenta / A si e a nuvem negra que sustenta" (5.21); even his famous leech simile might borrow its climactic motion ("se enche e se alarga grandemente") from a similar undulation in Lucretius ("trudatur et extendatur in undas").[89] To discern this Lucretian intertext, however, is to enter a Camonian moment of irony and skeptical ambivalence. For Camões's description of the waterspout frames a claim for the discovery of new knowledge and, therefore, for the achievement of the moderns over the ancients; it exalts the Portuguese mariners' (and poet's) daring attempt to surpass their classical fathers and (literally) chart new territory. "Se os antigos Filósofos, que andaram / Tantas terras, por ver segredos delas . . . / Que grandes escrituras que deixaram!" (If philosophers of old, who visited / So many lands to study their secrets . . . / What great writings they would have left us!) the poet writes, noting that only then would their work (like his) be "tudo, sem mentir, puras verdades" (all pure truths, without lying, 5.23). However, this claim to express an unvarnished, empirical truth that has been stripped of mythic accretion, hearsay, and superstitious belief is itself a frequent Lucretian refrain, one which in fact occasions the Roman poet's digression on waterspouts (6.424ff). Camões's use of Lucretius in defense of the moderns here serves multiple functions: it explains the natural phenomenon upon which Adamastor is based by drawing upon a mix of ancient text and modern observation, and it connects Camões and the Portuguese explorers to a long tradition of epistemological questing and intellectual demystification linked to *De rerum natura*. The gigantomachic confrontation between Adamastor and Gama thus comes to represent the clash between two modes of description (empirical, mythic) and two paradigms of explanation about the nature of the world. The giant represents the immense forces of the earth that the Portuguese must suppress and control, but paradoxically he is also a figure for an older, sedimented intellectual order, which the Portuguese—as new giants—must rise up to overthrow.

In a powerful analysis of the episode, David Quint argues that when the myth of gigantomachy here "is read as an allegory of the moderns' attempt to outdo the ancients and overthrow their authority, the Portuguese may claim victory where the mythical giants met defeat."[90] But Camões's seeming radicalism has a source and a trajectory. It evokes the Lucretian inversion of the gigantomachic myth, in which the rationalizing Epicurean giants, in their desire to understand the natural world and free themselves from the bonds of *religio*, seek to overthrow the mythic obfuscations of the Roman gods.[91] The *atrevimento* of the Portuguese—their transgressive crossing of intellectual and political frontiers, their assault on old forms of authority with new forms of understanding—is allied with the intellectual upheaval of Lucretius's giants. In this daring and equivocal move, shocking in its association of the Portuguese with the destabilizing effects of the "new" philosophies, the connection to Epicurean rationalization betrays an anxiety about the movement into a modern world.

To see the Portuguese in *Os Lusíadas* as the new giants winning a historic struggle over the bases of cosmic authority is also to understand why Adamastor's losses, both political and erotic, become Gama's successes within the economy of the poem. Several scholars have observed that the Adamastor episode is closely connected to the final cantos on the Ilha dos Amores.[92] Both offer mythic representations of colonizing strategies, but both also dramatize how the struggle for knowledge and control of the world in the sixteenth century was simultaneously political and metaphysical. In this context, Tethys's demystification of the pagan gods at the beginning of her cosmographic description ("fomos fabulosos, / Fingidos de mortal e cego engano") reflects a similar separation between orders of explanation emblematized by the waterspout of canto 5. Her dismissal of the pagan gods as convenient poetic devices may also owe something to Lucretius's dismissal of the Roman gods as mere allegories, though the preferred (new) order that follows for Camões is Christian rather than Epicurean.[93] And so, as Adamastor loses a nymph named Thetis whose body contains the promise of a new world order, Gama gains the ocean goddess Tethys in marriage, and with her, a glimpse of the cosmos in its entirety.[94]

This brutal symmetry, which is somewhat undermined by the poem's sympathy for Adamastor's loss and suffering, depicts the violent passage from one vision of world order to another, a passage that is both triumphant and shot through with anger and mourning. Camões's love for the ancient genre of epic itself, with its classical myths and allegorical retellings of historical events, collides with a modern world, skeptical of the utility of fictions and marvels. Fittingly, in *Os Lusíadas*, myth itself signals the transition from an older unity to the new, fractured vision of the whole.

Egalitarian Giants: Mutabilitie and Cosmic Desire

Chthonic giants are almost ubiquitous in *The Faerie Queene*. From book 1's Orgoglio, "a hideous Geant" born of earth and air, to the Egalitarian Giant of book 5, who measures with the cosmic elements, these gigantomachic figures are frequently used to connect the knights' immediate missions to larger cosmic concerns. This pattern reaches a climax in the *Cantos of Mutabilitie*, which thematize the gigantomachy, and by doing so crystallize the epic's exploration of political authority, cosmic process, and the emergence of a modern world.

"World" appears more times in the *Cantos of Mutabilitie* than anywhere else in *The Faerie Queene*, indicating a culmination of the poet's favorite subject.[95] The Titaness Mutabilitie's challenge to Olympian Jove and her demand for "the worlds whole souerainty" (7.7.16) signals a reexamination of the assumptions underlying the current order of the world. Here, Spenser addresses the philosophical problem of cosmic order and world-making through a myth about the struggle for political order and world dominion. The language of political struggle becomes a potent metaphor for the broader philosophical struggle to identify the laws of the cosmos; phrases such as "rule and dominion" (7.6.3) refer to the political authority of the two mythic figures, even as they stand for the force of cosmic dynamics. Condensed into this confrontation are also several other instances of the battle for world dominion: Jove's mythic war with Saturn and the Titans; the cosmic war between the principles of order and chaos, stability and change; the philosophic conflict between Lucretian Epicureanism and Christian Neoplatonism; and the gender battle between the female power of matter and the male principle of form.

Spenser repeatedly associates Mutabilitie with the Lucretian force of cosmic flux. But this heritage also explains the moral direction of her assault. Like Epicurus in *De rerum natura*, Mutabilitie is a Giant who dares to challenge the status quo, to lift mortal eyes up to challenge the might of the heavens, braving Jove's thunderbolt and vigorously questioning the myths of power which grant him dominion over the world. She too is a transgressor of boundaries: "all their statutes burst: / And all the worlds faire frame (which none yet durst / Of Gods or men to alter or misguide) / She alter'd quite" (7.6.5). The image of the "worlds fair frame" may in fact allude to the Lucretian *moenia mundi,* and the change she brings is distinctly Epicurean; it denies the eternity of the world ("good estate" and "happy state") to affirm cosmic decay and human mortality.

Mutabilitie's power also seems to evoke the Fall from an original paradise or Edenic golden age, revealing her to be an atheistic challenger to the narrative of Christian providential grace, much like Epicurus who challenges the idea of the Olympian gods.[96] Her plea to Nature begins with a subtle debunking of godhead as a matter of mere appearance: "And, gods no more then men thou doest esteeme: / For, euen the gods to thee, as men to gods do seeme" (7.7.15). In a move that echoes the opening book of *De rerum natura*, Mutabilitie, after explaining the composition of the world from the four elements, simply dismisses the Olympian gods—"How-euer these, that Gods themselues do call" (7.7.26)—as no more than personified aspects of the natural world.[97] This turn toward allegorical demystification is similar to Tethys's dismissal in *Os Lusíadas* and points to a strategic substitution of causal cosmological narrative for classical myth in the *Cantos,* marking them as self-consciously modern. [98]

The fear Mutabilitie inspires as well as her prodigious cosmic reach give allegorical form to an anxiety about the grim state of the contemporary world evident throughout *The Faerie Queene*'s fifth and sixth books. The epic failures with which these books conclude—Artegall's fall from favor and the eventual escape of the Blatant Beast—reflect a profound sense of dis-enchantment and disillusion, a crisis about where the centers of authority may now lie. If books 5 and 6 confront the political dimension of the issue with increasingly brutal, topical allegories, the *Cantos* return to the cosmic foundation: Mutabilitie now becomes responsible for the cosmic disorder in the Proem to book 5 and mirrors the rebellion of the Egalitarian Giant on a much grander scale.

Artegall's encounter with the Giant of the scales pits the knight of Jus-tice, the upholder of the political status quo, against another chthonic fig-ure who seeks to disrupt the established order by reevaluating the original creation of the world from its constituent elements. The Giant's political desire to "reduce" all things "vnto equality" (5.2.32) finds its expression in a cosmological experiment that literally shakes the foundations of the world as he threatens "all the earth [to] uptake." Artegall treats the Giant as though he were a Mutabilitie-in-the-making since he threatens to bring change where there has been none ("And mongst them al no change hath yet beene found . . . / All change is perillous, and all chaunce vnsound"). Both introduce transience into the world and thereby shatter the stability of an established world order that is ostensibly unshakeable. If the Egalitarian Giant wants to "weigh the world anew," according to Artegall, Mutabilitie actually alters "the worlds fair frame."[99] Moreover, the Giant's association with Ireland ties him ever more closely to the *Cantos* with their Irish setting and raises questions about the hierarchies of colonization within the moral economy of the cosmos.

Both Mutabilitie and the Egalitarian Giant appear to suffer the tradi-tional fate of the giants who rose up against the Olympians. But Spenser, like Lucretius, subtly reverses the moral implications of their uprising. Re-cent critics of the epic have been sympathetic to both giants' arguments and several suggest that Spenser's own sympathies may lie with Mutabilitie and the Giant rather than with Artegall and Jove.[100] It is the Giant who acknowl-edges "how badly all things present bee, / And each estate quite out of order go'th" and offers, as a remedy, an egalitarian ethics which may be the politi-cal corollary of Mutabilitie's ever changing view of cosmic flux: "Were it not good that wrong were then surceast, / And from the most, that some were giuen to the least?" (5.2.37). Consequently, natural processes of material transformation—the erosion of mountains and the overthrow of rocks into the "deepest maine"—come to symbolize necessary political revolutions: "Tyrants that make men subiect to their law, / I will suppresse, that they no more may raine; / And Lordings curbe, that commons ouer-aw; / And all the wealth of rich men to the poore will draw" (5.2.38). But this redistribution of power and wealth has a double function. While it registers real social misery and political unrest in England, it also suggests that change is necessary and perhaps inevitable. Such redistribution may in fact correspond to the natural cycles of elemental redistribution which Mutabilitie claims as the source of her authority and which she will celebrate in her pageant. Like Lucretius's Epicurean Giants, both of Spenser's giants initiate philosophic revolutions by speaking truth to power. Their rebellion represents a new kind of heroism that will pave the way to a new world order.

The fates of the two giants, however, could not be more different. In the political context of book 5, Artegall has little patience for such an assault on the state's authority, and his praise of the divinely created hierarchical cosmos as evidence against the Giant's claim feels distinctly defensive and uneasy. As if to acknowledge that this vision of a transcendental universe is lacking in persuasive force, Talus steps in to eliminate the naysayer with brutal efficiency. This destruction of the Giant, however, suggests that the creature's reasons are both convincing and subversive: only violent force can eradicate the danger he poses to the Elizabethan status quo.[101] The *Cantos*, however, move away from this politically expedient solution by replacing executive violence with a judicial hearing presided over by Dame Nature, a figure who represents the entire material world, as Pliny notes in the *Natural History*.[102]

Nature herself, as Anne Prescott has noted, is a "super-giantess" and thus materially linked to Mutabilitie, matter, and mortality.[103] A version of the earth, of the Magna Mater, and thus of the Lucretian Venus Genetrix, Nature is also genealogically linked to Mutabilitie, whom she calls her daughter. Though Nature's verdict is frequently read as a denial of Mutabilitie's suit, it achieves a paradoxical affirmation of the cosmic cycle and the ubiquity of transformation:

> . . . all things stedfastnes doe hate
> And changed be: yet being rightly wayd
> They are not changed from their first estate;
> But by their change their being doe dilate:

> And turning to themselues at length againe,
> Doe worke their owne perfection so by fate:
> Then ouer them Change doth not rule and raigne;
> But they raigne ouer change, and doe their states maintaine.
>
> (7.7.58)

Far from ruling against Mutabilitie, Nature naturalizes change, defining it as the very nature of things. She thus eases our passage into a modern world where knowledge is only partial and that harmonious view of the whole is always just out of reach.[104]

Promethean Worldmaking in Camões and Spenser

While Camões can conclude his epic with a glorious vision of a still whole world, such an apotheosis is not possible for Spenser. Instead, *The Faerie Queene* affirms its faith in human worldmaking, symbolized by Prometheus. Invoked as the creator of Elfe in book 2, Prometheus is also connected to Mutabilitie, as Jove intuitively understands (7.6.29); indeed, Spenser claims to have found her story, "registred of old, / In *Faery* land mongst records permanent" (7.6.2), a gesture that connects the *Cantos* to book 2 of *The Faerie Queene*. The Faery chronicle too opens with allusions to gigantomachy in the clash between Prometheus and Jove over the creation of man, and Spenser's sympathy for the Titan is palpable. For Prometheus emblematizes the Renaissance's own favorite image of its treasured ideals.[105] As an archetype for man-the-creator, Prometheus fashions himself and his world; his theft of divine fire for mortal humans becomes a symbol for man's discovery of his own rational intelligence. Symbolically, he holds out the promise of freedom from ignorance: in his more human guise, the Titan brings the arts of civilization, so that the knowledge he embodies is both poetic and philosophical. Thus, Prometheus becomes a figure for the epistemological quest itself—its promise, but also its transgressive nature, for the price of intellectual inquiry (the fire of the gods) is suffering.

Camões too invokes Prometheus in this vein, associating him with the transgressions of the Portuguese mariners. The Old Man of Restelo links the Titan's original sin of stealing the gods' fire to the daring voyages of exploration. It is Prometheus's fire which inflames the Portuguese heroes' desire to explore the world: "Trouxe o filho de Jápeto do Céu / O fogo que ajuntou ao peito humano / Fogo que o mundo em armas acendeu" (Prometheus stole the fire from heaven / Which rages in every human heart, / Setting the world ablaze with arms, 4.103). But this association only underlines the worldmaking nature of their voyage, and connects them to a successful, modern gigantomachy which the old order, figured by the Old Man, rightly fears.

While Camões carefully distances himself from both the Old Man and the Portuguese at the end of the fourth canto, allowing the Promethean imagery to stand as a double-edged celebration and condemnation of worldmaking, Spenser consciously identifies *himself* with the Titan. By identifying Prometheus as the maker of his Faery characters, Spenser aligns himself with the transgressive creator who dares to question divine decrees and reinvent the world. His poetic fictions, such as *The Faerie Queene,* thus become associated with the gift of Promethean fire; they are mytho-poetic images that may finally offer some promise of illumination in a world darkened by partial knowledge, intellectual incoherence, and seemingly fruitless philosophic questioning. A similar symmetry is the basis of Arthur Golding's commentary on Prometheus in his contemporary translation of Ovid's *Metamorphoses* (1567). Aiming to resolve the implicit incompatibility between Ovid's creation account and that of Genesis, Golding identifies Prometheus as the first poet ("first did images invent").[106] In Golding's account, the Titan's transgression is the animation of these images, the attempt to make an inanimate poetic world equivalent to the real world of God's creation. But herein lay the fascination of Prometheus for Renaissance poets: not only did this Titanic story about the origin of the world challenge the biblical account, it also showed how poetic myths might be valuable, indeed sufficient, for intellectual renewal. It is finally this Promethean vision of worldmaking—poetic, transformative, and transgressive—that characterizes Spenser's epic project and its vision of cosmic process in Mutabilitie.

CHAPTER FOUR

Cartesian Romance: Universal Origins and *Le Monde*

Early in 1632, a new book appeared in Florence presenting a *Dialogo*. Its frontispiece depicted three aged philosophers engaged in fervent debate; behind them ships stand at anchor in a port that opens onto a wider world. The title page advertises the author's fame: "Galileo Galilei Linceo, mathematico sopraordinario . . . e Filosofo, e Mathematico primario del . . . Gr. Duca di Toscana" (Galileo Galilei, the Lincean, extraordinary mathematician . . . and Philosopher and Primary Mathematician to the Grand Duke of Tuscany). Within a year, the author would be condemned for heresy by the Roman Inquisition, becoming perhaps the most famous martyr for modern science. All copies of his book would be seized and burned under orders from Pope Urban VIII. This public drama unfolding around Galileo's *Dialogue Concerning the Two Chief World Systems* garnered spectators across Europe. But the furor also occasioned a more private crisis for another philosopher-mathematician, the young René Descartes.

In 1632, Descartes was about to complete *Le monde,* whose working title echoed Galileo's. In November 1633, upon hearing the shocking news from Rome, Descartes hastily suppressed his work and fearfully wrote to a friend, the Minim Friar and staunch supporter of Galileo, Marin Mersenne:

> m'estant fait enquerir ces iours à Leyde & à Amsterdam, si le *Sisteme du Monde* de Galilée n'y estoit point, à cause qu'il me sembloit avoir apris qu'il avoit esté imprimé en Italie l'année passée, on m'a mandé qu'il estoit vray qu'il avoit esté brûlez à Rome au mesme temps, & luy condamné à quelque amande: ce qui m'a si fort estonné, que ie me fuis quasi resolu de brûler tous mes papiers, ou du moins de ne les laisser voir à personne.[1]

[having made inquiries a few days ago at Leiden and at Amsterdam, if Galileo's *System of the World* was available as yet, since it seemed to me that it was printed in Italy last year, I was told that it was in fact true, that it was burnt in Rome at the same time, and that he was condemned. This surprised me so much that I was almost ready to burn all my papers, or at least, not to let anyone look at them.]

Fearing persecution, Descartes would spend the rest of his life in the Protestant Low Countries, out of the Roman Inquisition's reach. And he would spend his career disguising and repackaging his early work in various forms, for he had learned Galileo's lesson well and was determined to avoid his fate.

THE WORLD AFTER GALILEO

Galileo and Descartes are often separated in histories of the New Science.[2] They never met, nor do they appear to have known much of each other in the turbulent early years of the seventeenth century. But the older Italian serves as a useful touchstone for the young French philosopher's ambitions and practice. While Galileo publicized his desire to reshape the world, claiming to remake the heavens by his discovery of new stars, Descartes shied away from public spectacle, deliberately manipulating the forms and genres of scientific dissemination to make his case without controversy. The specter of Galileo's fall haunts the Cartesian turn inward, away from the contested spaces of the external world and toward worlds within. The contrast between the philosophers' legacies could not be starker. Galileo is celebrated as the Copernican champion, a proto-modern mathematician and scientist who confronted questions of cosmic structure and world systems, challenging religious orthodoxy head-on; Descartes is celebrated as the champion of the *cogito* and inventor of the mind-body problem, a modern Augustine who defends God's existence through rational proofs. But these differences hide another story. Behind the coincidental convergence of their ideas in the early 1630s lurks a profound similarity: Descartes, like Galileo, was first and foremost a worldmaker.

Le monde, the treatise that Descartes suppressed in 1633, goes well beyond Galileo's *Dialogo* in its intellectual ambition and audacity. It is the first modern attempt to construct a systematic natural philosophy out of the corpuscular-mechanical approach to the natural world. Unlike other contemporary critiques of Aristotelianism, it reaches beyond antischolastic polemics to distill experimental scientific and mathematical findings into a wide-ranging theoretical paradigm.[3] Where Galileo critiques existing cosmic hypotheses, Descartes stakes his claim by producing a new "total theory"—the all-in-one modern replacement of ancient atomism, the Aristotelian organon, and Thomist syntheses—that offered a comprehensive explanatory framework for making sense of an expanding world. And Descartes was fully aware of the enormity of his intention. In a 1630 letter to Mersenne, he writes with hubristic flair: "je me suis resolu d'expliquer tous les Phaenomenes de la nature, c'est a dire toute la Physique" (I am resolved to explain all the phenomena of nature, that is to say, the entirety of Physics).[4] At a time when new philosophies were calling all in doubt, Descartes's conviction defies skepticism; Le monde aims to become the foundational modern scientific text to establish a coherent idea of the world.[5] It lays the groundwork for Descartes's reorientation of the relationship between God and world in the Meditationes, redefining for the later seventeenth century and beyond the encounter between religion and science.

Worldmaking and the Scientific Revolution

Descartes's worldmaking venture in Le monde marks a turning point in the long march toward the establishment of a Copernican world: it was the first attempt to craft a complete physics for a heliocentric vision of the universe. Responding to debates over the three world systems—Ptolemaic, Copernican, and Tychonic—it aspires to fulfill the almost apocalyptic "expectation of a new philosophy" which Stephen Menn has described as the defining characteristic of Renaissance philosophy.[6] In this way too, Descartes is Galileo's unlikely heir.

In The Assayer (1623), Galileo had defended himself against the charge that he had lamented "the calamity of our time for our not knowing the true and certain arrangement of the parts of the world."[7] But by the early 1620s, this lamentation was a common one across Europe. "The worlds whole frame," mourned John Donne in 1621, is "Quite out of ioynt . . . / And freely men confesse that this world's spent, / When in the Planets, and the Firmament / They seeke so many new."[8] In his First Anniversary, suggestively entitled "An Anatomy of the World," Donne makes the instability of the known world his theme. Alluding to recent astronomical observations, such as Galileo's discovery of the satellites of Jupiter, Donne complains that the world's integrity has been shaken: "'Tis all in pieces, all coherence gone; / All iust supply, and all Relation." Galileo's own view was not too distant from Donne's, though his tone suggests curiosity rather than distress: "the two systems [Ptolemaic and Copernican] being false, and that of Tycho null . . . [do] not blame me if like Seneca, I desire to know the true constitution of the universe . . . this is a lot to ask and I very much crave the answer."[9]

Le monde sought to satisfy that craving for explanation and thus to rebalance the world's frame. Like Galileo's *Dialogo*, which turned on the question of motion—of falling objects, of tides, of the earth itself—as a defense of the Copernican hypothesis, Descartes's treatise too centered on the twin themes of matter and motion, thereby connecting physics to cosmology. With a simple title—*Le monde*—Descartes signals his worldmaking desire, stretching the term "world" to encompass the earthly and the cosmic, making it a synonym for the universe as a whole.[10] Such a shift would eventually dictate a related, parallel shift in theories of mind and self, of human making and imagining.

The significance of *Le monde* thus depends on how we characterize the Cartesian project. Scholars continue to perpetuate a dichotomy, first established in the seventeenth century, between Descartes the metaphysician who demonstrates the existence of God and Descartes the natural philosopher who seeks a convincing and wide-ranging explanation of the natural world.[11] This clash between metaphysics and physics is even manifest in his archive: the major portion of Descartes's extant work (most of it unpublished in his lifetime) consists of scientific writings, which stand in stark contrast to the rather slim body of published philosophical essays that have become identified with him today.[12] But the unpublished scientific writings clearly inform the shape of the published Cartesian corpus: in 1633, Descartes suppressed not just his treatise but also the wide-ranging inquiries into scientific worldmaking that would undergird his thought for the next twenty years. To read *Le monde* against this background is to excavate the early meditations on the world that were the intellectual foundation for his later work on method, the *cogito*, and the nature of divinity. In the marriage of an innovative literary form to a new post-Copernican science, the early treatise suggests how Cartesian physics and metaphysics—questions about the nature of the world and of God—are closely related. It thus reorients the central questions of Cartesian philosophy away from the mind-body problem and toward what we may call the mind-world problem: is there a definitive, underlying order to the world and how can we know it?

Le monde emerged from a period of intense scientific and mathematical activity that engaged Descartes from about 1628 to 1633. These were years following his attempts to conceive of a universal mathematics, when he was occupied with problems in geometry and practical optics, the discovery of the sine law of refraction, and an interest in perceptual cognition.[13] The treatise began as an attempt to explain the phenomenon of parhelia, also known as sun dogs or mock suns, which occur when sunlight refracted through ice crystals produces brightly colored spots on either side of the sun. The Jesuit astronomer and mathematician Christoph Scheiner had observed a particularly striking appearance of these "multiple suns" at Frascati in 1629, and the conjunction of optics and astronomy clearly excited Descartes.[14] The question of mock suns may have opened a suggestive connection between individual perception and universal physical laws that made it emblematic of the challenges inherent in any attempt at scientific worldmaking. As an early link between mental images and the nature of the phenomenal world, Descartes's interest in parhelia also forecasts his own intellectual turn from physical and cosmological matters to psychological and metaphysical ones.

Descartes's interest in an atmospheric optical phenomenon rapidly expanded to take in all meteorological phenomena and then the whole physical world. In a 1630 letter to Mersenne, Descartes explained the growing scope of his project with excitement: "j'y veux inserer un discours . . . [qui] contiendra quasi une Physique toute entiere . . . la fable de mon Monde me plaist trop pour manquer à la parachever" (I want to include a discourse . . . which will contain almost a complete Physics . . . the fable of my *World* pleases me too much not to finish it).[15] Three years later, however, with the treatise almost ready for publication, Descartes suppressed it.

Fear of meeting Galileo's fate prevented him from publishing anything with unqualified pretensions to a new world-system, but the abandoned treatise and the concentrated scientific labor it entailed remained a touchstone for all his subsequent work. As though unable to suppress his early project entirely, Descartes provides a careful summary of its contents in the fifth part of the *Discours de la méthode*, while the scientific content of *Le monde* found its way into the accompanying *Essais*, notably in the *Météores* and the *Dioptrique*. Though Descartes seemed to turn away from scientific interests after 1637 and to move toward metaphysics, a shift that culminated in the publication of the *Meditationes* in 1641, natural philosophy was never far from his primary concerns. In January 1642, ten years after the suppression of *Le monde*, Descartes wrote to Constantijn Huygens with some glee:

Peut-estre que ces guerres scolastiques seront cause que mon Monde se fera bientost voir au monde, et je croy que ce serait dès a présent, sinon que je veux auparavant luy faire aprendre à parler latin; et je le feray nommer Summa

Philosophiæ, afin qu'il s'introduise plus aisément en la conversation des gens de l'École, qui maintenant le persécutent et taschent à l'étouffer avant sa naissance, aussy bien les Ministres que les Jésuites.[16]

[Maybe these scholastic wars will be the reason that my World will soon be seen by the world at large, and I believe that this could be as soon as now . . . and I will name it Summa Philosophiae so that it can be introduced more readily into the conversations of the Schoolmen, who are persecuting it now and are trying to suffocate it before its birth.]

Two years later, in 1644, Descartes published the *Principia philosophiae* consisting of material from the abandoned *Le monde* in revised rhetorical form: the literary structure and elaborate rhetorical devices of the earlier work now gave way to the question-and-answer format of a scholastic textbook.

Though *Le monde* itself never appeared in print in Descartes's lifetime, there are over thirty references to "mon Monde" in his correspondence. Almost twenty years after its suppression, in the *Conversation with Burman* (1648), he still appears to be discussing his first work and is reported to have confessed "that the few thoughts that he had concerning the world [*de Mundo*] are a source of greatest pleasure for him to look back on [*cum voluptate reminisci*]. He values them most highly and would not wish to exchange them for any other thoughts he has had on any other topic."[17] A now standard 1647 portrait by Jan Baptist Weenix depicts the philosopher holding a book bearing the inscription "mundus est fabula," a poignant reference to the work he was never able to publish.

The unfinished manuscript of *Le monde* was finally published posthumously in 1664 in an unauthorized version, before it was reissued in authorized form in 1677 by Claude Clerselier, Descartes's literary executor.[18] Even as the work presented new evidence for the philosopher's scientific innovations, its publication—over forty years after its conception—fueled debates over Descartes's suspected materialism, atheism, and other heterodoxies. But just what was so controversial about *Le monde*? Why does this marginal text demand a central place in a history of early modern worldmaking? To answer these questions, we must return to Galileo.

Fictions of the Cosmos

For Descartes, as for Galileo before him, questions about the nature of the world were best answered by engaging the speculative power of the intellect and by exploiting the blurry boundaries between the hypothetical and the real. *Le monde* lays out a framework for understanding the world in the form of a fable, a fictional narrative which invites the reader to witness imaginatively the creation of a new world (*nouveau monde*) according to the principles of Cartesian physics. Rather than logical proposition, Descartes exploits the energy of narrative unfolding, thereby suggesting that the world has an aesthetic order that could only be revealed through literary figuration.

But the power of literary figuration and hypothesis in the *Dialogo* were, in fact, also at the center of Galileo's 1633 trial.[19] The problem of hypothesis and its relation to the phenomenal world had been integral to debates over the Copernican model from the initial publication of *De revolutionibus orbium coelestium* in 1543, with its notorious preface by Andreas Osiander, who disavowed the truth-value of the work's cosmological hypotheses, to Christian Huygens's *Cosmotheoros* in 1698, whose hypotheses claim to reject fiction even as they engage in a kind of fiction-making.[20] As Frédérique Aït-Touati has recently shown, astronomy and hypothesis share a long history: in the seventeenth century, figures as diverse as Galileo, Kepler, Wilkins, More, Hooke, and Huygens capitalized on the overlap between fiction and hypothesis for imaginative exploration and thought experiment.[21]

Le monde belongs to this tradition but pushes its boundaries. The use of literary devices in a natural philosophical work was not in itself new since Galileo had used the dialogue, Froidmont the Lucianic voyage, and Kepler the dream vision.[22] *Le monde,* however, does not merely harness the epistemic potential of literary form; it interrogates the relationship between the fictional-hypothetical and the real, between imagined systems and material phenomena, as well as the nature of cosmic creation itself. In its pursuit of "the true constitution of the universe" it edges from cosmic hypothesis toward theological heresy. As Descartes would recognize, the affinities between the philosopher's intellectual recreation of the world and God's original creation were profound and unsettling. Not only did the new natural philosophy undermine the Thomistic Aristotelianism on which biblical theology had rested for centuries, it threatened to undercut the metaphysical foundations for human knowledge of the world and therefore of God himself.

In *Le monde,* Descartes argues that a fictional narrative can offer a new understanding of the ("real") world because of its epistemic force; in other words, fiction could reveal philosophic truth. At stake in this appeal to literary imagining is not only philosophy's traditional claim to universal certainty and truth (as opposed to the dubious phantasms of poetic invention) but also an entire theory of "the world." In his classic study of the Scientific Revolution, Alexandre Koyré describes the disorientation in terms of a two-fold loss: "man . . . lost his place in the world, or more correctly perhaps lost the world in which he was living and about which he was thinking, and had to transform and replace not only his fundamental concepts and attributes, but even the very framework of his thought."[23] *Le monde* is a powerful response to such loss. It seeks to recreate a "new world" in epistemological terms, filling the void left by lost certainties. By imagining the world's origin and development, Descartes offers a new, mechanistic creation story for the modern age.

Le monde can thus take its place amidst a tradition of worldmaking texts including scriptural expositions (such as Genesis and the hexamera of the Church Fathers), classical philosophy and poetry (as in Plato's *Timaeus* and Lucretius's *De rerum natura*), contemporary hexameral poetry (notably Du Bartas's *La sepmaine* and Tasso's *Le sette giorni del mondo creato*), and scientific theories (of Copernicus, Galileo, Kepler, Gassendi). By revising "the world" as a concept through the use of an imaginary model, Descartes intervenes in a rich tradition that had repeatedly tested the boundaries between poetry and philosophy. To recast philosophy in terms of fiction and suggest that such a fiction could pave the way to knowledge (*scientia*) was to tear down the traditional bases of philosophical value. More troubling from the Church's perspective, as Descartes himself knew, his project threatened the authoritative status of Genesis as the definitive explanation of the world, reducing it to the status merely of one more narrative competing for philosophical supremacy.

Descartes's foray into worldmaking thus deliberately sets up a subversive counterpoint between human and divine acts of creation, threatening to substitute God's original making of the world with a second, human act of imaginative construction. The natural philosopher in search of a grand cosmic theory faced a challenge parallel to that of the mapmaker who sought to reconstruct the world visually on the page. Both are compelled to take a god's-eye-view—a demiurgic epistemology—in order to encompass the whole. Inevitably, such a position raised profound metaphysical dilemmas, forcing a return from the cosmos back to the individual thinking self.[24]

This intellectual arc—from physical and epistemological considerations about the world to moral and metaphysical reflections on God and the thinking self—captures the shape of Descartes's career and also mirrors the trajectory of seventeenth-century thought. In the aftermath of *Le monde*'s cosmic reach, the *Discours de la méthode* returns to the form of the autobiography and essay even as it redefines how an individual can comprehend the world. The 1641 *Meditationes de prima philosophia,* often seen as the central expression of Descartes's mature philosophy, may also be usefully read against the cosmographic meditation and theodicy, alongside texts by Mercator, Montaigne, Milton, Pascal, and Kant.[25]

CREATING *LE MONDE*

Le monde contains one of the most seductive invitations in early modern philosophy: "Permettez . . . à votre pensée de sortir hors de ce Monde pour en venir voir un autre tout nouveau que je ferai naître en sa présence dans les espaces imaginaires" (Allow your thought to move beyond this world and venture to view another, entirely new one that I will bring into being in its presence within imaginary spaces).[26] Descartes lures the reader out of this world ("ce Monde") and into a new one by invoking a familiar binary—the Old World of Europe and the New World of America—transposing a commonplace of geographic expansion into the cosmic realm. A new Columbus, Descartes reveals a hitherto unknown vision of the world. But he is no terrestrial explorer concerned with empirical data and natural historical detail; nor is he cosmographer who reframes the world. His new world and the comprehensive picture of world order that it proposes will be brought into view by the sheer power of intellectual conception. Located in "espaces imaginaires," it cannot be reached through sense impressions of the phenomenal world.

Le monde's worldmaking project is thus positioned as a two-pronged one: it deconstructs the scholastic-Aristotelian basis of knowing the world through the senses and simultaneously offers a new method for knowing the phenomenal world through the intellect. The work's double beginning formally mimics these aims. The first five chapters serve as a systematic unmaking of the Aristotelian world, questioning the axiom that *nihil est in intellectu quod non prius in sensu* (nothing is in the intellect that was not first in the senses). This deconstructive move opens a conceptual space for Descartes's own Copernican, mechanistic philosophy, which counterintuitively argues for a vision of the world that belies its physical appearance.[27] The initial clearing of old epistemological foundations is then followed by a second, renewed beginning in the sixth chapter, where Descartes calls upon the reader to witness the birth of a new world as an epistemic thought experiment. The play between these two interlaced strands marks Descartes's initial foray into the central questions that will occupy his later philosophy:

is there an external world and how can I know its structure (form, essence, nature)? What is the nature of my mind as a cognitive being? What can I know with certainty? Does God exist and is knowledge of God necessary for knowledge of the world?

In this respect, *Le monde* becomes a counterpart to the *Meditationes* and offers an alternate entry point into Descartes's philosophy.[28] The two texts can even be considered inversions of each other. If one begins with physics and glances toward metaphysics, the other begins with metaphysics and touches on physical matters at the very end. Descartes starts with the phenomenal world in *Le monde* and uses it as springboard for questions about the mind and God; in the *Meditationes* he begins with the *cogito*, the thinking mind, and moves from this intellectual self to God and finally to the world beyond. Descartes's primary concern is thus the triangular relation between the thinking individual ("mind" or "cogito"), the material world ("body") within which she is situated and which she seeks to comprehend, and God, who gives both mind and world their metaphysical foundation. Descartes's concern is therefore not with the mind-body problem per se, but with what I have called the mind-world problem: the challenge of knowing the underlying order of the world, discerning whether such an order even exists, and discovering its relation to the divine. It is a concern at the heart of the worldmaking enterprise, and Descartes powerfully reorients it for modern philosophy.

Unifying the Whole

Descartes's confidence that the mind *can* know the world and God without resorting to sense impressions marks a radical break from the sixteenth-century worldmakers. He shifts the task of worldmaking away from a fundamentally synoptic operation—that is, the cosmographic accumulation and integration of parts into a whole—and reframes it as a kind of imaginative archaeology, the quest for a central principle that unifies the whole and which underlies everything.[29] By uncovering this principle through probing intellectual inquiry, the world's totality can be grasped in a single thought. To know the world outside, Descartes argues, we must delve into the mind.

Fundamental to this shift is Descartes's rejection of the Aristotelian view of knowledge that proceeds from particulars to universals—a process wherein sense impressions of the world are abstracted into universal ideas about categories and essences of things.[30] By insisting on the possibility of certain knowledge through the careful exercise of the intellect (his

"method"), Descartes dissociates himself from both the philosophical theology of the scholastics and from the Pyrrhonic skepticism of Montaigne.
Both these predecessors begin from the materiality of the external world
and bind the scope of human knowledge to it, though in markedly different
ways. For the scholastics, only God, the original creator, could comprehend
the entirety of the created cosmos and its ordered structure; human knowledge depended on partial information derived from interaction with that
world. For a skeptic like Montaigne, however, God is the ultimate, unknowable metaphysical other; to know the world is to know the self, to accept its
mutability and incertitude.

Descartes implicitly responds to both strategies, abandoning neither
world nor God as sources of knowledge, but by showing how the thinking
mind mediates between both and draws them into relation. For him, human knowledge of the world is neither fundamentally impossible nor the
sole province of the deity. Nor must it be resigned to the perspectivism of an
individual consciousness, as Montaigne suggests in the "Apologie." Knowledge of the world rests on the mind's ability to discern the inherent unity of
the whole, which derives from empirical truths established by God. In May
1632, Descartes wrote to Mersenne explaining this view:

> For the last two or three months I have been quite caught up in the heavens. . . .
> For although their distribution seems irregular . . . I have no doubt that there
> is between them a natural order which is regular and determined. The grasp of
> this order is the key and foundation of the highest and most perfect science of
> material things man can ever attain, for if we possessed it we could discover *a
> priori* all the different forms and essences of terrestrial bodies, whereas without
> it we have to be satisfied with guessing them *a posteriori* and from their effects.[31]

The emphasis here on moving away from *a posteriori* knowledge of the phenomenal world to an excavation of *a priori* principles that unify the whole
marks a strong rejection of earlier attempts to cobble together a synthetic
vision of the world out of its disparate parts.

It is worth pausing to consider Descartes's innovation here. Both the
scholastics and the skeptics deny our ability to comprehend the world's
totality and truly grasp its order. Behind this epistemic diffidence lurks a
profound suspicion of claims to universal knowledge as being made up, as
mere human conceptions that cannot be verified by any external measure.
The skeptical arguments of the First Meditation dramatize these concerns
as they seek to separate true knowledge from false impressions of the world.
Moving progressively through the dream argument to the idea of a deceiv-

ing God and finally an evil demon, Descartes harnesses the power of doubt to explore the conditions that might distort our apprehension of the world. In doing so, he measures the problem of conceiving the whole through its parts, of knowing the universe through the partial lens of the thinking self. The question he confronts is this: if we must only rely on partial information, how can we capture the underlying unity of the whole?

It is in response to this conundrum that Descartes effects a fusion between poetry and philosophy. The philosopher Stanley Rosen argues that the idea of the unity of the whole itself forms the basis of a fundamental kinship between poetry and philosophy despite the ancient terms of their quarrel. He asks, "What follows from the impossibility of an analytical or conceptual explanation of the unity of the whole? Either there is no whole, which is to say that the wholeness or unity of our experience is a perspective, that is, a poem. Or, at a somewhat deeper level of this response, it is we who must supply a cosmological myth of the whole with an accordingly rhetorical justification."[32] Writing of Plato here, Rosen locates the source of philosophic discomfort with poetic imagining in its paradoxical reliance on the latter. To articulate the unity of the whole is either to take Montaigne's perspectivist position (that we can only know the world through our individual experience of it) or to push even more deeply into poetic territory to "supply" a "myth of the whole." This latter response is Descartes's choice in *Le monde*.

It would be inaccurate to describe *Le monde* as a myth or a poem. But Rosen's astute distinction also explains why it has been so difficult to see worldmaking as central to the Cartesian philosophical project. Either we must accept that any attempt to describe the world is doomed to perspectivism (associated with poetry and individual subjectivity) or we must recognize cosmological narratives (myths of origin) as necessary and effective, if contingent, products of human artifice. It is almost commonplace now to relegate Descartes to the perspectival position and to assert that the modern subject first finds voice in his work.[33] In a widely influential interpretation, Jean-Luc Nancy argues that "le sujet, c'est celui qui a un *monde*: quelque chose qui est à sa disposition, quelque chose de prêt pour son usage. . . . Aujourd'hui, depuis Descartes, le sujet c'est le monde, et réciproquement" (The subject is one who has a world: something that is available to him, something ready for his use. . . . Today, after Descartes, the subject is the world, and vice versa).[34] But while this phenomenological perspective emphasizes the power of individual consciousness to shape a particular perception of the world, it does a disservice to the immensity of Descartes's vision. For his response is the latter, more deeply humanistic one.[35] Descartes's project consists in purposefully rebuilding a torn philosophical edifice with the recognition that the edifice itself may be a contingent artifact. To understand Descartes's imagined world in these terms is to take his emphasis on its fictionality seriously—to recognize that the world's unity may finally be best known through the intellectual imagination.[36]

Rosen's meditation thus sheds light on the challenges that faced Descartes as he sought to repair a broken unity and recover the world's order in the wake of Copernican, atomist, and materialist claims to truth. After the failure of "analytic and conceptual explanations" to offer a comprehensive vision of the world—a position we can associate with the cosmographic mode—a range of writers explored more metaphorically supple modes of imaginative speculation. The rise of utopian fictions in the seventeenth century, a hybrid genre with either a scientific or political thrust, is an example. It was used by writers as diverse as Bacon, Kepler, Winstanley, Harrington, Cavendish, and Boyle, and drew on travel narratives, philosophical treatises, and an imaginative desire to explain the whole. In *Le monde*, Descartes too turns to this mode, combining rigorous philosophic inquiry with poetic fiction-making to produce a comprehensive vision of world order through nontraditional philosophical means. *Le monde* consequently becomes the testing ground for his subsequent analyses of the *cogito*, the thinking mind that discerns itself and the world. Later in his career, he would chart the epistemic workings of this "intellectual imagination" in the *Meditationes*, seeking to ground it metaphysically in the creative potential of God's infinitude.

Relocating Descartes in this wider context also suggests why he returns to the ancient analogies between microcosm and macrocosm, self and world, only to recast the terms of their engagement once more. Like Mercator, who literally recreates the world on the page, or Montaigne, who imagines the fertile potential of humanly made forms, Descartes, the narrator of *Le monde* (as indeed, the Meditator in the *Meditationes*), harnesses the mind's ability to know the world by recreating it cognitively. And if the mapmaker's artisanal skill or the humanist's belief in the mind's generative force imitated and threatened to substitute the skills of the divine *artifex*, the Cartesian *cogito* now comes dangerously—and thrillingly—close to that divine creative power.

Imagining the World

Le monde conjures up a world in the making, taking the reader to a space beyond the reaches of biblical time, where we witness a second creation:

Permettez donc pour un peu de temps à votre pensée de sortir hors de ce Monde pour en venir voir un autre tout nouveau que je ferai naître en sa présence dans les espaces imaginaires. Les philosophes nous disent que ces espaces sont infinis et ils doivent bien en être crus puisque ce sont eux-mêmes qui les ont faits. Mais afin que cette infinité ne nous empêche et ne nous embarrasse point, ne tâchons pas d'aller jusques au bout; entrons-y seulement si avant que nous puissions perdre de vue toutes les créatures que Dieu fit il y a cinq ou six mille ans; et après nous être arrêtés là en quelque lieu déterminé, supposons que Dieu crée de nouveau tout autour de nous tant de matière que, de quelque côté que notre imagination se puisse étendre, elle n'y aperçoive plus aucun lieu qui soit vide. . . .[37]

[Allow your thought to move beyond this world and venture to view another, entirely new one that I will bring into being in its presence within imaginary spaces. The philosophers tell us that these spaces are infinite and they should certainly be believed because they have made [these spaces] themselves. But to prevent this infinitude from hindering and embarrassing us, let us not attempt to go out to the end; let us enter only so far as to lose sight of all the creatures that God made five or six thousand years ago; and after we have stopped in some established place, let us suppose that God creates anew so much matter around us that in whatever direction our imagination may possibly extend, it no longer perceives any place that is a void. . . .]

Descartes's invitation to enter into these contested spaces of the imagination reveals his intellectual sympathies more clearly than the later cagey and circumscribed arguments of the *Meditationes*. It suggests that we can know the true nature of the world—that is, its foundation in Cartesian physics— through a cognitive act of imagination rather than through a collation of sense impressions from the phenomenal world. This mental act inevitably leads to considerations about the origin and extent of the world, which touch on the nature of God, of creation (both divine and human), and of the mind, which imagines, conceives, and comprehends.

The key term is "espaces imaginaires": the intangible, mental spaces of the imagination, that storehouse of images in Renaissance theories of perception, which was on the verge of acquiring its more active, modern meaning as a shaping force rather than a repository.[38] Here, the imagination is a place which expands to take in the entire world, which for Descartes is a fluid universe completely filled with matter. Understanding the world, then, depends on understanding the nature of the imagination, which performs multiple functions: it contains the world (as a location), it is a means for comprehending the world (as an instrument), and it also helps to visualize and articulate the world (as a canvas or book).

This centrality of the imagination to Descartes's method has been long obscured by critical emphases on Cartesian rationalism, but it is vital to his worldmaking aims. Much of his early writing explores the functions of the imagination, frequently aligning it with poetic knowledge and expressing a guilty pleasure in its properties. In a letter to Guez de Balzac in April 1631, for instance, we get a rare glimpse into Descartes's imaginative life. "Je dors icy dix heures toutes les nuits," he reports "et sans que jamais aucun soin me réveille, apres que le sommeil a longtemps promené mon esprit dans des buys, des jardins, et des palais enchantez, où j'éprouve tous les plaisirs qui sont imaginez dans les Fables, je mesle insensiblement mes reveries du jour avec celles de la nuit." (I sleep here for ten hours every night and since no care ever awakens me, after sleep has for a long time led my spirit to wander in forests, gardens, and enchanted palaces, where I experience all the pleasures that are imagined in the Fables, I unconsciously mingle my daytime reveries with those of the night.)[39] Descartes, *rêveur*, is an unexpected figure, one who sleeps late and indulges in the pleasures of dream visions, freely mingling the philosophic thoughts of the day with the fabular, poetic delights of the nocturnal world. His evident enjoyment of literary figuration—imaginary forests, gardens, and enchanted palaces—combines here with his self-identification as a philosopher. These habits, however, appear to have been related to meditational practices in the seventeenth century. Peter Dear's suggestive connection between Descartes's meditations as exercises in "purposeful thinking" and the Jesuit Honoré Fabri's description of the *meditatio* as a method of study reveals that speculative reflection may have been an accepted philosophical practice that sought to harness the associative power of imagining for philosophic use.[40]

Through a series of early writings—including records of his dream visions, reflections on the poetic furor, and the rules for training the mind in the *Regulae*—Descartes had meditated at some length on the poets' ability to apprehend knowledge.[41] As early as 1616, he opens his Law Theses with an epigraph from Lucretius celebrating the power of poetry to expound philosophic doctrine.[42] Two years later, in the "two-imaginations note" of the *Cogitationes privatae*, his private notebook from 1618–19, Descartes describes poetry and philosophy as complementary faculties, observing that "there are more profound judgments in the writings of poets than of philosophers. The reason is that poets write through enthusiasm and the force of imagination: there are seeds of knowledge in us, as in flintstone, that are drawn out by philosophers through reason [and which] from the poets are shaken loose by the imagination and thus shine more brightly."[43] The philosophical reasoning feels labored in all senses beside the intuitive illumination provided by the poets' imagination. The context for these musings is especially significant. The *Cogitationes,* a loose collection of notes on problems primarily in mathematics and physics, record a period of intense and fruitful activity in practical science under the tutelage of Isaac Beeckman. Given Descartes's stated interest in developing a *mathesis universalis* at about this time, it is striking to find a note declaring the supremacy of poets to philosophers in no uncertain terms — Galileo had also explored the intuitive force of poetic argument.[44]

Such attempts to elaborate a cogent "method" are strongly marked by what Pierre-Alain Cahné has called the "nostalgie de la connaissance intuitive," the desire to attain knowledge with the immediacy and seemingly unmediated insight of a poet seized by enthusiasm.[45] This backdrop illuminates Descartes's appeal to the power of verisimilitude and fabular invention in *Le monde* and usefully complicates his apparent suspicion of fictions in the *Meditationes.* In *Le monde,* Descartes's deployment of the fable as an epistemic instrument — a tool that reveals knowledge — makes his treatise on mechanistic physics also a defense of the literary imagination. This defense depends on a strategic inversion of the Neoplatonic allegory of truth as a veiled figure, best known from Macrobius's influential *Commentary on the Dream of Scipio,* a work that itself presents a cosmographic meditation in the form of a dream vision. Macrobius argues that philosophers may use fabular narratives to veil truth and its mysteries, intentionally concealing them from the common gaze. But Descartes triumphantly collapses the distinction between the veil of allegory and the body of Nature: his fable will reveal truth more clearly, as though it were the naked body itself.[46] Not only does this claim unite philosophy and fiction; it suggests that fiction clarifies philosophic understanding.

Pointing to Descartes's poetic sensibility, despite his later claim in the *Discours* to have abandoned all humanistic study, Cahné highlights this fusion of poetry and philosophy: "Du poète au philosophe, la proximité est grande, et, dans sa manière d'imaginer l'impensable, en toute necessité, Descartes se situe lui-même dans la familiarité du poète" (The proximity between the poet and the philosopher is great, and in his manner of imagining the unthinkable, Descartes necessarily situates himself close to the poet).[47] Descartes draws on the speculative leap of imagination associated with the poet precisely when philosophical proposition proves inadequate. World-making demanded this leap into the space of the unthinkable, forcing the Aristotelian philosopher to abandon the traditional bases of knowledge and propose a new all-encompassing vision of the cosmos that would be built from first principles. Confronted with this challenge, Descartes invokes the fabular language of the poet.

At the very end of the fifth chapter of *Le monde*, immediately before his claim to create a new world, Descartes pauses to reflect on his method in terms that evoke the Macrobian *fabula* once again:

[Je] serai même bien aise d'y ajouter quelques raisons pour rendre mes opinions plus vraisemblables. Mais afin que la longueur de ce discours vous soit moins ennuyeuse, j'en veux envelopper une partie dans l'invention d'une fable, au travers de laquelle j'espère que la vérité ne laissera pas de paraître suffisament, et qu'elle ne sera pas moins agréable à voir que si je l'exposais toute nue.[48]

[I would be happy to add several arguments to make my opinions more plausible [*vraisemblable*]. But so that the length of this discourse seems less boring to you, I want to cover one part in the form of a fable, through which I hope that the truth will not fail to appear clearly, and that it will be no less pleasing to see than if I revealed it in complete nakedness.]

Descartes's careful juxtaposition of arguments, opinions, and fables here culminates his longstanding interest in the potentials of the poetic imagination—what he describes as the "intellectual imagination" in the *Regulae*—to discern the clear and distinct ideas that lead to true knowledge. Significantly, Descartes describes both *Le monde* and the *Discours de la méthode* as fables, linking his ambition to reimagine the world and the foundations of universal knowledge to the adage "mundus est fabula" (the world is a fable) attributed to Pythagoras, and thus to the origins of philosophy itself.[49] The allusion informs his 1647 portrait, where the philosopher appears as a modern Pythagoras initiating a new beginning for philosophy. But what Descartes actually provides is an ironic twist on the Pythagorean model: the ancient philosophical belief that the sentient world is an insubstantial fiction gives way to the Cartesian version in which the insubstantial fable becomes the explanatory foundation of the world's true order.[50] With this inversion, Descartes dismantles the ontological opposition between the real and the imaginary and reorients questions about the origin and structure of the world away from metaphysics and toward physics. As fictional and real blur in *Le monde* and subsequent works, Descartes's imagined world heals the traditional chasm between a *logos*-centered philosophy and a *mythos*-centered poetry.[51]

As the space within which Descartes's new world unfolds, the "espaces imaginaires" connect the new Cartesian physics to a tradition of philosophic worldmaking that relies on imaginative reconstruction. As the allusions to

Pythagoras and Macrobius suggest, Descartes frames his new philosophy as a cosmogony to rewrite a foundational text in that tradition: Plato's *Timaeus*, the dialogue that presents an entire metaphysically grounded theory of world order. The only Platonic work known to the Latin Middle Ages, the *Timaeus* is the original worldmaking text in the classical tradition, offering a theory of the whole that touches on natural philosophy, geometry, mathematics, medicine, metaphyics, theology, and epistemology. For the early moderns, its similarity to Genesis was compelling, and it provided an alternate framework to Aristotelianism, making it possible to imagine other systems that could explain the nature of the visible world.[52] But Timaeus's vision of the world is explicitly founded on a "likely tale" (*eikos logos*), a plausible but ultimately hypothetical account of cosmic form narrated in the form of a cosmogony.[53] Descartes may have borrowed the double beginning of *Le monde* as well as the use of cosmogonic narrative as thought experiment from his classical model. And while he departs from the metaphysical distinctions so dear to Plato, his own mechanistic conception of the world presents similar problems of verisimilitude to those that plague Timaeus, and which result in a kind of "scientific poesis."

Both Timaeus's *eikos logos* and Descartes's fable set forth worlds conjured through imaginative power in order to transcend the epistemological divide between sense impressions and their underlying order (whether Platonic forms or Cartesian corpuscles), which is ultimately intangible. Both use verisimilar narratives to unfold systems of explanation that make sense of the visible world. By turning to Plato, Descartes draws on the very philosophic strategies that Aristotle had rejected and uses them to liberate philosophy from the grip of Aristotelian scholasticism. In *Le monde*, Descartes transforms the imagination itself into a potent weapon in a philosophic battle—a matter crucial to the argument of the Sixth Meditation, where the affirmation of the imagination as a cognitive function is the first step toward the mind's apprehension of the world ("body").

Meditations on Infinitude

Descartes continues to be described as the early modern philosopher who "married Platonic metaphysics to Democritean physics," endowing both with an Augustinian twist.[54] This intellectual mixture itself situates him within a tradition of early modern worldmaking, for one of its markers is the attempt to combine Platonic and Epicurean (or more broadly speaking, materialist or atomist) paradigms. Descartes's union of these two seemingly incompatible philosophical systems is not in itself new. But his work departs from previous efforts in its programmatic confrontation with the "God question" that was inevitably raised by the clash between a demiurgic metaphysics (Platonic) and an atheistic vision of the natural world (Democritus, Epicurus).

Descartes's early work, particularly *Le monde*, registers his attempt to think through the thorny question of God's relation to the world. These attempts often stumble into dangerous theological territory, revealing fissures in the Cartesian system that Descartes will seek to shore up in the *Meditationes* and the *Principia*. But it is precisely in these moments, when Descartes seems to have dropped his guard, that we glimpse the subversive potential of his philosophy—something his contemporaries would quickly seize upon, much to his own chagrin.

One such fissure is in fact the *espaces imaginaires* of *Le monde*, which allude to a vexed question in philosophy and theology. A technical term in scholastic philosophy, "imaginary space" referred to the possibility of extra-cosmic space, that is, the existence of a void beyond the fixed spheres. The finite world of Aristotelian physics denied any such possibility, but throughout the Middle Ages and the early modern period the question drew fierce debate. Descartes's evocation of the term in *Le monde* therefore connects his project to another contentious strand of early modern worldmaking—do we live in a closed world or an infinite universe?[55]

A potentially heretical issue, the question could only be discussed in hypothetical terms. Extra-cosmic, possibly infinite, space occupied a conceptual no-man's-land between fictional projection and philosophic proposition. By situating his new world—supposedly fictional and heuristic, but eventually real—within this slippery mental space, Descartes engages a set of questions that plagued would-be worldmakers: is the cosmos finite, infinite, or indefinite? How is the nature of the (created) world thus related to God, its creator?

These were thoroughly discussed topics in the scholastic tradition. In the sixteenth century, such reflections on infinitude, well-known from medieval commentators, culminated in the Jesuits' Coimbra commentaries on Aristotle's *Physics*, a series of texts that had formed the backbone of Descartes's education, with a chapter entitled, "What is Imaginary Space?"[56] Buried in this chapter, we find a compelling discussion of the use of fictions and mental conceptions to determine questions about the nature of God and the world. Here, the world's finitude is contrasted to the infinity of God, but there was also a long tradition that questioned the basis for such absolute distinctions. By the mid-seventeenth century, theories of worldly finitude were under pressure—not only from astronomers and natural philosophers whose observations of celestial phenomena such as the supernovae of 1572 and 1604 raised questions about cosmic expanse, but also from theologians for whom assertions of the world's finitude seemed to place limits on God's omnipotence.

Descartes's own position on this issue is far from clear and scholars continue to debate it.[57] Though he is at pains in his correspondence to distance himself from claims for an infinite world, he never explicitly denies this belief. Descartes firmly holds that the world is "indefinite," that is, our finite human intellect cannot comprehend its limits and therefore cannot determine its nature. God alone is "infinite," while the *cogito* is clearly finite. It is this apprehension of the gulf between finite self and infinite deity that becomes the basis for the proof of God's existence in the Third Meditation; significantly, Descartes never explicitly addresses the nature of the world.[58] Such fine distinctions leave open the possibility—both in *Le monde* and in the Third Meditation—that the world *could* be infinite. We simply do not have the cognitive ability to know with certainty. Descartes here opens an epistemological dimension (can we know the extent of the world?) to what had been a primarily metaphysical question (what is the nature of the world?), and thereby sidesteps any potentially troubling conclusion. But he strategically does not foreclose the possibility of cosmic infinitude.

Noting the "opportune consequence" of such a move, several scholars have argued that Descartes's distinctions are driven by political caution.[59] Alexandre Koyré famously declared that the distinction between the infinity of God and the indefiniteness of the world "was a pseudo distinction, made for the purpose of placating the theologians."[60] But even if we accept counterarguments that claim a theoretical necessity within Cartesian metaphysics for such careful distinctions, it is difficult to escape the suggestion of the world's infinitude—a suggestion that prompts troubling questions about the nature of God and the deity's position within the cosmic structure. Descartes's own repeated statements in his correspondence betray prudential concerns with regard to the topic. In the well-known 1647 letter to the French ambassador Pierre Chanut addressing Queen Christina's concerns that he held the heterodox theory of an infinite world, Descartes invokes Nicholas of Cusa as an intellectual predecessor, arguing, "Cardinal Cusa and several other doctors have supposed the world to be infinite without ever being censured by the Church."[61] Descartes's comment here is either incorrect or deliberately misleading, for Cusa's discussion in *De docta ignorantia* clearly distinguishes two senses of infinitude that are differently applicable to God and the cosmos.[62] By eliding the two, Descartes has it both ways once more—he simultaneously distances himself from the taint of heterodoxy even though he actually does not disavow the possibility of cosmic infinitude.

While a similar circumspection is evident in other letters from the late 1630s onward, *Le monde* engages with the issue more openly.[63] By playing on the multiple meanings of *espaces imaginaires*, Descartes explicitly takes on theological matters. If he can procreate his new world (literally, *je ferai naître*) within the fertile spaces of the creative, shaping imagination (*espaces imaginaires*), the very emergence of that world challenges God's creative omnipotence and control of the world. Descartes thus highlights his own self-consciousness about the implications of his creative act: making a new world in what may be the infinite spaces of the universe sets up a comparison between Descartes's imagined literary creation and God's creation of the world.[64] A daring and provocative gesture, it shifts the agency of world-making from an original creative deity to contingent human thinkers who must use the creative imagination to understand the phenomenal world.

Descartes may have been thinking back to these problems, and perhaps even to the specific Coimbra commentary, when he jokingly notes, "The philosophers tell us that these spaces are infinite and they should certainly be believed because they have made [these spaces] themselves." The connection, simultaneously parodic and profound, between physics, cosmological model-building, and fiction obliquely reveals how Descartes seeks to displace the Ptolemaic-Aristotelian system: by shifting the grounds of philosophic explication and engaging the problem of creation itself—creation of the world as well as of theories about the world.

SUBSTITUTING GOD: ON HUMAN AND DIVINE WORLDMAKING

Le monde shows affinities to both classical and Christian cosmologies. But its real target is nothing less than a systematic revision of Genesis.[65] In structural terms, *Le monde* and the related *Traité de l'homme* form a work that closely resembles the Genesis account of cosmic and human creation.[66] Descartes's presentation, however, juxtaposes competing visions of divine and human worldmaking, implicitly voicing skepticism about both the accuracy and the explanatory power of the biblical account in its Thomist-Aristotelian iteration. In a clear allusion to the first words spoken by the creator-God in

Genesis 1:3 ("Fiat lux. Et facta est lux"), *Le monde* opens with a revisionary theory of the nature and properties of light, and the fictional Cartesian cosmogony is followed by chapters describing the formation of stars, planets, and comets. Though unfinished, the celestial physics of *Le monde* was meant to be followed by an account of human physiology in the *Traité de l'homme*, thereby establishing a mechanistic account of the creation.

In December 1630, Descartes wrote to Mersenne confessing his ambition: "Je vous diray que je suis maintenant aprés [*sic*] à demesler le chaos, pour en faire sortir de la lumiere, qui est l'une des plus hautes & des plus difficiles matieres que je puisse jamais entreprendre; car toute la physique y est presque comprise" (I will tell you that I am now about to dispel chaos in order to bring forth light, which is one of the most sublime and most difficult tasks that I might ever undertake; for almost the entirety of physics is included therein).[67] The echo of Genesis 1:4 suggests that Descartes had made the connection between his new physics and Genesis before he actually set it out in narrative form. Yet commentators have shied away from attributing full responsibility to him for the boldness of this move; in a tense religio-political environment, a radical rewriting of Genesis was probably not a prudent plan.[68] Descartes, however, makes the connection explicit in another letter to Mersenne from the same period:

> I have come to the description of the birth of the world, in which I hope to include most parts of Physics. And I will tell you that rereading the first chapter of Genesis, I discovered not unmiraculously [*non sine miraculo*] that all of it could be explained so much better according to my thoughts, it seems to me, than by all the ways in which the interpreters have explained it, something I have never hoped for until now. But now, after having explained my new Philosophy, I propose to show clearly that it agrees with the verities of faith much better than that of Aristotle."[69]

This is an astonishing statement, particularly in light of the Church's condemnation of Descartes's works in 1663 and the accusations of atheism that would plague the Cartesians. To offer an explanation of a miracle (the creation *ex nihilo*) in naturalistic terms was itself a potentially blasphemous claim; the casually added "non sine miraculo" suggests an ironic consciousness of the theological implications of his undertaking. But Descartes's assertion here also provides evidence for his intention to rewrite the foundations on which natural philosophical and theological doctrines were based—not to discredit the Mosaic account but to reconcile the seeming opposition between science and scripture, soon to be made prominent by the Galileo affair. To do this, Descartes would have to rethink the relationship between God and world.

In God's Image

Le monde exploits a crucial slippage between Descartes as the creator of the *nouveau monde* and God as the *possible* creator of the same world. While Descartes initially tells us that *he* will create a world in the *espaces imaginaires* ("je ferai naître"), he subsequently modifies this to imagine God creating a new universe ("supposons que Dieu crée de nouveau"). Despite invoking God as a prime mover, Descartes enters the vexed theological debate over the creation *ex nihilo*, which, following Thomist doctrine, was one of the central mysteries of Christian theology.[70] By setting up a parallel between himself as a worldmaker and God's creative omnipotence, he activates an implicit analogy between poetic invention ("l'invention d'une fable") and the great miracle of Genesis 1—both create worlds by making "something" from "nothing." But the analogy has a double-edged effect. It demystifies the creation account of Genesis even as it elevates the Cartesian intellectual imagination to quasi-divine status. And it introduces a deliberate ambiguity into the nature of creative process itself. It is far from clear whether Descartes might be intentionally questioning the authority of scripture and church dogma, or whether he, like Plato before him, is attempting to make the creation rationally intelligible through the use of a persuasive parallel. Like the question about infinite space, which threatens to depose God from the center of the cosmos, Descartes's move in *Le monde* threatens to substitute a divinely anchored vision of the world with an uncertain but equally powerful story told by a human maker.

This heterodox strain in Descartes's writing draws on a particular literary trope: the image of the human mind encountering the prospect of an infinite world. Alluding to Montaigne, Cicero, and Plato, Descartes fuses contrasting traditions about the agency of the world's creation to emphasize his own claim for a conceptual equivalence between human and divine making. This insistence on the mind's creative power also signals the turn in his later writings from the world to the thinking self.

Descartes claim for mind's ability to construct a new world in *Le monde* builds on Montaigne's exploration of the same theme in "Des coches," the late essay on the discovery and conquest of the Americas. Montaigne's engagement with an Epicurean vision of the cosmos in that essay, where the mind makes and explores infinite worlds, haunts Descartes's own foray into

worldmaking.[71] Where Montaigne, drawing on Cicero, indulged in a potentially atheistic picture of a human mind shaping the forms of the phenomenal world, Descartes now returns to investigate the nature of worldmaking
itself by allowing an original divine act to slip into human hands.

Only too aware of this problem, Descartes follows the cosmogonic account in *Le monde* with an ingenious disclaimer that avoids confronting the
historicity of Genesis directly but which specifically invokes divine authority
for his own imaginary world. At the end of the sixth chapter, he slyly claims
that "[Au] lieu que, pouvant distinctement imaginer tout ce que j'y mets, il
est certain qu'encore qu'il n'y eût rien de tel dans l'ancien monde, Dieu le
peut toutefois créer dans un nouveau: car il est certain qu'il peut créer toutes
les choses que nous pouvons imaginer." (Since, everything that I set out here
can be distinctly imagined, it is certain that even if there were nothing of this
kind in the old world, God could, nevertheless, create it all in a new one:
for it is certain that he can create all the things that we can imagine.)[72] As
a fiction, Descartes's new world pretends to function merely as a counterfactual hypothesis ("qu'il n'y eût rien de tel dans l'ancien monde"), but in
fact its status as a possible world which God *could have* created suggests that
the "ancien monde" of the schoolmen, derived from Genesis, is not the only
epistemically sound, divinely sanctioned theory of world order.[73]

Such a craftily defended alternative to scholastic philosophy effectively
destabilizes the Thomist-Aristotelian interpretation of the Genesis account
and suggests how a mechanist, Copernican science could indeed be compat-
ible with theological truths.[74] As Descartes would write in the "Letter to Di-
net" that accompanied the second edition of the *Meditationes,* theology and
the new natural philosophy did not have to be at odds: "As far as theology is
concerned, truths can never be in conflict with one another, and it would be
impious to fear that any truths that philosophy discovers could be in conflict
with those of the faith. Indeed, I maintain that there is no religious matter
which cannot be equally well or even better explained using my principles
than by using those commonly accepted."[75] Descartes's exploration of the
relationship between human knowing and divine making in *Le monde* suggests how his new system could claim divine sanction. If the *possible truth* of
his fiction was being grounded in God's potential ability to create anything
that we can imagine, to question the verity of his imaginary world would
then be to question the infinite creative power of God himself.[76]

But Descartes is fully conscious of the dangerous instability introduced
into his philosophy by the use of heuristic fiction-making in *Le monde*. His
return to the scene of creation in the *Discours de la méthode* carefully omits
the troubling analogy between himself and God as creators of new worlds:

[Je] me résolus de laisser tout ce monde ici à leurs disputes, et de parler seulement de ce qui arriverait dans un nouveau, si Dieu créait maintenant quelque
part, dans les espaces imaginaires, assez de matière pour le composer, et qu'il
agitât diversement et sans ordre les diverses parties de cette matière, en sorte
qu'il composât un chaos aussi confus que les poètes en puissent feindre, et que,
par après, il ne fît autre chose que prêter son concours ordinaire à la nature, et
la laisser agir suivant les lois qu'il a établies.[77]

[I resolved to leave all of this world to their disputations, and to speak solely
of what would be present in a new one, if God created right now somewhere
in the imaginary spaces, enough matter to compose it; and that he animated
the diverse parts of this matter differently and without a fixed order so as to
compose a chaos as confused as that which the poets could invent, and that,
thereafter, he did nothing other than lend his ordinary concurrence to nature,
leaving her to follow the laws which he had established.]

The conceptual structure of Le monde remains intact, but Descartes avoids
any ambiguities about creative agency. He removes the formal parallels to
Genesis even as he retains the three central ideas that form the basis of
his challenge to scholastic Aristotelianism: the notion of a transcendent,
omnipotent God who creates a universe composed entirely of inert matter, the emphasis on the laws of nature established and guaranteed by an
immutable God, and perhaps most important, a *developmental* account of
world-formation moving from chaos to order.[78] The narrative development
of the imaginary world from chaos transforms the mysteries of Genesis into
a properly explanatory account of the creation.[79]

But this explanatory move masks the central shift of authority in Descartes's philosophy: creation occurs because of the laws of nature acting
on inert matter, *not* through the active, interventionist agency of God. One
of the key aspects of Descartes's physics—and his later metaphysics—is,
in fact, the rather distant, intellectually central but spiritually inaccessible
figure of God, who seems to be a rational principle rather than a vital force
of divine action. There is no Platonic demiurge with an artisanal touch in
Descartes's imagination. The world can no longer be physically shaped,
contained in a book, or on a globe; its immense totality can only be imagined into existence through the exercise of the thinking mind, the *cogito*.

A World without God?

It is easy to see why Descartes would hastily suppress Le monde, with its
formal imitation of Genesis, after the condemnation of Galileo. More interesting, however, is the Cartesian afterlife of Le monde's foray into reimagin-

ing the original creation of the world. Descartes would never again directly place himself alongside God as a worldmaker, but in later works, particularly the *Meditationes,* he would try to resituate God with regard to the created cosmos, seeking to establish a legitimate basis for human knowledge of the world. If Cartesian philosophy has a "religious flavor," as John Cottingham observes, it is for this reason: God underwrites Cartesian physics, giving it a strong metaphysical basis; but this appeal to a transcendent deity simultaneously frees the Cartesian Meditator to uncover the geometric, mathematical structure underlying the natural world.[80]

Such a double-edged appeal to the divine has also contributed to the bifurcation of much contemporary Cartesian scholarship, dividing it between defenders of a devoutly Catholic, metaphysical Descartes and a rational, secular, scientific Descartes.[81] But these disagreements may ultimately emerge from a paradox within the Cartesian corpus. Like the double appeal to *both* God *and* the human imagination in *Le monde,* Descartes's vision of the world's foundation remains troubled by the question of agency: does the order of the world finally derive from a human or divine maker? Or more precisely, what is the relationship between human knower and prior (divine) making?

In an extraordinary reflection on the linked histories of philosophy and theology, the late Pope John Paul II touches on this very question, attributing the modern, post-Enlightenment turn away from God to Descartes:

> In the pre-Cartesian period, philosophy, that is to say the *Cogito* ('I am thinking') or rather *Cognosco* ('I acquire knowledge') was subordinate to *esse,* which was considered primary. For Descartes, by contrast, *esse* appeared secondary, while he viewed the *Cogito* as primary. This . . . marked the decisive abandonment of what philosophy had been hitherto, particularly that of St. Thomas Aquinas. . . . [For Aquinas] God as fully self-sufficient being (*Ens subsistens*) was considered as the indispensable support for every *ens non subsistens,* for every *ens participatum,* that is to say, for every created being, and hence for man. The *Cogito ergo sum* carried within it a rupture with this line of thought. The *ens cogitans* (thinking being) thus became primary. After Descartes . . . all that is *being*—the created world, and even the Creator, is situated within the ambit of the *Cogito,* as contents of human consciousness.[82]

While this analysis of the Cartesian *cogito* and its metaphysical foundations is problematic in many respects, John Paul's passionate demonization of a perceived Cartesian self-sufficiency highlights two key aspects of Descartes's philosophy and its centrality for a history of worldmaking and modernity.

First, it points to an instability within Descartes's metaphysics about the relative creative capacities of human and divine makers that we have glimpsed in *Le monde*. John Paul's accusation that Descartes subsumes "the created world, and even the Creator" into human consciousness goes beyond the familiar accusation of perspectivism. It argues that Descartes simply absorbs the creative capacity of the deity into the *cogito*, thereby reversing the traditional priority of divine creation over human imitation.[83] This claim is particularly interesting because it is not borne out by any of Descartes's texts, though it is a key element in attacks on Cartesianism from the seventeenth century onward. But the demystification of the divine that John Paul attributes to Descartes and his supposed turn toward the fragile but powerful potentials of human making *is* a familiar element in worldmaking texts throughout the early modern period, from Mercator to Milton.

In the Third Meditation, Descartes is at pains to ground the *cogito*'s ability to know in a beneficent creator-God, just as the narrator's vision of the new world in *Le monde* is also rooted in divine sanction. But it is difficult to shake the sense that by using God as the metaphysical foundation for his physics, Descartes simply "kicks God upstairs," thereby separating God from the workings of the world even as he grounds claims to knowledge of the world in God's continuous creation and conservation of it.[84] John Paul's argument touches precisely on this seeming paradox between a God-centric world and a world devoid of divine intervention. It gets to the heart of how Descartes manages to establish a theologically acceptable metaphysics for the new Copernican science: Cartesian metaphysics simultaneously acknowledges the role of God but is also able to distance thorny questions about divinity from natural philosophy's investigation of the world. Descartes's philosophy from *Le monde* onwards strategically revises the relationship between the natural world and the divine.

Second, and equally significant, John Paul's assertion that Descartes breaks with Aquinas goes beyond the classic observation that Cartesian philosophy rejects Thomist scholasticism and points instead toward another theological bugbear. In the *Meditationes*, Descartes pointedly rejects the cosmological arguments for the existence of God, proposed most prominently by St. Thomas, and, after the causal arguments of the Third Meditation, adopts a version of the ontological argument, first proposed by St. Anselm, in the Fifth Meditation. This move has several implications. Most obviously, it sharply decouples knowledge of the world from knowledge of God. By rejecting *a posteriori* arguments for the existence of God, Descartes takes a step toward defusing any potential theological problems created by a new materialist and Copernican physics. The claims of the new science thus cannot have any purchase on theological truths. By reorienting philosophical theology away from cosmological arguments and toward ontological ones, Descartes is still able to argue for existence of a creator-God who underwrites the laws of nature—without appealing to the senses—by using his rational intellect. The result is an elegant metaphysics that is perhaps radically threatening to traditional theology, as John Paul would argue as recently as 2005. Descartes affirms the presence of a transcendent God who provides the metaphysical foundation for knowledge and the world; but this deity is also strategically removed from the world itself in both physical and ethical terms.

As Stephen Gaukroger notes, mechanism required a "particularly unyielding conception of God's transcendence," redirecting philosophy away from divine and human engagement with the phenomenal world and toward the mind's relation with a transcendent God who creates empirical truths. This cognitive process demanded an initial withdrawal from the sense impressions of the world before it could initiate a return to that world of embodiment, a move that Descartes describes in the Sixth Meditation. There, Descartes finally seems to elide nature and divinity: "There is no doubt that everything that I am taught by nature contains some truth. For if nature is considered in its general aspect, then I understand by the term nothing other than God himself, or the ordered system of created things established by God."[85] In this moment, God is at once absorbed into the world and quietly removed from it.

Atheism and Censorship

Descartes's vision of a transcendent God attracted the ire of both Protestant and Catholic authorities. From the battle with Voetius and trial of Utrecht in the early 1640s to the condemnation of his works by the Congregations of the Holy Office and the Index in 1663, the polemic against Descartes's philosophy turned on the sense that Cartesian mechanism sought to separate God from the world. Voetius's accusations against Descartes were, in fact, built on the suspicion of atheism that grew out of the argument for a transcendent God. Martin Schook's *Admiranda methodus* compares Descartes to Lucilio (Giulio Cesare) Vanini, famously burned at the stake in Toulouse in 1619 for his heresies. Schook argues that Descartes had set aside traditional proofs for the existence of God (that is, the cosmological argument) and instead advanced proofs so weak that readers might deny the existence of God altogether—a tactic that Mersenne, among others, had attributed to

Vanini.[86] Descartes's discrediting of *a posteriori* proofs of God's existence, his assertion of an innate idea of God, and his suggestion that God is *causa sui*, all derive from his powerful separation of God from the phenomenal world.

Similarly, the evaluation of Descartes's works by the Holy Office was initiated by the internuncio in Louvain who complained of a renewal of the errors of Democritus and Epicurus—in other words, of suspected atheism. In this case, however, the looming threat concerned the impact of Cartesian physics on the doctrine of transubstantiation. In his censure, Stephanus Spinula, one of the two Roman qualificators assigned to evaluate Descartes's work, distills these concerns down to the implications of Descartes's transcendent, non-interventionist portrayal of God, as he notes among his *animadversiones*:

> First, that God from the beginning created, at one and the same time, all the material beings that now exist, have existed, and will exist. From this it follows that there is no generation of material substances or new production of material qualities. . . . Second, that there is no prime matter, nor any material substantial form. . . . Fifth, that the universe of corporeal substances has no limits and is delimited by no place; and that all possible worlds in fact exist.[87]

These physical theories have a direct bearing on the Genesis account and, as Spinula clearly recognizes, challenge literal interpretations of Genesis upon which the hybrid of Christian cosmology and Aristotelian physics rested. He is quick to notice that Cartesian physics hints at an infinite universe, even though he extends its claims in the un-Cartesian direction of a plurality of worlds. The decentered nature of the cosmos, on Spinula's rendering of Descartes, emerges narratively from an initial separation of God from the world at the original moment when God created everything "at one and the same time."

All five objections that Spinula raises are already present in the early summary of *Le monde* that Descartes had inserted into the *Discours de la méthode*; only one (the fourth) actually concerns heliocentrism and the motion of the earth, which are traditionally noted as grounds for the condemnation. Indeed, Spinula's censures are based partly on a passage in the *Principia* that is almost identical to discussions of divine agency in the *Discours* and *Le monde*:[88]

> Ce peu de suppositions me semble suffire pour m'en faire servir comme de causes <ou de principes,> dont je déduiray tous les effets qui paroissent en la nature, par les seules lois cy-dessus expliquées. Et je ne croy pas qu'on puisse

imaginer des principes plus simples, ni plus intelligibles, ni aussi plus vraysem-
blables, que ceux-cy. Car bien que ces lois de la nature soient telles, qu'encore
mesme que nous supposerions le Chaos <des Poëtes, c'est à dire une entiere
confusion de toutes les parties de l'univers,> on pourrait tousjours demonster
que, par leur moyen, cette confusion doit peu à peu revenir à l'ordre qui est à
present dans le monde, & que j'aye autrefois entrepris d'expliquer comment
cela auroit pû estre.

[These few assumptions seem to me to be sufficient to serve as the causes <or
principles> from which all the effects observed in our universe would arise in
accordance with the laws of nature set above. And I do not think it is possible
to think up any alternative principles for explaining the real world that are
simpler, or easier to understand, or even more probable. It may be possible to
start from primeval chaos <as described by the poets, i.e. a total confusion in
all parts of the universe> and deduce from it, in accordance with the laws of
nature, the precise organization now to be found in things; I once undertook to
provide such an explanation.]

(*Principia*, 3.47)[89]

Descartes gestures back here to the following moment in *Le monde*, a kind
of primal scene for Cartesian philosophy:

Car Dieu a si merveilleusement établi ces Lois qu'encore que nous supposions
qu'il ne crée rien de plus que ce que j'ai dit et même qu'il ne mette en ceci
aucun ordre ni proportion, mais qu'il en compose un chaos le plus confus et le
plus embrouillé que les Poètes puissent décrire: elles sont suffisantes pour faire
que les parties de ce chaos se démêlent d'elles-mêmes et se disposent en si bon
ordre qu'elles auront la forme d'un Monde très parfait et dans lequel on pourra
voir non seulement de la lumière, mais aussi toutes les autres choses, tant géné-
rales que particulières, qui paraissent dans ce vrai Monde.[90]

[For God has so marvelously established these Laws that, if we suppose that
he created nothing more than what I have said and that sets nothing in any
order or proportion, but that he composes from it a chaos as confused and as
muddled as any the poets could possibly describe; these laws are sufficient to
ensure that the various parts of this chaos disentangle themselves and arrange
themselves in such order that they have the form of a perfect world in which we
can see not only light, but all other things, general as well as particular, which
appear in this real world.]

(*Le monde*, 6)

The repetition of terms, phrases, and entire clauses from one text to the
next should alert us not only to the conceptual novelty of Descartes's theory
of creation but also to the strategic role that God plays. While the *Principia*

carefully omits an explication of God's precise role in the creation, *Le monde* fills in the gaps: God only creates that primal chaos and establishes the laws of nature. These empirical laws then become the underlying order according to which matter transforms itself into "un Monde très parfait." Descartes's allusion here to the chaos of the poets may allude to Ovid and Lucretius, but more importantly it connects the creative capacities of human writers to the creative agency of God, suggesting that human reflection and imagination may capture the truth of divine making. Later, in the *Principia,* Descartes adopts a cautious, conditional tense ("nous supposerions . . . on pourrait"), clearly disentangling the real from the hypothetical. If the exuberant worldmaking of *Le monde* demanded a productive fusion of human and divine, by the 1640s the philosophical textbook that is the *Principia* had to cover its potentially heretical tracks.

CARTESIAN ROMANCE

The contemporary theological response to Descartes's philosophy suggests that arguments surrounding Cartesian science, unlike the debates over the work of Galileo and Kepler, were not limited to natural philosophical questions such as heliocentrism, the motion of the earth, the infinite extent of the universe, or the more contentious theory of vortices in a fluid cosmos. They also repeatedly return to the audacity of Descartes's conception, its hypothetical expression, and its implications for the Mosaic account of divine creation. Newton would note, with a measure of disgust, that it is "unphilosophical to seek for any other Origin of the World, or to pretend that it might arise out of a chaos by the mere laws of Nature."[91] And further: "[It] is not to be conceived that mere mechanical causes could give birth to so many regular motions. . . . This most beautiful system of the sun, planets and comets, could only proceed from the counsel and domination of an intelligent and powerful being."[92]

Newton is concerned here not with the principles of Cartesian mechanism *per se* but with Descartes's "unphilosophic" desire to apply those principles to an explanation of the world's origin, which would eventually displace the Genesis account. Like Voetius and Spinula, Newton's objection too stems from Descartes's vision of a transcendent God removed from the world; in the Newtonian account, natural philosophy approached natural theology through its *a posteriori* confirmation and celebration of God's design. Thus for Newton—and his colleague Robert Boyle—any attempt by natural philosophers to engage or comment on scripture, and thus to produce a new theory of the world, was philosophically suspect. Paolo Rossi explains this dilemma succinctly: "For Newton, the biblical account tells real facts in a 'popular' manner. But if these same facts could be explained 'mechanically' or 'scientifically,' would not the biblical account necessarily end up as just an 'allegory' or a metaphorical description of a different process of events? Would not that narration come to be dissolved into a different narration?"[93]

This is, of course, precisely what the imaginary world of Le monde had to accomplish if Descartes was to establish an entirely new vision of the world and its order. The epistemological status of Genesis had to be subtly altered from factual narrative to probable fiction or "allegory," so that the fictional Cartesian account could then become its natural philosophical explanation. To achieve this delicate balancing act, Descartes had to draw on the authority of scripture while simultaneously disentangling it from the dogmas of scholastic interpretation. His solution is to harness the potential of divine creative agency for his own philosophical project.

By making the suggestiveness of fictional verisimilitude central to the worldmaking enterprise, Descartes decisively revises philosophy's rules of engagement with the world. To know the world, he suggests, is to recreate it in the intellectual imagination. Such an emphasis on Descartes's investment in literary form challenges the traditional portrayal of Cartesian thought as hostile to the poetic imagination and responsible for the demise of poetry's claims to truth.[94] And yet, even in his lifetime, Descartes's philosophy was frequently attacked for its literariness, its rhetorical efficacy, and its emphasis on a kind of figurative language that was powerful and insidious enough to disseminate a new philosophical paradigm without reprisal.[95]

Once again, Galileo presents a useful counterpoint. Where Galileo writes a dialogue, examining and destroying arguments against Copernicanism, Descartes strategically offers a fable. If Galileo, the lover of Ariostan romance, is compared by his most recent biographer to Don Quixote rushing at full tilt into the windmills of the Inquisition, Descartes too would be condemned, by no less than Blaise Pascal, as a Don Quixote trapped in a "romance of nature."[96] Such comparisons hint at the problem of "Cartesian romance" which plagued the followers of Descartes for much of the late seventeenth and eighteenth centuries, and trouble easy characterizations of the scientific revolution, the rise of empiricism and rationalism, and the diminution of poetry's place in modern intellectual life. Instead, Descartes's reorganization of the world's frame through the fictional form of Le monde suggests that to understand the modern reconsideration of the relations between mind and world, we may have to recover a submerged narrative about worldmaking and the poetic imagination in the seventeenth century.

Christiaan Huygens astutely describes the problem in a 1693 letter to Pierre Bayle:

> Mr. des Cartes avoit trouvè la maniere de faire prendre ses conjectures et fictions pour des veritez. Et il arrivoit a ceux qui lisoient ses Principes de Philosophie quelque chose de semblable qu'a ceux qui lisent des Romans qui plaisent et font la mesme impression que des histoires veritables. . . . Il devoit nous proposer son systeme de physique comme un essay de ce qu'on pouvoit dire de vraisemblable dans cette science en n'admettant que les principes de mechanique. . . . Nonobstant ce peu de veritè que je trouve dans le livre des Principes de Mr. des Cartes, je ne disconviens pas qu'il ait fait paroitre bien de l'esprit à fabriquer, comme il a fait, tout ce systeme nouveau, et a luy donner ce tour de vraisemblance qu'une infinite de gens s'en contentent et s'y plaisent.[97]

> [Mr. Descartes has found a way to have his conjectures and fictions taken for truths. And so, people who read his *Principles of Philosophy* experience something similar to those who read delightful romances and it has the same effect as seemingly true stories [*histoires veritables*]. . . . He should have given us his system of physics as an essay so that we could distinguish what was plausible [*vraisemblable*] in this science, and not admitting anything other than the principles of mechanics. . . . Notwithstanding this tiny bit of truth that I find in the book of Mr Descartes's *Principles*, I do not disagree that he has made apparent his gift for building, as he has done, this entire new system, and has given it this impression of verisimilitude [*ce tour de vraisemblance*] that an infinite number of people are satisfied and delighted by it.]

Huygens's complaints are directed toward matters of literary form rather than substantive matters of physics: he is perturbed that Descartes's philosophical treatise offers the same intellectual pleasure and narrative drive as romances or fictional histories. But it is also this effect of *vraisemblance* that makes Descartes an accomplished system-builder.

The great Cartesian novel, Huygens suggests, is built upon a fusion of literary and philosophical concepts of plausibility, which create a persuasive, comprehensive vision of the world, even if its physical theory is suspect. This fluidity between fact and fiction, and the troubling mélange of philosophical truth and aesthetic pleasure, is of course an intentional effect that Descartes labored to create. Thus, Huygens's critique becomes a powerful acknowledgement of Descartes's successful project of philosophical revision and sheds light on its key features. Indeed, Huygens resorts to the very literary analogy that Descartes himself uses when he compares the work

to "romans qui plaisent." In his preface to the *Principia,* Descartes advises
the reader to "lire ce livre . . . d'abord tout entier ainsi qu'un roman" (to
read through the entire work, as one would a romance).[98] This advice for
reading the *Principia*—a mammoth work in four parts with over five hundred and forty individual theses—seems a peculiar one, but it suggests how
Descartes may have understood the epistemic function of his work. Though
the *Principia* is the scholastic textbook version of *Le monde* stripped of all its
fictional trappings, Descartes still wants the work to be read for narrative
rather than propositional logic; the rhetorical and epistemic thrust of the
book, he seems to suggest, will be revealed through its overall plot rather
than its individual arguments.

A similar analogy, made by Robert Boyle in 1674, explains Descartes's
vision. "The Book of Nature," writes Boyle, is "as a well contriv'd Romance
[where] the parts have such a connection and relation to one another . . . that
the mind is never satisfied till it comes to the end of the Book."[99] The complex interconnectedness of romance narratives, composed of diverse, seemingly unrelated details, provides a pointed parallel to the activity of natural
philosophy which also seeks to establish the complex interconnectedness
of seemingly unrelated phenomena in the natural world. As an expansion
on the "book of nature" *topos,* the narrative analogy suggests that interconnectedness is the *condition* of the natural world, whose signs reveal a densely
woven but coherently ordered whole—if only we keep reading. For both
Descartes and Boyle, the rhetorical form of narrative fiction becomes a potent means of knowing because it mimics the narrative of nature itself.

The image of Descartes's ideal reader eagerly devouring the *Principia* as
a romance serves as a useful gloss on *Le monde,* his only extensive venture
into the realm of narrative fiction. The fabular form of this work has long
troubled historians of philosophy and science, who have argued either that
this early foray into literary imagining reflects the lack of adequate ma-
terials for a metaphysically grounded natural philosophy or that it serves
purely tactical ends by hiding Descartes's Copernican sympathies.[100] But
such claims reduce the rhetorical complexity of *Le monde*'s fabular structure
to the status of a straightforward natural philosophical essay and miss its
epistemological inventiveness. Philosophical texts, like romances, must fas-
cinate, tease, and grip the intellect by making their subjects visually present
and instantly persuasive—particularly when their goal is to explain and
encompass the entire world.

This underlying literary logic of worldmaking emerges to the fore in
the seventeenth and eighteenth centuries through debates over the relative

merits of Cartesian "romance" and Newtonian "realism."[101] When Newton famously declares "Hypotheses non fingo" in the *Principia mathematica,* he almost certainly has in mind Descartes's contrary assertion in the *Principia philosophiae,* also at 3.47, that "the falsity of these suppositions does not prevent the consequence deduced from them being true."[102] The conflict cuts to the heart of Cartesian epistemology and its reliance on the power of the poetic imagination. Newton's determined attempt to distance his project from the taint of feigning thus becomes an attempt to extricate natural philosophy not merely from what he perceived to be a methodological mistake but more importantly from crucial epistemic problems introduced into it by Descartes.

As Newton well understood, Descartes uses fictional elements in his philosophy to erase the ontological difference between various categories of thought and make possible an entirely new set of philosophical connections. The imaginary world of *Le monde* for instance claims to help distinguish the real world as it materially exists from a heuristic model used to explain a new mechanistic philosophy—but, in effect, it both exposes the difference between philosophical system and material world and collapses it. We can only understand the order of the (real) world, it seems, through philosophical systems that are self-created and self-imposed.[103] While Descartes himself was not a relativist (or atheist) in any sense, the narrative exposition of *Le monde,* and the strategic reorientation of God's relation to the world in the *Meditationes,* open a crack in philosophy's imagining of the cosmos which would subsequently take a radical turn in Hume's attack on worldmaking in the *Dialogues Concerning Natural Religion* (1779).[104]

The potent mixture of fiction and scientific hypotheses in Descartes's work necessarily drew him outside the realm of philosophy as traditionally understood and placed him in the heterogenous company of worldmakers, as Newton, Boyle, and a host of other philosophers and commentators recognized. Thus, when in 1721 Matthew Prior fulminated against "system-makers and world-wrights," he may well have had Descartes and his followers in mind.[105] It is a revealing historical irony that the word "worldmaking," now a privileged and hotly debated term in philosophy, originated as a pejorative term in the seventeenth century to denounce people—most notoriously Descartes—who advanced hypotheses about the creation of the world.[106]

CHAPTER FIVE

"This Pendant World": Creating Miltonic Modernity

"Thus thou hast seen one world begin and end," the archangel Michael tells Adam, at the beginning of the final book of *Paradise Lost*, pausing momentously "betwixt the world destroyed and world restored."[1] Awestruck and appalled, the first man is silenced by the grand spectacle of creation and destruction. Until this point, he has restlessly questioned what he sees, transforming the middle books of the epic into an extended dialogue on the nature of the universe. In this silence, in the long pause that opens the final book poised between old and new worlds, Milton reveals not only an unfolding biblical drama but also a historical transition of global, and yet deeply personal, dimensions.

Milton himself must have felt both the dumb-struck wonder of Adam as well as the triumphant revelatory power of Michael. His life spanned the better part of a century that historians repeatedly identify as the bridge to a recognizably modern world, defined variously in terms of scientific advancement, socio-economic upheaval, and political revolution. And his long career, from the early poetry of the 1640s and the notorious regicidal pamphlets of the 1650s to the fallen, "evil days" of the Restoration, was entwined politically and intellectually with many of those changes. This proximity to major historical events of his time gives Milton's writings—particularly the retrospective later works—a unique doubleness, a Janus-like quality that balances a belated, nostalgic humanism that would not have been out of place in the sixteenth century with an early intellectual liberalism that would have been at home in the eighteenth. Poets before Milton had described the world, but it had been a finite and perfectly crafted entity, comfortably limited in scope. The sublime sweep of the narrator's gaze across the created universe in *Paradise Lost*, from the "dark Abyss," the realm of Chaos and eternal Night, to the empyreal heaven, the plurality of planetary systems and "this pendant world," now carries a thrilling, double power: a new humbling apprehension of the uncontrollable vastness of space and time, but also the individual poet's triumphant ability to contain it all within his imaginative grasp.

REINVENTING THE WORLD

Milton's universe teems with worlds. Raphael sails through "worlds and worlds" when he glides down from Heaven on a sunbeam; Satan sees "innumerable stars, that shone / Stars distant, but nigh hand seemed other worlds / Or other worlds they seemed . . ."; Uriel reminds us that God "hath bestowed / Worlds" on Adam and Eve.[2] If empirical investigation alone could establish the subject of a poem with any certainty, *Paradise Lost* would be a poem about the world: the word "world" appears one hundred and thirty times in the epic, more often than either Adam or Eve, and cedes place in terms of frequency only to God and Earth.[3] The adjectives most frequently applied to the world in Milton's poem are familiar ones— "new," "new created," "other," or "another"— calculated to recall Europe's encounter with America.[4] And like the new geographic discoveries of the early modern period, the worlds of *Paradise Lost* are novel and unfamiliar. They lurk at the edges of the known in the vast sea of infinite cosmic space, like the Hesperian isles "famed of old"; and like the islands of the Mediterranean or the Americas, they invite travelers and theorists to make sense of their origin and order.

Profuse and multiple, the worlds of *Paradise Lost* thus map a new cosmic and intellectual consciousness where knowledge of "the world" is never complete, never finite. As the scope of "world" blurs into the boundless universe, there are always new worlds being created, new knowledge to be discovered. Writing in the wake of Montaigne, Galileo, and Descartes, who struggled variously with the skeptical problem posed by this rapidly changing world-picture, Milton inherits the challenge of comprehending the whole, of seeing "things invisible to mortal sight." His acute awareness of this epistemological limitation—the need to capture the world's totality despite the fundamental impossibility of the task—fuels his epic quest to explore the implications of worldmaking.

Milton goes beyond his literary predecessors, Camões and Spenser, who had tested the worldmaking potentials of the epic through allegory and classical myth. *Paradise Lost,* by contrast, reimagines the *urtext* of European worldmaking itself as Milton defiantly takes on the subject of Genesis—the topic that even Descartes had abandoned with *Le monde.* Genesis presented a dangerous invitation to would-be worldmakers, for it promised cosmic renovation on an unprecedented scale. Like the cosmos whose creation it described, the status of Genesis as the ideal, unshakeable union of scientific knowledge and salvation history was increasingly under assault in the seventeenth century. To shore up the subject of Genesis was effectively to reinvent the world. Milton was thus not merely drawing on scripture to buttress the claims of his epic and his own poetic fame; rather, his epic sought to remake the world and reimagine its sacred origins for a post-Cartesian era.

Genesis and Its Authors

The plot of *Paradise Lost* and its origin in the opening chapters of Genesis have become so thoroughly naturalized within European literary history that it is difficult to reconstruct the shock of readers upon its publication. Reacting violently to Milton's intellectual audacity, readers labeled him an atheist, suggesting that the poem's heterodox treatment of the foundational text of Christian theology was an act of destruction rather than creation.[5] Even the admiring Andrew Marvell, in his dedicatory poem to the second edition, compares Milton's achievement to Samson's destruction of the Temple: "(So Sampson groped the Temple's post in spite) / The world o'erwhelming to revenge his sight."[6] Such ferment reflects the gravity of Milton's choice: to rewrite Genesis in the late seventeenth century was to enter pointedly into debates about the nature of the world, to take a stance with respect to the "new philosophies," and to intervene in the vexed matter of God's engagement with his creation.

Like Montaigne, Spenser, and Descartes before him, Milton takes the measure of man and the universe and finds the old relations of order and proportion wanting; unlike them, however, he directly confronts the implications of a changing world-picture for religious faith. *Paradise Lost* asks how the new global consciousness and cosmic view affect Genesis, the foundation of Christian moral philosophy, natural history, and cultural teleology. How did a post-Columbian and post-Galilean perspective on the world, with its skeptical investigation into the nature and order of things, its acknowledgement of cultural heterogeneity and difference, affect the ancient tale of the divine making of the world?

By the 1660s, when *Paradise Lost* was largely written, the familiar outlines of the Ptolemaic *oikoumene* with its concentric celestial spheres had definitively given way to a Copernican, mechanistic cosmos and to now well-established commercial and geo-political interactions from the Americas to the Far East. Newton had made his key discoveries in differential calculus; Boyle conducted experiments on the properties of air and described them at the newly chartered Royal Society; the English East India Company established its first trading posts in Masulipatam, Surat, and Madras; and John Locke was drafting the Fundamental Constitution of the Carolinas. Against this backdrop, the subject of Genesis was a potentially incendiary one because it touched on two problems at the heart of seventeenth-century thought.

The growing persuasiveness of naturalistic explanation associated with the New Science was challenging literal interpretations of Genesis and leading many to demand a new synthesis between the book of nature and the book of scripture. Already, at the beginning of the century, Galileo would invoke St. Augustine's commentary on Genesis to defend himself from threats of heresy and to argue that Copernicus "understood very well that if his doctrine was demonstrated it could not contradict the properly interpreted Scripture."[7] If the biblical account of origins, properly interpreted, was to provide an effective foundation for a modern world-system, it required powerful reimagination and retelling in response to the empirical observations of natural philosophers. More fundamentally, however, the breakdown of old systems of world order and their gradual reestablishment not by divine but by human means in the form of new scientific, political, and ethical theories and narratives raised troubling questions about the very necessity of a concept of deity to understand the world. To examine Genesis in the light of a new world-picture was therefore to ask the almost unutterable question about the nature and existence of God and his relationship to human making.

Paradise Lost stages this dilemma as a clash of worldmaking narratives, from Satan's colonizing assault on Eden to various characters' retellings of the Genesis story in their quest to understand how the world was crafted. This preoccupation with the literal act of making the world—the entire cosmos, but also the many "worlds" within it—marks Milton's sharp rejection of the epic's favorite theme of imperial desire, which now literally becomes a Satanic practice. In its place, the poet fully embraces the cosmological turn initiated by Camões and Spenser. The poem repeatedly juxtaposes competing creation accounts: the opening invocation in book 1, Uriel's and

Raphael's strikingly different descriptions in books 3 and 7, Satan's shocking denial of divine creation in favor of self-begetting and Abdiel's furious insistence on created being in book 5. These culminate in a speculative dialogue on world systems that Milton dramatizes in book 8 between Adam and Raphael. Taken together, these narratives produce a strongly skeptical approach to the matter of creation itself—and thus to the question of God's relation to the world—as they point to a fundamental mystery at the world's origin. For Milton, even more than for Descartes, skepticism about our ability to understand the world empirically through the senses—"the unspeakable desire to see, and know" (3.662)—leads to a powerful affirmation of faith in the divine, in "the things invisible to mortal sight." At the same time, such epistemic uncertainty heightens the importance of humanly made fictions of the world—fictions, like *Paradise Lost* itself, which give the decentered, amorphous "new world" of early modernity a renewed coherence, order, and harmony. The poem thus revels in a newfound understanding of world order as humanly and poetically created, even as it longs to defend belief in an absolute, divinely established order of things.[8] Caught between human and divine makers, *Paradise Lost* becomes a powerful investigation of *how* new world-systems are produced, where they originate, and why we need such narratives of overarching order when faced with a transition to modernity.[9]

Milton's epic partakes in the double-edged promise of the post-Cartesian moment as it explores the ramifications of continuing to believe in Genesis. While a sweeping poetic imagination with a cosmic gaze is everywhere in evidence, Milton's defense of a divine creator repeatedly highlights both the power and the metaphysical isolation of the human thinker who dares to deny a prior model. Though he delights in the power of his own artfulness, he is painfully aware of the risk involved in human creativity, a sensation that Eugenio Garin famously described as "standing on the edge of an abyss."[10] To assert "the metaphysics of man the creator" was, as the early humanists knew well, to challenge the priority of the divine Maker. It required embracing the fear of cosmic loneliness, so resonantly evoked by Pascal, who would write of the new mechanistic cosmos in the late 1650s, "le silence eternel de ces espaces infinis m'effraie" (the eternal silence of these infinite spaces terrifies me).[11] In its engagement with Genesis, *Paradise Lost* addresses this tension between divine and human creations, confronting the skeptical dilemma of partial knowledge and fallen language as well as the emotional need for a stable, divinely sanctioned foundation. The poem's theodicy is therefore inextricably connected to the problem of worldmaking—whether the original creation of the world, or the poet's modern act of re-creating a new world system.

Cosmos and Hearth

The poetic form of *Paradise Lost* yearns to mimic the divine order of the world. Perhaps more than any other early modern text, the poem zooms through dizzying changes of scale, from the wide vistas of the cosmos to the precise enumeration of Eden's flora, from the wild ringlets of Eve's hair to the dramatic collision of inchoate matter in the boundless sea of Chaos. Each detail seems to celebrate Raphael's evocation of the intricate *scala naturae*, the tightly structured hierarchy of the created cosmos whose elements are bound together by "one first matter all." But in its attention to form—cosmic, poetic, political, philosophic, and theological—*Paradise Lost* struggles with the ethical consequences of a deeply interconnected view of world order and the implications of such changes of scale for quotidian living.

Milton's epic has been celebrated as both the first to measure the "void profound" of modern cosmic space and the first to focus on bourgeois domestic matters. It juggles the opposing attractions of home and world, the intimate circle of Adam and Eve "imparadis'd in one another's arms" with the lure of "spaces incomprehensible" in the vast universe. To connect these two extremes of scale, Milton returns to the ancient analogies between self and world, redefining the relations between human action and cosmic knowledge. Writing almost a century after Mercator and Montaigne, Milton struggles with a distinctly modern tension between what the geographer Yi-Fi Tuan describes as "cosmos and hearth," the "urge to stay in our burrow, to find sustenance in familiar things" and the "open spaces with their promise of a different, more challenging future."[12] Despite its nostalgia for prelapsarian harmony, an imagined past in which the world was perfectly ordered, *Paradise Lost* chronicles the final breakdown of the old link between microcosm and macrocosm as Nature groans at the Fall, at the severance between human and natural worlds. But in this cleavage, Milton sees the opportunity for a new, imaginatively forged basis for human connection with the world: Adam and Eve must find the "paradise within" as they look out at "the World . . . all before them." *Paradise Lost* thus suggests how worldmaking in a potentially infinite universe leads back to the circumscribed self.

Much has been written about Milton's reaction to the New Science and the peculiarly hybrid cosmology of his epic, which seems to move awkwardly between the closed spheres of a Ptolemaic world and the open cosmic panoramas of a Copernican universe.[13] But worldmaking for Milton is less a matter of cosmological hypothesis and more a matter of cosmic understanding. To understand the world is to grasp one's place within the whole—a particularly challenging goal if the whole is never entirely in view. *Paradise Lost* dramatizes this difficulty, asking how individuals might make ethical choices when the grounds of knowledge are insecure and the frame of the world feels unstable. In Raphael's turn away from the speculative pleasures of cosmology in book 8 to the domestic imperatives of "thyself and thy being," Milton does not reject the New Science; he marks an important copula between natural and moral philosophy. For even as the apocalyptic desire to uncover the absolute truths of creation and stable world order fails in *Paradise Lost*, Milton forges a fragile new ethics of living within a modern world of contingency.

GAZING ON THE COSMOS

As early as 1631, in his postgraduate years at Cambridge, Milton was dreaming of the cosmos. In "At a Vacation Exercise," the poem that concludes his Sixth Prolusion, the young poet imagines how "the deep transported mind may soare / Above the wheeling poles, and at Heav'ns dore / Look in, and see each blissful Deitie."[14] This vision of a human gaze lifted above the convexity of the earth and traveling through the cosmos is a familiar one. It looks back toward the tradition of the cosmic dream vision derived from Macrobius's *Somnium Sciopionis* and widely imitated in early modern cosmographic texts. And it anticipates, in its transgressive desire and creative confidence, the flight of the narrator "through utter and through middle darkness" to Heaven in book 3 of *Paradise Lost*.[15] As in his later epic, the early poem charts a voyage through the elemental spheres as it seeks to "sing of secret things that came to pass / When Beldam Nature in her cradle was." In Milton's hands, however, this central cosmographic trope of exploration undergoes an important change that parallels a larger shift in attitude toward the discovery of universal knowledge in the seventeenth century. Terrestrial travel and reportage alone were insufficient and no longer novel; to grasp the entirety of the world—now aligned with the vaster scale of the cosmos—was to travel, like Scipio, through the universe. Spatial extension demanded an intellectual pursuit through time, a desire to return to the original moment of the world's creation, much as Descartes had sought to do by imagining a second creation in *Le monde*.

Questions about creation and the nature of the world in *Paradise Lost* are punctuated by long views of the cosmos, as Milton juxtaposes cosmic travel with investigations into the structure of the universe. The poem interlaces the desire to "look in, and see" with attempts to excavate underlying systems of order, from the end of Book 2, where Satan looks out into the uncharted "secrets of the hoarie deep," his first assessing gaze over the created universe in book 3, and his subsequent dialogue with Uriel on the world's creation, to Raphael's parallel journey to Eden that spans the entire universe and Michael's final triumphant revelation of world history to Adam in books 11 and 12. This narrative strategy fuses three strands of early modern worldmaking: the cosmographic descriptions and travel narratives of the sixteenth century, the mapping impulse fueled by the new cartographic technologies, and the quest, usually associated with the new science of the seventeenth century, for a new physics that could explain the "laws of nature" that governed the world. All three come together in the epic's strategic allusions to Galileo and the telescope.

Each invocation of the astronomer in the poem's similes is a prelude to a visionary description of the expanding, multiple worlds of the cosmos.[16] The poem's repeated insistence on wonder at the discovery of new worlds— "new lands," "other worlds," "the universe," "shining globes," "imagined lands," "worlds and worlds"—occur alongside the Galileo similes, associating his observations with a wider interest in the changing shape of the world. In book 1, Satan's shield is compared to the moon, "whose orb / Through optic glass the Tuscan artist views / . . . to descry new lands, / Rivers or mountains in her spotty globe" (1.287–91). In a meditation on discovery, expansion, and changes of scale, Galileo's observations of the moon here are analogous to new world discoveries, even as they project terrestrial exploration outward into the reaches of the cosmos. The simile of celestial exploration here parallels and prefaces the fallen angels' own exploration of Hell, the new realm that lies on the edge of the cosmos, and which, like a dark double, anticipates the fallen human world. As a figure which literally telescopes space *and* time, the "optic glass" is reminiscent of the "long and prospective glass" of "At a Vacation Exercise" and marks a double moment of cosmological and spiritual instability: it suggests a contemporary breakdown of the old (Aristotelian-Ptolemaic) systems of order, just as it reflects back on the original metaphysical breakdown of cosmic harmony initiated by Satan's conquest of the Edenic world.

Satan's journey through the cosmos also culminates in a Galilean analogy. As the fiend travels "Amongst innumerable stars, that shone / Stars distant, but nigh hand seemed other worlds, / Or other worlds they seemed, or happy isles," eventually landing on the sun, he is compared to "a spot like which perhaps / Astronomer in the sun's lucent orb / Through his glazed optic tube yet never saw" (3.565–90). Once again associated with a change of scale, the image of the "optic tube" is responsible both for the liberation of the cosmic imagination, which now can see so many "other worlds," and for the collapse of the boundary between the supralunary and sublunary worlds—a collapse brought about, among other things, by the discovery of sunspots. Satan's association with sunspots suggests a reconfiguration of the link between the invisible, spiritual realm and the visible, material world, for the simile hinges on a seeming, but ultimately false, likeness between temporal and spiritual matters.[17] Galileo himself was a strong proponent of the separation of theology and science, of the invisible ways of God from the visible book of nature; the envious Satanic perspective on the cosmos, itself a parody of God's view over his works, accentuates this gap. Indeed, the simile may also glance at Galileo's *Letter to the Grand Duchess Christina* (published in 1636, shortly before Milton's travel to Italy), which lays out a position on the relation between scripture and science perhaps akin to Milton's own in *Paradise Lost*.[18]

Milton's final—and only named—reference to Galileo highlights the central epistemological challenge facing human worldmakers: the insurmountable chasm between an omniscient view of the world's totality and the inevitably partial human gaze. As Raphael "sails between worlds and worlds" speeding down to Eden on a sunbeam, his gaze over the cosmos is unimpeded: "no cloud, or, to obstruct his sight, / Star interposed" (5.257–58). To emphasize this clarity, Milton presents a deceptive simile: Raphael's angelic view is unlike the nocturnal gaze provided by "the glass / Of Galileo" which "less assured, observes / Imagined lands and regions in the moon," or even worse, the limited vision of the "pilot from amidst the Cyclades / Delos or Samos" who "first appearing kens / A cloudy spot" (5.260–68). Here, the enhancement of human vision by the invention of the telescope is revealed to be still partial, still "less assured."[19] It can enable a new glimpse of hitherto unknown worlds in the cosmos, but it cannot comprehend it.

Several of the poem's key themes repeatedly cluster within the Galilean similes: cosmic travel narratives, speculative visions of "other worlds," the epistemological problem posed by limited human vision, and the spiritual and metaphysical consequences of a shift of worldview. Taken together, these signal Milton's transformation of the epic into a genre that reflects on poetry's role in worldmaking. It is no accident that these images occur within similes, poetic tropes that invite and highlight the work of the imagination in analogic thought. In *Paradise Lost,* the poetic imagination absorbs and underwrites the system-building initiatives of mapmakers and astronomers like Mercator and Galileo; it alone can provide the limpid vision of the cosmic whole.

In this, Galileo becomes a "cryptic self-portrait" for Milton himself, a likeness that the poet ultimately surpasses.[20] Maura Brady argues that the problem with the telescope lay in the limitedness of the view it offered: one could never have a complete view of the cosmos (which Satan and Raphael are both given in *Paradise Lost*) when looking through the "optic tube."[21] While there is undoubtedly a moral injunction to be drawn from this epistemic limitation (and Raphael will make it in his admonition to Adam to "be lowlie wise" in book 8), it is precisely this gap between telescopic fixity and the supernatural gaze that provides the artist—both Tuscan and English—with the gift of epistemic openness, precipitating the flight of imagination. For dramatizing the difference between the two epistemological viewpoints is Milton's purpose: painstaking empirical investigation, which reveals visually verifiable pieces of the whole, must be combined with the precarious inductive leap of the speculative imagination in order to produce a view of the big picture, the cosmic panorama.

By connecting his own poetic practice, which soars and flies through the cosmos, with specific technologies of seeing and representing the world (cosmographic, cartographic, astronomical), Milton overgoes predecessors such as Camões and Spenser, who stopped short of thrusting the epic beyond the Virgilian dialectic of cosmos and imperium. In *Paradise Lost,* Milton transforms culminating images of cosmographic rapture, such as the global survey offered to Vasco da Gama by Tethys in the final cantos of *Os Lusíadas,* into a sustained series of narrative episodes that engage and surpass his major epic antecedents. The world is made visible in its entirety through the power of the poetic imagination.

Celestial Cartographers: Poetry and Cosmic Mapping

If earlier writers, such as Montaigne and Spenser, were fascinated by a cartographic imagination, Milton expands that probing cartographic gaze into the cosmos. In *Paradise Lost,* the travels of Satan and Raphael through the landscapes of the universe resemble the travels of New World voyagers and their favorite epic predecessor, Ulysses. Navigational metaphors familiar from discovery narratives are transferred to the skies as Milton tests the vocabulary of terrestrial mapping against the new imaginative frontier of cosmic space. Such a move was a staple of the *seicento* Italian epic tradition in the wake of Galileo: Marino's *Adone,* for instance, includes a prophecy about Galileo's telescope.[22] But Milton extends these narrow associations of Galileo and Columbus as analogous intellectual pioneers into a broader rumination on the consequences of an expanding world. When the narrator imagines "gentle gales / Fanning thir odoriferous wings" and whispering "whence they stole / Those balmy spoils" as Satan sails toward Eden, Milton translates Camões's lyric cosmography onto a galactic and spiritual plane: Satan is like "to them who saile / Beyond the Cape of Hope, and now are past / Mozambic . . ." (4.156–61). The enjambment and the contracted "Cape of Hope" suggests both Satan's spiritual and cosmographic location; he has crossed an invisible historical and moral boundary from which there will be no turning back, much like the Portuguese in their confrontation with Adamastor, the spirit of the Cape. The terrestrial conquests of the sixteenth century seem to anticipate the spiritual challenges and celestial expansions of the next century.

Such imagistic transfers, which collapse the boundaries between terrestrial and celestial planes, are also a characteristic feature of seventeenth-century natural philosophy and suggest Milton's complicated alignment with its aims. Catherine Gimelli Martin has argued, for instance, that Milton's allusions to Galileo may have a source in Francis Bacon's *Descriptio globi intellectualis,* where the philosopher praises Galileo's "optic instruments" as being akin to "skiffs and barks" that "have opened a new commerce with the phenomena of the Heavens"; in the same text, Bacon, like Milton after him, compares stars to "so many small and invisible islands" in a "sea of ether."[23] Bacon's language here has a Camonian bent and evidently draws on the same stock of literary tropes as Milton. But unlike the philosopher, the epic poet is at pains to separate the literal, political consequences of such images from their metaphoric, metaphysical thrust: commerce and conquest are diabolic activities in *Paradise Lost,* while contemplation and admiration are properly angelic. Satan and Raphael consequently offer two versions of the discovery voyage, *in malo* and *in bono,* reflecting morally opposed attitudes to the cosmic vision laid out by the divine worldmaker. The human pursuit of *scientia,* Milton suggests, is caught between these extremes.[24]

Against this context, Milton's allusions to Galileo take on a new force, for they act as glosses to *both* angelic and diabolic missions. If Galileo is the poem's emblem for the new science, it is neither dismissed as Satanic deception nor celebrated as a privileged means to spiritual understanding.

Instead, Milton counterpoints his superhuman characters' wide-angled perspective on the entire cosmos to the partial view of the human astronomer-mathematician, who, even with the enhanced vision of his new instrument, can only discern tantalizing glimpses of extramundane places and details that hint at the greater order of the universe. In his double association with Satan and Raphael, Galileo thus points to both the pitfalls and the potentials of the new natural philosophy. He becomes a charged emblem of the human desire to know the world, of human worldmakers who are heroic in their ambition but doomed by their material and epistemic limitations.

Like Milton, Galileo too yearns to "look in, and see," and thus the allusions to the telescope activate the poem's thematics of sight and insight, blindness and true vision. The doubling of stargazing with New World exploration implies a suggestive parallel between maps and telescopes as instruments that facilitate vision. Milton invokes the power of the telescope to play on a new understanding of spatial distance, where distant objects seem close and places hitherto unseen can be made visible.[25] This magnification and diminution of worlds extends the epistemological revolution wrought by the new cartography that had made rapid changes of scale ubiquitous. Galileo's invention of the telescope opened the entire cosmos to similar scrutiny: it revealed new objects and distant places to the human gaze, but as a consequence the mapmaker's dilemma of synthesizing so many new pieces of information into a coherent picture—and the consequent leap from empirical detail to speculative vision of the whole—became the dilemma of the astronomer.[26] Into this space between the individual observations of celestial motions and the desire for a synthetic picture of the entire system, the poet enters.

While the importance of romance narratives and fabular imagining for early representations of the New World is well established, it remains little noticed that a similar symbiosis between fiction and science animated the Copernican revolution in astronomy.[27] If maps and cosmographies of the sixteenth century reveal the indebtedness of European explorers to classical and medieval stories, the new celestial mapping projects of the late sixteenth and seventeenth centuries reveal just how deeply the new astronomy drew on practices of fiction-making. As the use of new instruments—the cross-staff, quadrant, and the telescope in particular—revealed a stellar world previously unknown and fertile for discoveries, the potential for the intermingling between the real and the imagined was especially strong.

Early celestial maps demonstrate this fusion of aesthetics and measurement. As voyages to the Southern hemisphere revealed new stars and stellar patterns, new constellations were created to corporealize areas of stars that had been recorded but which were noted as unformed or outside the boundaries of the classical Ptolemaic constellations.[28] In 1589, for instance, Petrus Plancius added the Southern constellations Crux and Triangulus Antarcticus on a globe, expanding these again in 1597 to twelve new constellations based on observations from the Keyser-De Houtman voyages.[29] But in addition to arranging the new stars into constellations named mostly after biblical and natural historical subjects, Plancius also added unempirical embellishments, most notoriously Polyphilax, a completely fictional constellation.[30] Even Johann Bayer's 1603 *Uranometria*, the first star atlas and the first to draw a fairly accurate map of the Southern hemisphere, succumbs to the temptation of fiction. Its map of the Southern constellations reveals an aesthetic desire to shape new constellations, a pleasure in naming and mythic representation, alongside the astronomer's strict eye for position and measurement. At the extreme of such ventures is Julius Schiller's 1627 star atlas, *Coelum stellatum christianum,* which implemented a radical reworking of the constellations, transforming them into figures from the Bible: the twelve apostles replace the twelve signs of the Zodiac.[31] A revision of Bayer's atlas, the conceptual thrust of the work, which found many sympathizers, draws its motivation from the power of naming and narration. To reformulate the constellations and reimagine the relation between stars would be to excise the ancients from the skies and reclaim the heavens for Christianity—a project not unlike various seventeenth-century calls for a science that would reveal the glory of God.

These works, along with other important milestones such as Hevelius's 1647 *Selenographia,* the first atlas of the moon, provide an important backdrop for Milton's cosmic voyagers. Scholars have long acknowledged Milton's deep geographic knowledge and passion for maps, evident in *Paradise Lost*'s many topographical surveys and catalogues of places, but it is worth noting that mapbooks in the seventeenth century were not confined to terrestrial cartography.[32] The first sign of the expansion of "world" to include both the heavens and the earth emerges in maps and atlases, which, already by the late sixteenth century, began to include inset celestial maps or cosmological diagrams.

As early as 1584, the Cologne edition of Ptolemy's *Geography* opens with a set of celestial maps, while the first maps of the Southern constellations appear as insets on Petrus Plancius's famous world maps of 1592 and 1594.[33] By the late seventeenth century, world atlases inevitably contained celestial data: Andreas Cellarius's important star atlas, *Harmonia macrocosmica*

(1660), for instance, was published as the eleventh volume of Jan Jansso-
nius's *Atlas novus*, an earlier edition of which Milton may have owned, and
which he inquires about in a letter to Peter Heimbach.[34] The monumental
eleven-volume Blaeu *Atlas maior* (the other atlas which Milton discusses
in his correspondence), published in the 1660s, similarly announces its
combination of earthly and celestial concerns (*solum, salum, coelum*) on its
extended title pages. Like Cellarius, its intended movement toward a "harmonia macrocosmica" finds expression in a discussion of the Ptolemaic,
Tychonic, and Copernican hypotheses as a prelude to its maps of the earth.
Together, these suggest that the earth can only be understood from the epistemic perspective of the cosmos. Likewise, John Seller's 1672 world map
includes a map of the moon and cosmological models alongside a double
hemispheric map of the earth. As late as 1696, the Jaillot-Sanson *Atlas nouveau*, a work that set the foundation for eighteenth-century cartographic
projects, repeats this trope: the work opens with a *Planisphaerium coeleste*
that juxtaposes an elaborate star chart with the Ptolemaic, Tychonic, and
Copernican models of the cosmos.[35]

These works—and Milton's possible exposure to them—point to the existence of a visual basis for the doubling of cartography and astronomy in
Paradise Lost. They highlight the poet's worldmaking desire by locating it
within a contemporary corpus of texts that sought to encompass the world
and reimagine the relations between the earth and the heavens. Not only
did maps and atlases bring the terrestrial and celestial realms into visual
juxtaposition, they also raised (often explicitly) troubling theoretical questions: Where were the boundaries of the universe? Where was God located?
How could we know?

New Science and the Hexameral Epic

Paradise Lost responds to this crisis of order and epistemic boundlessness.
Milton's cagey refusal to choose between the chief world systems as he celebrates the "new spatiality" occasioned by Copernicanism thus may not
indicate intellectual backwardness or nostalgia. Rather, it emphasizes the
processes of knowledge-making rather than on their final (inevitably uncertain) product. The epic's engagement with the new cosmology therefore
should not be relegated to a unidirectional engagement with cartographic
or philosophic sources. It participates in a dynamic contemporary interplay between discourses and genres which sought to comprehend the world
through their unlikely interconnections, often producing peculiar intel-

lectual hybrids. A good example, which Milton most likely knew well, is Henry Hexham's English translation of Mercator's *Atlas* (1636). Here, the poet would have found interpolated quotations from Josuah Sylvester's *Divine Weeks* sprinkled into Mercator's treatise on the creation of the world by his translator. Almost half a century after the original publications of the *Atlas* (1595) and Du Bartas's *La sepmaine* (1578), both early worldmaking texts that sought to connect natural philosophy, cosmography, and theology, the two are fused into a hybrid work that purports to offer a new view of the world. In Hexham's renamed atlas, *Historia mundi*, poetry, cosmography, and technical science coexist and even enhance each other; Milton's later use of Galileo in *Paradise Lost*'s poetic maps of the cosmos thus seems to mirror and invert Hexham/Mercator's co-option of Sylvester/Du Bartas.

Milton himself was deeply influenced by Sylvester's translation of Du Bartas's creation epic, which remained popular in England in the seventeenth century, and he sought to surpass it in *Paradise Lost*.[36] But its citation in Hexham's atlas reveals the extent to which the cosmological and hexameral epics of the Renaissance were embedded in the emerging modern discourses of cartography, astronomy, and natural philosophy. Milton's immersion in these disciplines connects *Paradise Lost* to a contemporary intellectual context as well as to a literary tradition of engaging with scientific discourse. Not only does the epic demonstrate a mastery of early modern worldmaking strategies and genres, it thoroughly digests and rewrites its two great predecessors in this venture: Du Bartas, and most importantly, Dante. Both poets are the objects of parody in the Paradise of Fools, an episode that derives both from *La Sepmaine* and the Limbo of the *Inferno*. Both are dismissed, however, for their celebration of a Ptolemaic world of crystalline spheres—a cosmology and world-picture that *Paradise Lost* decisively replaces with dissolution of the concentric shells into the unimpeded vistas of cosmic space.[37] Milton's allusion to "the Tuscan artist" in book 1 may point to his substitution of a Dantean cosmology with a Galilean one—as some scholars have observed, the terms of the allusion fit both figures— thereby suggesting another complex doubling of poetry with science.[38] Both Tuscan artists sought to describe the cosmos, though in quite different terms; and yet, Galileo's first known public lectures were a geometric explanation of the traditional geography of Dante's hell.[39]

Such multilayered allusions tease out the relations between poetry and science. Milton argues that the two are necessarily interlinked in the worldmaking endeavor, for poetry must draw on the details of scientific observation, even as science depends on the poetic imagination to produce that cosmic "view from above." But the fusion of empirical eyewitness report-age and the imagined odysseys of *Paradise Lost* creates a powerful tension which underlies the epic's claim both to describe and to question a new world order. The gaze of Satan or Raphael across the created worlds is in effect a map of Milton's universe. But in giving his fictional characters these blueprints of the whole, Milton acknowledges that such synthetic visions of world-systems—the basis of worldmaking—can only be produced as works of imagination. This is a double-edged game; while the poem explains cosmic order and relation at length, it simultaneously doubts this knowledge by disclosing its human origins, thus producing a deep-seated thread of skepticism in the poem. It is this thread that Milton will exploit as he inter-rogates the matter of Genesis.

WRITING OF GENESIS

Creation is a central preoccupation in Milton's poetry, a theme that long predates the conception of *Paradise Lost*. His earliest foray into the mysteries of genesis is also his first major work in English, a poem not coincidentally written on his twenty-first birthday and precociously marking his coming of age as a bardic, Christian poet.[40] The "Ode on the Morning of Christ's Na-tivity," the lyric that opens Milton's 1645 *Poems,* imagines a skillful Platonic *artifex* who brings order from chaos:

> Such Musick (as 'tis said)
> Before was never made,
> But when of old the sons of morning sung,
> While the Creator Great
> His constellations set,
> And the well-balanc't world on hinges hung,
> And cast the dark foundations deep,
> And bid the weltring waves their oozy channel keep.[41]

This vision owes much to the Augustinian tradition of grafting the classical creation account in Plato's *Timaeus* onto the opening chapters of Genesis, and is a thoroughly conventional, though beautifully resonant, portrayal of the world's beginning. Yet, this foundational emblem of an artisanal divine maker stands in stark contrast to the elemental wildness of "At a Vacation Exercise," written shortly thereafter: there Milton imagines a seemingly eternal clash between the forces of order and disorder, as he tells of "how green-ey'd Neptune raves, / In Heav'ns defiance mustering all his waves."[42]

In these lines, as in the contemporaneous, "Naturam non pati senium," Milton evokes—only to ward off—the spectre of a Lucretian cosmos characterized by the collision of atoms and the eventual death of the world. Standing against this dark vision of a cosmos shaped by chance and eventual dissolution is the glorious vision of Christian-Platonic creation in the "Nativity Ode."

Here, it is God and not "Beldam Nature" who creates the world; raving Neptune can no longer muster "all his waves" because God bids "the weltering waves their oozy channel keep." This repeated image of the ocean threatening to overwhelm the shore reprises a familiar early modern motif for the menace of uncontrollable disorder encroaching on the fragile order of civilization: it appears in the preamble of Montaigne's essay on the cannibals, in Spenser's Mutabilitie Cantos, and most famously, perhaps, in Shakespeare's *Sonnets*. For Milton, the elemental struggle is always reminiscent of the creation and destruction of entire worlds; the encroaching of chaos on creation becomes a sign for the disintegration of world order and the need for cosmic renewal.[43]

In "Naturam non pati senium," for instance, the poet insists on the fixity of the universe, guaranteed by an omnipotent Creator-God, as an antidote to the poem's first half, which captures the terror of Lucretian flux and universal decay. Against the crumbling floor of heaven and the crashing, colliding poles of the earth ("sono dilapsa tremendo / Convexi tabulata ruant, atque obvius ictu / Stridat uterque polus"), images that will return in *Paradise Lost* as the cosmographic consequences of the Fall in book 10, stands a Miltonic unmoved Mover: "At Pater omnipotens fundatis fortius astris / Consuluit rerum summæ, certoque peregit / Pondere fatorum lances, atque ordine summo / Singula perpetuum jussit servare tenorem" (But by founding the stars more strongly the omnipotent Father has taken thought for the universe. He fixed the scales of fate with sure balance and commanded every individual thing in the cosmos to hold to its course forever).[44] Similarly, in the "Nativity Ode," the geometric precision and stillness of the Creation stands against the flight of the pagan spirits and polymorphous gods, who represent the idolatrous worship of the natural world rather than of its creator. Christ's birth, which marks human salvation and the redemption of a corrupt, decaying world, can thus only be compared to the original act of divine creation.

This drama between cosmic flux and fixity, which animates so many of Milton's early poems, draws on a vocabulary, well established by 1630, that pitted a godless, materialist cosmos (often identified with Epicurus or Lucretius) against the harmonious symmetries of Christian-Platonic forms—a dynamic already visible in the work of Montaigne and Spenser. By the mid-seventeenth century, this contrast became the organizing philosophical lens through which the struggle to establish a new system of world order that addressed the shortcomings of Scholastic Aristotelianism unfolded. Philosophers as diverse as Gassendi, Descartes, Hobbes, Henry More, and Newton inevitably turned to both Plato and Lucretius, whose Renaissance reception was already deeply entwined.[45] Many sought a middle way that would combine features of both philosophies into a viable explanation of the nature of the world. Milton too participates in this large-scale philosophical thought experiment.

If the early poems echo this conflict of systems, *Paradise Lost* explores the possibilities for their integration even as Milton marks their fundamental metaphysical differences. Thus, the Platonic creation of book 7, achieved with golden compasses, is contrasted to "The secrets of the hoary deep, a dark / Illimitable Ocean without bound, / Without dimension, where length, bredth, and height, / And time and place are lost . . ." (2.891–94). Here, the epic reprises the counterpoint in the early poems between the chaotic exuberance of Neptune and Nature and the artifactual solidity of the "well-balanc't world." But now the realm of Chaos also becomes the poem's heterotopic space of uncreation, the zone of anti-genesis which seems to demand an act of creation, that "secret thing" which Milton so longs to see, to narrate, and, perhaps, to reproduce. By bringing such questions about the world's original creation to the forefront of his epic, Milton does what none of his contemporaries had dared to do explicitly: he tests the effect of the new philosophic paradigms of world order on interpretations of Genesis, the sacred text whose claims to revealed knowledge surpassed the human reasoning of pagan philosophers. *Paradise Lost* participates in the wider revival of interest in interpretations of Genesis over the course of the seventeenth century as knowledge about the creation and proof of its occurrence became integral to struggles over new forms of religious, political, and scientific order.

Milton was not alone in his desire to lay bare the details of creation, to penetrate beyond the cryptic, suggestive images of Genesis 1–2 and imagine a world taking form. Histories of biblical interpretation routinely list the sudden flowering of new exegeses following the Reformation (the Genesis interpretations of Calvin and Luther were particularly influential) and the trend continued well into the early eighteenth century.[46] The seventeenth century saw the publication of a slew of texts on various aspects of Genesis alone—a departure from the Reformation emphasis on the New Testament, especially the Pauline epistles—influenced partly by a better linguistic command of the Hebrew Bible and Septuagint and partly by ongoing doctrinal battles and the pressures of new geographical and scientific discovery.

By the 1650s, literal interpretations of Genesis were being strongly contested and reimagined in allegorical, accommodationist, or even hypothetical terms; to write a poetic account of the biblical creation story in this period was to engage provocatively in a contentious debate over the basis of a legitimate world order authorized by the act of creation itself. A census of pamphlets concerning Genesis interpretation in the seventeenth century reveals over two hundred items published in England alone, with the majority falling after 1640. It is likely that the numbers are similar for the Continent, particularly in the Netherlands and Germany.[47] The text of Genesis unites cosmic order and human political order in a single causal chain, and many of these works combine natural philosophic or political concerns with biblical exegesis.[48] Recent critical attention to Milton's heresies and the radicalisms of various seventeenth-century sects tell a similar tale: the status of Genesis was anything but stable by the mid-seventeenth century.[49]

The existence of these works, obviously in dialogue with each other and reflecting the concerns of a wide-ranging readership, indicates a shift in modes of biblical interpretation while it hints at some reasons *why* such a shift may have occurred. By the Enlightenment, biblical scholarship had become clearly divided into two camps: biblicism or systematic theology, now practiced mostly by church-affiliated theologians who considered scripture a living text along typological lines; and the nascent historical-critical school, divided between the rationalists and the orientalists, both of whom saw the Bible as a product of historical contingency and denied it a privileged epistemological position as revealed truth.[50] If these methodologies became crystallized by 1753, when Jean Astruc first advanced what was to become the documentary hypothesis of the sources of Genesis, early lines of division are already apparent in texts of the later seventeenth century, when the genre of biblical exegesis became sharply divided between the systematic works of church divines and the *annotationes* of learned laymen (of which Milton's *De doctrina christiana* is an example).[51] This tendency toward generic separation signals a shift in the purpose of exegesis: biblical interpretation, particularly Genesis interpretation, was no longer confined primarily to doctrinal matters and their national political consequences; it was now urgently implicated in legitimizing (or denying) changing cultural practices such as commercial speculation, imperial expansion, scientific inquiry, and (following the peace of Westphalia) international relations. Moreover, the rise of radical sects in England, with their emphasis on the rational exegesis of scripture, eventually led to a questioning of the very notions of Godhead and revelation. It was becoming clear that contemporary historical events—whether the political turmoil of the English Revolution or the philosophical upheaval associated with the anti-Aristotelianism of the New Science—demanded new interpretative strategies for scripture and a revised understanding of its authority in order to maintain structures of religious belief.

When Milton turned to writing his epic in the 1650s, he was also completing *De doctrina christiana*, which like similar commentaries of the period contains more references to Genesis than to any other book of the Bible, excepting the Psalms and Proverbs.[52] Taken together, *De doctrina* and *Paradise Lost* situate Milton at the heart of the transformation in religious belief occasioned by a new sense of the world. Both are poised between the doctrinal and political battles of midcentury, of which Milton himself was a veteran, and the rational turn in biblical exegesis associated with the Socinians and Deists.[53] Both works struggle to reimagine biblical texts in order to mitigate the contingencies of a world perceived to be somehow new, even "modern." However, despite considerable recent scholarly attention to Milton's heresies and singular theology, the implications for *Paradise Lost* of the vast cultural project to reinterpret Genesis still bear analysis.

Milton's choice of Genesis as his subject may have itself been a revolutionary and radical gesture.[54] His analysis of the fissures within Genesis, several of which he had discussed in a theological vein in *De doctrina*, reveal a philosopher-poet who brings a skeptical gaze to the story of origins even as he seeks to assert faith in its account of the nature of things. By staging multiple narratives of creation within the poem, Milton exhaustively explores various versions of the Genesis tale. But he does so in order to give the problem of world order itself a point of origin. *Paradise Lost* takes the opposing Lucretian and Platonic narratives of the world's creation and locates their double origin within Genesis itself, suggesting that this central cultural-philosophical conflict is an original—and foundational—aspect of the sacred text.

This is an astonishing move in the history of early modern worldmaking. In Milton's hands, poetry becomes an epistemological instrument, seeking the truth within cosmic narratives—of the Bible and of the classical corpus—by imagining and retelling them. Poetic narration for Milton, like Galileo's telescope, reveals inconsistencies and fissures that question and dismiss the old cosmologies as it seeks an authoritative account of the world's creation for the modern age. And it is poetry that produces the effect of cosmic order, even as it discloses an original uncertainty, a primal skepticism, about the world's origin within the Bible itself. Almost counterintuitively, poetry becomes the chief instrument for the renovation of faith against the rational pressures of a nascent scientific modernity.

Versions of Creation

In the opening invocation to *Paradise Lost*, the poem's ostensible subject (the Fall of man from Paradise) is overshadowed by a deeper concern with the act of creation itself. The scene engraved in the minds of the poet and his readers is not the transgressive "mortal taste" but the emergence of the world, and of the poem itself, into being. Twice within the space of fifteen lines, the narrator asks to be told about the moment of creation: "Sing heavenly Muse . . . / In the beginning how the heavens and earth / Rose out of chaos . . ." (1.6–10) and again, "Instruct me, for Thou know'st; Thou from the first / Wast present, and with mighty wings outspread / Dove-like satst brooding on the vast Abyss / And mad'st it pregnant" (1.19–22). This insistence on recovering a true account of the event will recur over and over again through the poem. In his desire to "look in, and see" the world's origin, the narrator pursues a specific method reminiscent of Galileo and familiar to any reader of seventeenth-century natural philosophical texts: he calls upon the first eyewitness to the original event and conducts a complete interrogation.

The entire epic can in fact be read as a compilation of several eyewitness accounts of the creation, for this emphasis on an empirical ratification of what exactly happened in the beginning surfaces repeatedly, becoming a primal scene in the poem's own development. Milton reproduces this encounter—where an observer (here, the heavenly Muse) is questioned about the moment of creation by a belated interlocutor (here, the narrator)—at several points in the epic. But at each stage, the interlocutors and the answers are different. Thus, a similar tableau concludes the second book as Satan interviews the Anarch of Chaos; it frames the third book with the narrator's invocation to light and the final conversation between Satan and Uriel; at the end of book 4, Eve asks Adam about the nature of the creation; at the end of book 5, Satan and Abdiel dispute the moment of creation; at the beginning of book 7, Adam asks Raphael to describe the creation of the world; at the beginning of book 8, Adam persists with his inquiry into the nature of things; it is only at the beginning of book 12 that Adam is finally silenced as he watches one world end and another begin. These competing narratives argue that Milton, like many of his contemporaries, may not, in fact, have had a clear monolithic vision of the Creation in mind. He may have intended to interrogate the illusion of just such an orthodoxy.

The uniqueness of Milton's insistence on the different versions of creation, and their importance for a reading of *Paradise Lost,* becomes clear in comparison to the long tradition of hexameral literature. Against the post-classical Christian epics of Marius Claudius Victor, Dracontius, and Pseudo-Hilary, Dante's *Commedia,* and the more recent *La sepmaine* of Du Bartas or *Le sette giorni del mondo creato* of Torquato Tasso, Milton's strategy of questioning witnesses to the creation is unexpected and unsettling. Where Dante, Du Bartas, and Tasso are all careful to subordinate their poetic ability to the majesty of the creative deity, cautiously averting their imaginative vision instead of demanding to know, creation for Milton becomes a site for philosophic questioning. The narrator's demands, Satan's questions, Eve's and Adam's questions, even Raphael's questions all center on the fact of creation and its consequences. Knowledge of the creation is not self-evident. It must be searched out through laborious investigation, figured (among other images) in Satan's voyage through chaos toward "this pendant world."

The novelty of this strategy becomes apparent in the contrast between Milton's proliferating creation accounts and the single unified narrative of Du Bartas, who clearly identifies the agent and the act and energetically dispels any doubts that the reader might experience. The translation by Josuah Sylvester renders Du Bartas's narrator in this cautious way: "This Trinitie (which rather I adore / In humbleness, then busily explore) / In th'infinit of Nothing, builded all / This artificiall, great, rich, glorious Ball."[55] The English "artificial" exactly captures the French emphasis on "artifice" and fixes beyond any doubt the nature of God as the *artifex,* the one who makes. Metaphors of building abound in the *Divine Weeks,* which celebrates artisanal labor: God is compared to carpenters, weavers, potters, and even a shipbuilder. Du Bartas explicitly describes the Creator in terms of manual work—the world bears the marks of his *laboring* ("porte de son *ouvrier* empreinte") and not merely "the Builder's beauty." The Huguenot Du Bartas's commitment to Reformed theology leaves no room to question who exactly effected the creation: he affirms the doctrine of creation *ex nihilo* by a unified Trinity, and at each turn anticipates the arguments of skeptical ancients (notably the atomists) and calls on them to account for their unbelief and godlessness.[56] The declarative impulse of Du Bartas's poem is mirrored in Tasso's contemporary *Sette giorni del mondo creato*, where the Catholic poet, writing at the height of the Counter-Reformation, invokes the Trinity and the creation of a new world, once again emphasizing fashioning ("al fabricar del novo mondo").[57] Here again, the poet does not presume to demand or question; he prays that God might make the earth and sky, which are brought forth purely by his goodness.

Attempts have been made to trace the creation accounts in *Paradise Lost* to particular sources in the rich tradition of hexameral literature. But Milton differs strongly in tone from such poets as Du Bartas/Sylvester and Tasso.[58] His closest predecessor is perhaps Dante, who closely questions Beatrice about the cosmos in *Paradiso* 28; but that dialogue is singular, complete, and perfectly ordered, much like the crystalline beauty of the Ptolemaic universe, which the poet is privileged to see. In *Paradise Lost,* however, the apostrophic invocations to each narrator of the creation is demanding and interrogative in spirit: "Sing," "Say first," "Instruct me," "Brightest seraph tell," "But more desire to hear . . . / The full relation. . . ." That the resulting tales are internally inconsistent further outlines Milton's departure from the tradition in which he ostensibly situates his epic.

By juxtaposing several partial reports, Milton insists on a prismatic view of an original event that remains distant and inaccessible, demanding that we distinguish the (imagined) event from its representations and recollections. But taken together, Milton's multiple accounts of the creation effectively question the biblical text itself, stripping it of the epistemically privileged status traditionally accorded to scripture. The gesture is as radical as it is courageous. Milton confronts the possibility that we have no certain evidentiary basis to believe that the creation occurred in any particular manner. Some accounts may be just as valid as others.

From the poem's opening lines, Milton highlights a fundamental fissure within the Genesis account itself that arises from the well-known distinction between Genesis 1 and 2, now described as the Priestly and Jahwist versions.[59] While Genesis 1 (P text) offers an abstract account of a creation summoned into being by the mere agency of divine speech, Genesis 2 (the Jahwist or J text) describes God as a craftsman, fashioning the creatures with skill and painstaking care. Biblical philology emphasizes the distinction between the Hebrew *bar'a* (to create) used extensively in Genesis 1 and *yatsar* (to mold into a form, to fashion), characteristic of Genesis 2.[60] The difference of emphasis is significant enough even to be seen as contradictory: an instantaneous rising up at the command of the Word versus a vital impregnation and prolonged gestation that gradually shapes matter. The contrast is analogous to the difference between the Lucretian and Platonic versions of cosmic creation that distinguish between spontaneous genera-tion and laborious making, which Milton also explores at length. The chal-lenge was to reconcile these paradigmatic oppositions in the classical and biblical cosmogonies into a single, unified vision. At stake was the founda-tion of world order itself. By healing the breach, the exegete or poet could determine whether the world was fundamentally chaotic, unpredictable, and contingent or whether it was a thing made, circumscribed, stabilized into a *kosmos*.

The extraordinary efficacy of biblical exegesis from the early church onward and the triumph of Aristotelian poetic theories in the sixteenth century had produced such an influential synthesis of the two visions of creation that it has often been difficult even to see the original fault lines. Yet even Augustine worried profoundly about the problem of Genesis, writ-ing two interpretations early and late in his career, and returning to it in both the *Confessions* and the *City of God*.[61] While the problematic differences within Genesis are now explained away by the philological argument of the documentary hypothesis, which asserts that the P text was a much later addition to an original J text, neither Augustine nor Milton had recourse to such an elegant textual solution—nor, one suspects, would they have been satisfied with it. For the contrast between P and J raises a fundamen-tal problem about understanding the world with direct consequences for human action and belief. Was there a divine order in the natural world, infused into it at the moment of creation and eclipsed at the time of the Fall, that could be recovered by the faithful? Or, what seemed all too plausible in the late seventeenth century, was there in fact no such order, no benevolent deity structuring the course of history and time, but only chance encounters in which man made his own meaning? For the seventeenth century, the twin theories of creation were thus also the vexed foundation of worldmak-ing efforts—a foundation based not on a specific theoretical perspective but on *questions* about origin which are authorized, and suspended, in no less a text than the Bible.

Milton uses this tension between creation accounts to produce a strong skeptical undertow, an effect particularly evident in the three creation nar-ratives of books 3 and 7. In these passages, three different speakers—the narrator, Uriel, and Raphael—offer three different, hybrid versions of the creation to three different auditors (the reader, Satan, and Adam). In each of these versions, we are presented with variations on the themes introduced by the opening proem: creation by manual fashioning or creation by self-generation, both of which are brought together majestically in the grand narrative of book 7. In the invocation to book 3, creation is compelled by the "voice / Of God," clothed by "holy Light," but takes place as a spontaneous generation: "The rising world of waters dark and deep, / Won from the void and formless infinite" (3.8–12). Uriel, an original eyewitness, expands this version: "at his Word the formless Mass, / This worlds material mould, came to a heap," but the power of the Word now becomes invested with the qualities of an artifex, combining the two versions of creation into one: "Each had his place appointed, each his course / The rest in circuit walles this Universe" (3.708–21). Raphael's extended creation account in book 7 returns to these earlier retellings and fully juxtaposes Platonic geometry ("He took the golden Compasses . . . / . . . to circumscribe / This Universe, and all created things . . .") with vitalist infusion and spontaneous organization: "His brooding wings the Spirit of God outspred, / And vital vertue infus'd, and vital warmth / Throughout the fluid Mass . . . / . . . then founded, then conglob'd / Like things to like . . . / And Earth self ballanc't on her Center hung" (7.211–42).

Vital inconsistencies in these lines make it impossible to treat the sum of these narratives as a smooth, uninterrupted recollection of the primal event.[62] In the proem to book 3, it is light, "the bright effluence of bright essence increate" which invests the "rising world" with a mantle—an action that looks very similar to the outspread "brooding wings" of the Spirit of God in book 7. In the later version, however, there is no light and creation seems to take place within "the Darkness profound," as the Son rides *out* of heaven and into the din of lightless chaos. The problem of agency is even more complicated by Uriel's account in book 3, in which cosmos is simply called forth from chaos by "his voice"; here, the creation of light is preceded by the taming of "confusion," never mentioned in the book's invocation where the creation of light is depicted, as in Genesis 1, as the first active gesture of the creating deity (it is also unclear who exactly Uriel identifies as the creator). Uncertainty about the creative agent is compounded by uncertainty about method, as the accounts offer contradictory statements about the nature of "the formless mass / This worlds material mould," its activation and its shaping.[63] Some of this can be explained away with reference to the long tradition of exegeses on the Pentateuch, which disputed the exact details and order of the creation in fine detail. One might, for instance, argue for a first and second creation separated by the fall of the rebel angels, treated in books 3 and 7 respectively—a view sanctioned by the exegetical tradition but nowhere stated in *Paradise Lost* itself. More recently, however, Elizabeth Sauer has pointed to the difference between the creation narratives of books 3 and 7 as an instance of the epic's multivocality but hesitates to pursue the logical consequences of such clamor for a reading of the poem as a whole.[64] Milton's contradictory creation narratives, in fact, call attention to their mutual inconsistency and self-consciously emphasize the contradictory sources from which they arise, both within Genesis and beyond.

The juxtaposition of spontaneous generation and artifactual making in these passages can be traced to specific Platonic and Lucretian images; but while the former (the compasses, the hierarchical harmony of celestial bodies) are well-established Christian-Platonic tropes, the presence of the latter has been surprisingly neglected. Milton draws quite extensively on the fifth book of *De rerum natura*, which describes the birth of the world in a series of organic metaphors.[65] Lucretian matter is energized, moving, and self-propelling, and Milton adapts those images of an animate cosmos whirling toward coherent form. But there are also quite precise echoes: Uriel's explanation of how the remaining elemental particles "in circuit walles this Universe" (3.721) seems to render a version of the Roman poet's *moenia mundi*; "The Earth self-ballanc't on her Center hung" (7.242) echoes the Lucretian "terram . . . / in medio ut propulsa suo condensa coiret" (5.484–86); the repetition of "paresque / cum paribus iungi res" (5.437–38) finds translation in "conglob'd / Like things to like" (7.239–40); even the "black tartareous cold Infernal dregs" (7.238) have their Lucretian analogue in the "subsedit funditus ut faex" (5.497).[66] The density of Milton's allusions here, and their metaphoric intermingling, produces a philosophic fusion through poetic practice; the hybrid accounts of creation simultaneously disclose their metaphysical fractures and dissonances even as the poetry itself works to heal and harmonize them. The effect is twofold: Milton introduces a skeptical voice that repeatedly returns to the uncertain origin of the world, but at the same time, in its effort to produce aesthetically satisfying narratives of creation, the poem itself stands in for the stable order and cosmic symmetry that the narrator craves.

Paradise Lost thus offers a surprising perspective on the "crisis" occasioned by the new science on religious belief. Historians of science have long argued that the so-called clash of science and religion in the early modern period is a post facto construction of Enlightenment historiography, but have been hard pressed to explain just how skepticism and faith could so readily coexist. Milton's engagement with the effect of the new philosophies on Genesis suggests that the poetic imagination may have offered a resonant third way. *Paradise Lost* suggests how poetry—as a form of human worldmaking that imitated a presumed divine original—could unite religious faith, rationalistic exegesis, empirical method, and a changed understanding of world order to produce a new kind of skeptically grounded faith, one which acknowledged uncertainty but triumphantly asserted its own creative potential.

The Heresy of Self-Begetting: Theodicy and Worldmaking

Milton's emphasis on the power of human making repeatedly threatens the singular omnipotence of divine creation as the poet becomes a mortal *doppelgänger* for the creative deity. From the poem's opening invocation, the thematics of the world's creation are entwined with the poet's own creation of his poem, which in turn seeks to reveal the world in its cosmic entirety. Both poet and world are literally inspired and infused with the creative agency of the same "spirit."[67] This doubling of poem and world establishes a complex analogy between poet and god, and though Milton is mostly careful to subordinate human constructions to divine ones, over the course of the epic the mortal poem increasingly seems to stand in for knowledge of an original creation that can never be completely verified.

The poem's anxiety about this doubling emerges in its tentative investigations of a what a cosmos stripped of divine creative agency might look like: Milton raises the specter in the Paradise of Fools in book 3, and even more sharply in Satan's confrontation with Abdiel in book 5. Here, Milton tests the hypothesis that there was no defining moment at which the world's creation occurred and no particular original creator:

> That we were formd then saist thou? and the work
> Of secondarie hands, by task transferd
> From Father to his Son? strange point and new!
> Doctrin which we would know whence learnt: who saw
> When this creation was? rememberst thou
> Thy making, while the Maker gave thee being?
> We know no time when we were not as now;
> Know none before us, self-begot, self-rais'd
> By our own quick'ning power . . .
>
> (5.853–61)

Satan's insurrection is complete when his repudiates God as a creator by denying his own creatureliness and demoting the fact of divine creation into a mere theory.[68] But Satan's skeptical questions highlight the epistemological blank at the center of any world theory or universal system of order: there is no certain knowledge about the nature of the world's creation, its overarching structure, or who (if anyone) composed it. The only thing certain is our own consciousness of our "quick'ning power," our perceptions of ourselves and the world around us. If Satan sounds suspiciously like Descartes, it is perhaps no accident, for the radical skepticism of this passage bears a strong resemblance to the First Meditation.[69] But where Descartes used the skeptical process as a means to establish God as an innate idea, Satan stops short of pursuing that step; instead, Milton allows his question to remain unanswered—Abdiel's impassioned reaction is more an assertion of God's creation than a systematic response to Satan's insidious argument, even though the question had been debated at length since the publication of the *Meditationes* in 1641.

Instead, Milton distills the conflict between Satan and Abdiel into a clash between an atheist and creationist account of the world's foundation, a clash that mimics once again, the confrontation between Lucretius and Plato. Abdiel's Christian-Platonist image of Son is a familiar and comforting one, well known from the writings of Augustine and thoroughly developed in the medieval church. This anthropomorphic creator who makes the objects of the world with his hands is a reassuring figure whose motivations are comprehensible through analogies of the passionate artisan or the protective parent. He creates the foundation of the world, establishes its structure and systems, endows it with a purpose. By contrast, Satan's vision of initial emergence is shadowy and intangible, threatening the authority of Genesis with what Regina Schwartz has called the "heresy of self-begetting."[70] The unbounded energy in Satan's vision of self-making, of thrusting forth into existence "by our own quickn'ing power," subtly denigrates the image of unseen "secondarie hands" with its emphasis on manual labor; it contains the seed of the Romantic celebration of individual creative genius over laborious artisanal industry.

Satan's heresy has profound roots that seventeenth-century debates over science and theology merely revived. It draws on an intertwined tradition of skepticism and atheistic cosmologies, well known from classical sources such as Lucretius as well as from within the Old Testament itself. *De rerum natura*, for instance, responds directly to the theory of divine creation with a series of naturalistic arguments designed to demonstrate that there is no need for a concept of deity to explain the workings of the world (5.154–86).[71] When Satan reimagines Genesis 1 by eliminating the agency of the divine Word and attributing instead to matter or nature the will-to-create, he closely echoes Lucretian rhetoric.[72] Like Lucretius, Satan argues that existential liberty can only be achieved when tales of divine creation and intervention in the world are exposed as myths (*religio,* literally "that which binds").[73] In connecting Satan's heresy here with Lucretius, Milton takes his place in a long tradition of Christian defenders stretching back to Lactantius, who, in *De ira dei,* identifies the defeat of the "formidable argument" of Epicurean atheism and its denial of divine creation as central to any theodicy.[74] One key reason for this is that *De rerum natura* implicitly counters Platonic and biblical belief about how the particular things of the empirical universe had been fashioned.[75] Early modern writers retrospectively viewed Lucretius's early articulation of an anti-creationist position as a template for later debates that emerged when increasingly convincing empirical explanations by the new science challenged traditional concepts of divine agency.[76] The Roman poet's searching questions, here voiced by Satan, articulate a new interest in how precisely the universe had come into being.

The demand for proof of divine creation was, however, not simply the product of contamination by atheistic classical texts: similar questions find expression within the Bible itself. Thus, alongside the Lucretian echo in Satan's speech, Milton juxtaposes the Voice in the Storm from the Book of Job: "Where wast thou when I laid the foundations of the earth? . . . Who hath laid the measures thereof, if thou knowest? . . . Knowest thou it because thou wast then born?" (Job 38:4-21). As the skepticism of Lucretius collides with God's terrifying demand for faith in Job, Satan's speech in book 5 becomes a privileged locus for some of the key conflicts precipitated by the challenge of worldmaking: the clash between knowledge and unknowability, between doubt and belief, between divine creative power and the shaping force of the human imagination. At first glance, the Job allusion seems a characteristic Miltonic irony at the expense of Satan, since the memory of the biblical text with its cosmic grandeur undercuts Satan's claim to self-creation. But the passage remains a dilemma for biblical commentators, for the magnificent power of God's speeches in Job 38-41 never directly answers Job's anguished questions about the meaning of his existence and the just order of the world (Job 3-28); the hidden God of the Hebrew Bible refuses to offer any consolation by affirming the existence of a just world-system.[77]

This psychic trauma has produced a significant tradition of interpreting Job as one of the most skeptical books of the Bible, a text that raises questions that cannot be answered by rational exegeses, but merely by the sublimity of God's vision of the creation and what Paul Ricoeur has called an "unverifiable faith."[78] The Book of Job—a poem of worldmaking that parallels Genesis itself—identifies the deity by his creative power.[79] But as Carol Newsom notes, the God of Job, unlike the God of Genesis, seems to identify most with images of chaos and sublime terror: creation here is far from a vision of steady craftsmanship and comforting control; creation teems with an unmasterable violence against which Job's parochial, human desire for order is tragic and pitiable.[80] Set alongside Lucretius, the Book of Job adds to the chorus of voices that dispute the Genesis vision of cosmos, doubt its elegant symmetries, and struggle to reconcile its promise of divinely sustained order with a conflicted existence that seems governed by contingency rather than providence. Milton's introduction of this text into his epic at the very abyss of theistic denial also reveals the extent to which the theodicy of *Paradise Lost* is deeply interwoven with its investigation of worldmaking: the justification of the ways of God to men is finally tied to evidence of a coherent system of world order.

Beyond the superficial clashes between theistic and atheistic cosmologies, *De rerum natura*, the Book of Job, and *Paradise Lost* all explore a deeper, seemingly unshakeable human desire for reassuring narratives of origin and order, for consolation in the face of mortality, destruction, and loss. Despite its professions of liberty from the constraints of religion, Lucretius's epic is haunted by a melancholy for the inevitable endings of worlds; yet, at the same time, it epitomizes a courageous imaginative stand against the forces of entropy. In its trial of faith, Job reveals an existential need for belief in the divine that arises out of questions about the individual's place in the cosmos. By combining these voices, Milton probes the relationship between ideas of order, divine agency, and the socio-cultural need for organizing narratives. To suggest, along with Lucretius and Satan, that traditional systems of order were merely post facto human constructs was to offer an exhilarating sense of liberation—but as the book of Job suggests, to deny a divine creator was also to take on a terrifying metaphysical isolation.

Folded into Satan's defiant claim then is not only an individual uprising against an oppressive God but also an attempt to make sense of the need for divine authority in establishing systems of cosmic and earthly order. Milton knew that if Genesis was to remain a persuasive basis for world order, it would have to offer answers to the Satanic questions; by incorporating them into the texture of his epic, he transforms poetry itself into a powerful, ongoing commentary on the ancient and contemporary challenges to believing the biblical account of creation. Poetry holds together the skeptical tensions of competing voices, and its own formal symmetries perform a reassuring vision of cosmic order. By exploiting poetry's own creative potential, which simultaneously imitates and challenges divine creativity, Milton captures the contradictory desire to retain a notion of individual imaginative freedom without the price of cosmic loneliness.

FROM COSMOS TO HEARTH: A COSMOGRAPHIC MEDITATION

If *Paradise Lost* celebrates the "rapture of space," it also remains committed to affirming the singular importance of place, the particular, circumscribed habitus of human community and endeavor.[81] The first mention of cosmic structure by a human character in the poem is also an attempt to bind individual and cosmos in the ancient tradition of the microcosm: in book 4, Eve's love lyric to Adam celebrates the perfect interconnection between human and habitat. For Eve, "Adam *is* Eden"—"With thee conversing I forget all time, / All seasons and their change, all please alike" (4.639-40). In the triple repetition of "all," Milton signals the oneness of person and place; the diurnal repetitions of Eve's song perform an eternal recurrence, while its chiastic poetic form imitates the perfect orderliness that binds the mutable and heavenly spheres. At this precise moment of cosmic correspondence, Eve breaks the charmed circle to ask about the world beyond: "But wherefore all night long shine these, for whom / This glorious sight, when sleep has shut all eyes?" (4.657-58). The centripetal energies of the poem's inward, domestic turn are threatened by the centrifugal lure of the starry skies above.

The memory of Eve's question clearly underlies Adam's astronomical dialogue with Raphael in book 8, and the juxtaposition of her song and question returns in the Satanic dream she recounts in book 5. Eve and her question are in fact central to the poem's confrontation with a changing worldview, for it is through Eve that Milton filters the effects of the separation of microcosm from macrocosm, of the book of nature from the book of scripture. It is in the gap between Eve's lyric, itself a poetic vestige of an older era, and her "Galilean" question, which looks ahead to the new science, that *Paradise Lost* situates its deepest questions about the individual's place in the cosmos.[82] This is not only because Eve is the ostensible agent of the Fall, but because she binds together the spatial poles of cosmos and hearth. She is simultaneously the generative center of the domestic epic and the curious speculator who questions the received order, probing at the poem's fissures of doubt. And it is Eve who is the first human to be granted a vision of the cosmos—albeit a Satanic one—reminiscent of Scipio's dream vision in Macrobius's *Somnium*.[83]

Through Eve's shifting relationship with Eden and Adam, Milton signals the effects of a large-scale cultural re-centering, as the links between self and world are reoriented away from the ordered correspondences still evident in Mercator's *Atlas*. Eve holds together the poem's two seemingly separate subjects—cosmological exposition and heroic action—and initiates an inquiry into the connection between epistemology and ethics, abstract cosmic order and particular human relationships. Eve's song and question in book 4 thus frames book 8 and symbolically marks off the cosmographic section of the poem from the human drama to follow. As the book of final cosmological speculation stands between the majestic account of creation (book 7) and the separation of human and divine harmonies at the Fall (book 9), Milton pauses to offer a final cosmographic meditation, a reflection on the implications of Eve's response to the world around her. Book 8 strategically inverts the trajectory of Eve's thoughts in book 4: it opens with Adam's repetition of Eve's question to Raphael, initiating a crucial dialogue on the human ability to know the world, and it concludes with the angel's urging the first couple back to the closed domestic circle of Eve's song: "Sollicit not thy thoughts with matters hid, / Leave them to God above . . . / . . . joy thou / In what he gives to thee, this Paradise / And thy faire Eve . . ." (8.167–72). The distance between astronomy and domesticity measures the broader intellectual shift in the period from cosmological speculation to the ethics of self-regulation, from the distant stars above to the ethics of partnership and what Kant would later call "the moral law within."[84] Book 8 thus opens a safe space within the poem for speculation and doubt, for the explicit hesitations, skepticisms, and anxieties of creaturely conjecture, as the poem gradually leaves behind the desire to see the grand vision of cosmic creation and turns instead to human struggles of ethical choice, to "the better fortitude / Of Patience" (9.31–32).

Milton highlights this shift from the epic narration of Creation to the skeptical dialogue about cosmic structure in his revision of *Paradise Lost* from the ten-book edition of 1667 to the twelve-book edition of 1674.[85] The 1674 text clearly distinguishes between the creation account, now a neatly self-contained unit, and the epistemological uncertainties of the subsequent book, which takes as its subject the temptation to imagine, analyze, and speculate about self and world. This new division emphasizes the cosmic dramas of creation and destruction that occupy the central third of the poem (books 5-8) and permits a new beginning within the epic (marked by the invocation to book 7) as the Son's new creation of the world coincides with the beginning of the second half of the poem.[86] Structurally, then, the 1674 text calls even more attention to acts of creation than the 1667 edition had done; it sets up a counterpoint between two versions of worldmaking—divine fashioning in book 7 and human conjecture in book 8—and invites a meditation on the relationship between the two. The result suggests a return from the unknowable reaches of the world back to the intimate spaces of the self.

A Dialogue on World Systems

Adam's reaction to Raphael's glorious creation poem in book 7 is a surprising one. It provokes a moment of doubt that threatens to undercut the entire Ptolemaic-Aristotelian cosmology upon which orthodox interpretations of Genesis were based:

> something yet of doubt remains . . .
> When I behold this goodly Frame, this World
> Of Heav'n and Earth consisting, and compute,
> Thir magnitudes, this Earth a spot, a graine,
> An Atom, with the Firmament compar'd
> And all her numberd Starrs, that seem to rowle
> Spaces incomprehensible . . .
> . . . reasoning I oft admire,
> How Nature wise and frugal could commit
> Such disproportions . . .
>
> (8.13–27)

Adam's puzzlement over the seeming disproportion of subordinating the immensity of the heavens to "this Earth," which seems no more than "a spot, a graine / An atom" (later described again as a "punctual spot"), gestures toward the breakdown of an anthropocentric vision of the world and a new acknowledgment in the late seventeenth century of a decentered cosmos, once imagined by Nicholas of Cusa. The repeated description of the earth as a "spot" echoes an ancient Stoic disdain adopted by the mapmakers of the previous century, most famously cited from Seneca's dismissal of earth as "hos punctum," a trivial pinpoint in relation to the cosmos as a whole.[87] But even as this textual trace connects Adam's questions to the long tradition of the Stoic *kataskopos,* their scientific import draws on heated contemporary debates about diurnal rotation and geocentrism; the clash between ancient and modern sources foregrounds once more the uncertain place of human beings in what appears to be an infinite cosmos.[88]

Adam's condition here closely resembles that of the seventeenth-century intellectual. His "reasoning" leads him to doubt, despite Raphael's explanation of the creation and the angelic fall (that is, despite the revelation offered by scriptural sources, especially Genesis), and he wants further clarification for the troubling "disproportions" for which the angel's tale does not quite account. He is caught between admiration for the cosmic order that Raphael has outlined and a suspicion that its elements do not cohere, between a desire for an anthropocentric vision of cosmic purpose and an apprehension that such desire is no more than a fantasy. Adam's doubt thus articulates the tension between the new philosophies and traditional interpretations of Genesis, by taking up one of the classic questions in the philosophy of physics and religion, "the problem of the imperfection of a world, itself created by a perfect God."[89] If the Aristotelian universe was perfect— flawless, complete, and limited in scope—and mirrored the divine creator, how could the evident infelicities, asymmetries, and "disproportions" (such as sunspots), increasingly disclosed by new scientific instruments and experiments, be reconciled with a benevolent, perfect God?

The matter is central to histories of early modern science. Figures such as Bacon argued that such a desire for correspondence between God and world was no more than an Idol of the Theater; at the other extreme, Newton would claim that God's continuous involvement in his creation was only discernible through the imperfections of the world.[90] The imperfect-world problem with its anti-creationist thrust could also be traced back to Lucretius—and was routinely done so over a long period, as numerous citations by Gassendi, Boyle, Charleton, Newton, Bentley, Blackmore, and Hume attest.[91] In book 2 of *De rerum natura*, the poet bases his climactic argument against a theory of divine creation on evidence of worldly imperfection: "nequaquam nobis divinitus esse creatam / naturam mundi: tanta stat praedita culpa" (the nature of the universe has by no means been made for us through divine power: so great are the faults it stands endowed with).[92] In book 8, Milton too seems to make this connection, emphasizing the philosophical allusion to Lucretius and the ancient atomists as Adam compares the earth to "an atom," an unmistakable allusion to the Epicurean cosmology (and perhaps the new mechanics).[93]

Imagining the earth as an insignificant atom in a universe of "spaces incomprehensible," Adam invokes the vastness of the infinite universe as well as the terror of transience that permeates so much of Lucretius's poem.[94] But Milton's analogy suggests a way out of the agony of doubt: the atomic earth is not just miniscule and insignificant; it is also indivisible (in the etymological sense of *atomus*, "a particle incapable of further division"), singular, at once an integral whole and an elemental particle in a greater totality. At the very moment of its decentering, the earth in *Paradise Lost* becomes analogous to the individual self; as Adam ("red earth") phonetically echoes atom, Milton restores a fragile, complex correspondence between microcosm and macrocosm based on partiality, uncertainty, and doubt rather than on visions of cosmic wholeness.

The ensuing dialogue on world-systems places *Paradise Lost* within a wider philosophic frame of worldmaking efforts rather than against particular scientific debates over the Ptolemaic and Copernican systems.[95] Book 8 participates in the well-established genre of the cosmic dialogue reaching back to ancients such as Plato and Lucretius, medieval sources such as Macrobius and Dante, and contemporaries such as Galileo, John Wilkins, and Alexander Ross.[96] As Raphael weighs the Ptolemaic world and Copernican universe against each other, the early modern struggle to identify the world's true order is poised in a still moment of dispassionate consideration.[97] Scholars have been frustrated by Milton's refusal to endorse one or the other system explicitly, but this depiction of competing world-systems as plausible though ultimately improvable hypotheses may be meant to highlight the utility of hypothesis itself. Such a reflection within has a dual function: it refers outward to contemporary natural philosophic debates, but it also highlights the poem's own deepest fears about the ultimate knowability (and demonstrability) of cosmic order and thus of the goodness of God. The aporia produced by Raphael's refusal (or inability) to tell "whether heaven move or earth" (8.70) while simultaneously asserting the omniscience of "the great architect" responds to this fear, for the angel recognizes the question underlying Adam's astronomical dilemma as one concerning the place of human beings in the universe, and thus in relation to the divine.

The implicit analogy between scientific hypothesis and poetic imagination—a trope used repeatedly in the period—consequently threatens to implicate the epic itself in Raphael's warning to "admire" God's works rather than "to try / Conjecture" (8.75–76). For the extended comparison of the Ptolemaic and Copernican systems is ultimately an extended description of worldmaking in action: it is the desire to "model heaven / And calculate the stars," to "wield / The mighty frame, how build, unbuild, contrive / To save appearances" (8.79–82).[98] The slippage of agency from deity to human is crucial in this passage. While God, "the great architect," actually wields "the mighty frame," humans can only argue, speculate, and "contrive / To save appearances." But Raphael's gentle mockery also contains an empathetic acknowledgment of the intellectual and emotional need to reconcile the external appearances of the world with our understanding of how it is structured. The unceasing task of reconciliation, of asking "what if" and playing out thought experiments, seems to lead only to a burgeoning number of plausible options. The result is confusion, error, and cultural anxiety, for there is no known thread that leads out of the labyrinth of speculations, hypotheses, and other philosophical fictions. At best they offer temporary solace, but the cycle of doubt and longing begins again. And it is in anticipation of (and in reaction to) this stalemate with its attendant conflicts and despair—a portrait, albeit partial, of the intellectual climate in early modern Europe—that Raphael warns Adam to "be lowlie wise" (8.173).

Through Raphael's speech, Milton acknowledges the epistemological blank at the world's beginning, accepting, after eight books of restless interrogation, that we cannot know the truth of the whole. There is a tragic grandeur in this acknowledgement, which stems not from any lack of intellectual courage but from the sheer impossibility of the task. The inclusion of Raphael's warning is not so much a marker of piety or anti-scientific prejudice as it is of Milton's heroic, if inevitable, failure to uncover the origins of world order. Raphael's recital of world-theories is also a reminder that all such systems may ultimately be no more than speculative hypotheses that offer consolation. Their power lies precisely in their function as therapeutic fictions. If the devils of book 2 are doubles for a fallen humanity, they have already discovered that philosophic ruminations are a "pleasing sorcerie" that can "charm / Pain for a while or anguish."[99] There is no harm in these intellectual games, Raphael seems to suggest, as long as we recognize their insufficiency and do not mistake them for truth.

Such reasoning, however, cannot but affect the interpretation of Genesis, as it is retold in *Paradise Lost* but also as Milton understands its status amid the swarm of worldmaking narratives. Raphael's recourse to rhetorical "what-ifs" following the ostensibly authoritative account of creation in book 7 inevitably raises doubt about the finality of his own tale and leaves the issue of creation glaringly unresolved. To entertain the possibility that Genesis may not entirely reflect the truth of things is not, however, to dismiss its utility or continuing importance. By questioning Genesis and revealing its lacunae, Milton can affirm greater faith in it as a narrative of world order because the process of interrogation allows us to recognize the need for such narratives of order, though they—like *Paradise Lost* itself—may ultimately be no more than idols.

The Paradise Within

Though *Paradise Lost* features a miniature Galilean dialogue on the two chief world systems, the epic is finally concerned with earthly matters. "Like Copernicus's and Kelper's astronomy," Catherine Gimelli Martin notes, "Milton's cosmos remains earth-centered but not actually Ptolemaic."[100] It is this post-Copernican anthropocentrism—profoundly different from the ringing affirmations of Pico's famous *Oration on the Dignity of Man* (1486)—that Raphael articulates when he finally urges Adam to turn away from the cosmos and back to the hearth:

> Sollicit not thy thoughts with matters hid,
> Leave them to God above, him serve and feare;
> Of other Creatures, as him pleases best,
> Wherever plac't, let him dispose: joy thou
> In what he gives to thee, this Paradise
> And thy faire Eve; Heav'n is for thee too high
> To know what passes there; be lowlie wise:
> Think onely what concernes thee and thy being;
> Dream not of other Worlds, what Creatures there
> Live, in what state, condition or degree,
> Contented that thus farr hath been reveal'd
> Not of Earth onely but of highest Heav'n.
>
> (8.166–78)

Readers have frequently felt that Milton somehow betrays his own revolutionary desire to "look in and see" in this passage, abandoning the pursuit of knowledge for a pessimistic anti-intellectualism. But Raphael's speech here recalls the movement of Montaigne's "Apologie de Raimond Sebond," tracing a similar trajectory that investigates numerous alternate theories of world order, but which eventually concludes by embracing the partiality of human experience. Like Montaigne, Raphael emphasizes human limitation, without access to a god's-eye view or to an understanding of true cosmic order; all we can know is what we see, what we conjecture, what we experience through our own being. And like Montaigne's turn away from the seductions of *kataskopos* and his embracing of the mutable world and self, Raphael too gently leads Adam away from kataskopic desire toward a humbler—and perhaps richer—knowledge of himself, his partner Eve, and the world that they inhabit and shape together (Eden).

This (re)turn to the local concerns of the individual self in Milton, as in

Montaigne, represents a potent response to the dilemma of skepticism, and marks the split between natural and moral philosophy—knowledge of the natural world and knowledge of the self and of human action—that gradually develops over the late seventeenth and eighteenth centuries. Raphael's speech, in fact, takes its place in a long line of early modern texts, beginning with the "Apologie," which struggle with the consequences of separating epistemology from metaphysics.[101] Both Montaigne and Milton want to have it both ways: they vividly portray the terror of skeptical despair, even as they long to celebrate the power of human fictions (whether understood as "philosophy" or "poetry"). Both adopt a fideist position because their vision of the skeptical abyss propels them to choose between nihilism or belief. But their belief, which seems to shine all the more strongly for the experience of labyrinthine doubt, derives from understanding the importance of human speculation and the comfort of coherent world-systems, even if they are merely made up.

In book 8, Milton decisively breaks with the temptations of speculative cosmology and establishes a new focus on human action, suggesting that a new world-system might well be constructed through individual ethics rather than natural philosophy. At the end of the "Apologie" Montaigne affirms faith in God as an eternal, immutable entity, urging an appreciation of human limits: "For to make the handful bigger than the hand, the armful bigger than the arm, and to hope to straddle more than the reach of our legs, is impossible and unnatural. Nor can man raise himself above himself and humanity; for he can see only with his own eyes, and seize only with his own grasp."[102] The similarity of this sentiment to Raphael's "Heav'n is for thee too high / To know what passes there; be lowlie wise / Think onely what concernes thee and they being" (or to the narrator's anxious "know to know no more" in book 4) suggests Milton's programmatic reorientation of worldmaking away from abstract conceptual questions and toward the tangible business of establishing an ethical order. This turn toward ethics is strongly evident throughout the final books of *Paradise Lost*, which explore the ramifications of the Fall as a descent into history with the consequent necessity for moral choice and action. The Fall also marks a decisive shift of focus within the poem (as it does in Genesis) away from the cosmic vistas of worldmaking to the theater of human action, suggesting that the realm of religion is finally human moral conduct rather than philosophical explanation.

The emphasis on self-analysis and self-regulation, which Adam and Eve suffer in their painful growth from Edenic innocence to post-lapsarian experience, is also characteristic of the final book of Montaigne's *Essais*, where we are told in no uncertain terms:

> Of the opinions of philosophy I most gladly embrace those that are most solid, that is to say, most human and most our own. . . . Philosophy is very childish to my mind, when she gets up on her hind legs and preaches to us. . . . It is an absolute perfection and virtually divine to know how to use our being rightfully. We seek other conditions because we do not understand the use of our own, and go outside of ourselves because we do not know what it is like inside.[103]

Paradise Lost charts a similar movement. It signals the abandonment of speculative philosophy for the stern discipline of ethical choice through symbolic internalizations of the world. As early as book 4, the nihilistic Satan tormented by the consequences of his actions knows that the poem's spatial vistas are in fact psychological ones: "Which way I flie is Hell; my self am Hell / And in the lowest deep a lower deep / Still threatning to devour me opens wide, / To which the Hell I suffer seems a Heav'n" (4.75–78). Satan, however, seeks to fold the cosmos into himself and is ultimately unable to abandon what John Gillies describes as the "rapture of space" for the "particularity of place."[104] It is for this turn away from the seductive but intangible infinity of space for the attainable domestic pleasures and solvable quotidian dilemmas of the garden that Raphael advocates so strongly. Only in the final lines of the poem will Adam and Eve come to understand its import as the "paradise within" is balanced by the world of choices outside:

> The World was all before them, where to choose
> Their place of rest, and providence their guide:
> They hand in hand with wandering steps and slow,
> Through Eden took their solitary way.
>   (12.646–69)

Richard Bentley, Milton's notorious eighteenth-century editor, famously objected to the final line with its emphasis on solitude, claiming that it contradicted the earlier promise of a guiding providence. But Milton knew that the Fall brought with it a recognition of cosmic loneliness, the realization that human beings make their own choices and their own fictions — even the dream of Eden and the promise of providential intervention. For the first time in the poem, Adam and Eve — and we, along with them — face that recognition. Milton's dense cast of characters all fall away, leaving only individuals who must now confront a vast, fractured whole and make it meaningful once more.

Epilogue: From Cosmography to Cosmopolitanism

Antwerp, once more, ca. 1590. A new map is for sale. Unsigned and lacking even a printer's mark, it takes a curious form: a world map inside a motley cap. The outline of a jester's head and sphere-topped mace dominate the image. But where his face should be, we find the world. Human features have been replaced by the geometry of oceans and continents. And through a visual play of perspectives, the Fool and the World have become enmeshed, inseparable and interdependent.

This copperplate engraving of unknown provenance has become one of the most widely reproduced images of the early modern world, capturing in a single frame the literal expanse and metaphoric breadth of the period's vision of the "world" itself.[1] Janus-like, the map looks back toward medieval morality plays and warnings against worldly temptations—and forward to the post-Columbian world. The mathematical acumen of Ortelius's latest plate signals the disenchanted, instrumental rationality of Weberian modernity. But the teasing Erasmian motley cap draws the image toward an imaginatively sophisticated humanism grounded in classical and Christian metaphysics. Poised between philological humanism and the new science, the engraving reimagines the relationship between an archetypal self and reconfigured global space. Just twenty years after the landmark publication of Ortelius's *Theatrum orbis terrarum,* the Fool's Cap Map (fig. 18) satirizes, deconstructs, and paradoxically celebrates the act of creating the world *as* an image. It acknowledges the making of a modern world picture as a self-reflexive, knowing, and thoroughly problematic endeavor.

If the story of the world's emergence into modernity is one of unveil-

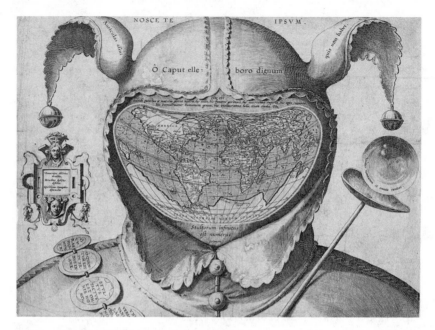

FIGURE 18. "Cordiform World Map in Fool's Cap" (ca. 1590). Courtesy of the New-
berry Library, Chicago (call no.: Novacco 2F 6).

ing, of stripping away the unshakeable belief in a stable, divinely established order of things, the tale of its reinvention as a modern object is one
of skeptical reenchantment and human fashioning. The map reflects on the
shifting relations between self and world, *artifex* and artifact, in ways that
still resonate over four hundred years later. It is no accident that this image
has become a favorite emblem for contemporary scholars seeking to link
the early modern past with the postmodern present.[2] For even as the image
foregrounds one of the most significant intellectual shifts of the period—the
remaking of the term "world"—it introduces a new keyword that signals
the future of worldmaking endeavors.

Hidden in a cartouche on the left is a cryptic Latin legend that announces
the advent of a new philosopher: one Epichthonius Cosmopolites, whose
name marks the emergence of the cosmopolitan ideal into modern European vocabulary.[3] Whether a pseudonym for the engraver, or the identity
of the fool, the name hints at an allegorical program. The Latinized faux-
Greek pseudonym is an erudite humanist joke, a play on words that juxtaposes *epichthonios*, a favorite Homeric epithet for mortal humans meaning
"terrestrial" or "earthly ones," with *cosmopolites*, "citizen of the world," the

tag that recalls Diogenes's famous declaration.[4] The Fool's Cap Map captures a moment of transition, even as it anticipates the transformation of the worldmaking project itself, from visions of cosmopoesis to the ethics of cosmopolitanism.

Symbolizing an expansive, Homeric sense of humanity, Epicthonius Cosmopolites registers the pull between local particularity and universal aspiration. This tension, already at the core of the Greek ideal of the cosmopolitan, acquires new energy in the wake of the worldmaking project that had reconceived the relations between human and world. Despite the desire to transcend the bounds of the particular and embrace the cosmos, the *cosmopolites* remained tied to the *polis,* the city-state. In a newly modern world, Epicthonius is a reminder that alongside a novel apprehension of global affiliation and an unprecedented ability to reach beyond the local, the renovation of the world brought with it a recognition of human limitation. We remain grounded in the particularity of our bodies, our faces, ourselves. The image is aptly entitled "Nosce teipsum" ("know yourself"): the cosmopolitan World-Fool suggests that knowledge of the self is bound up with knowledge of the world and vice versa.

Epichthonius Cosmopolites challenges contemporary theorizations of globalization and cosmopolitanism, concepts encoded within his name and evoked in the engraving. In recent years, cosmopolitanism has come to be seen as a byproduct of a globalized world, variously associated with the moral and political obligations of states toward foreigners, with the mobility afforded by global empires, with the mandate of international institutions to act transnationally, and with the individual citizen's capacity to cross national and cultural boundaries.[5] Epichthonius Cosmopolites, however, connects the term to a broader meditation on the philosophical bases for human connection and universal feeling, suggesting some four hundred years before Anthony Appiah that cosmopolitanism "shouldn't be seen as some exalted attainment: it begins with the simple idea that in the human community, as in the national community, we need to develop habits of coexistence: conversation in its older meaning, of living together, association."[6] Contemporary cosmopolitanisms have struggled to integrate yearnings for global and universal connection with skepticism about transcending national and local identifications. The Fool's Cap Map suggests how these concerns might begin with the challenges of the worldmaking project that has been the subject of this book.

To return to Epichthonius Cosmopolites, then, is to discover the seeds of modernity's global future in the experiments of the sixteenth and seventeenth centuries. It suggests how the struggle between the particular and the universal finally finds its apotheosis in that hybrid fusion of self and world. Epicthonius suggests how a worldmaking project that begins with Mercator's quest for a *cosmopoeia* could evolve into the eighteenth-century cosmopolite who could move through the world as though it were his home. As an ethically motivated vision of the world contained within the self and emerging from the individual subjectivity of a particular human maker, the Fool's Cap Map thus presents a tantalizing glimpse at an alternate geneal-ogy for a cosmopolitan modernity, one whose roots lie in an Erasmian six-teenth century rather than merely following from the post-Kantian think-ing of the Enlightenment.

The integration of fool and world is a final fitting emblem for *The World-makers*. Beyond traditional tropes of microcosm and macrocosm, this book, like the map, has sought to explore a "project of mediation" between the global and the local, the vast abstraction of the world (for the map itself is a scheme, an abstraction) and the circumscribed particularity of the persons who inhabit it.[7] Their reciprocity suggests that the two poles are not in opposition despite appearances to the contrary. Rather, they are mutually enabling and depend upon each other for intellectual coherence. Though it has often been argued that the individual modern subject comes into ex-istence after Descartes through the separation of self from world, images such as the Fool's Cap Map are a reminder that this sensation of separa-tion was itself derived from the worldmaking project. As these pages have argued, to understand the world as a thing made by human ingenuity and imagination is to reconceive the relation of the knowing human self and the world at large. No longer is the self merely subordinate to the world; it simultaneously creates the very world of which it was also a part.

Showcasing and questioning the desire to capture the world in a single gaze, the World-Fool troubles the seeming transparency of the map and draws both its maker and viewer into the frame. It pays homage to new-found worldmaking prowess. But it also relentlessly undercuts it. The Fool's Cap Map juxtaposes admonishments against worldly vanity and folly taken from Ecclesiastes and the Stoics against Ortelius's 1587 edition of the already iconic "Typus orbis terrarum" (fig. 9). The slightly sinister bells on the jester's cap even reflect the world map in miniature strokes, reducing its geometric precision to allegorical parody—the feat of world mastery is seemingly no more than empty jingling. The Latin quotation across the Fool's head underlines this paradox. Based on Pliny's *Natural History*, it reminds the viewer, like Adam in *Paradise Lost*, that from the perspective of the entire universe, the earthly world is no more than a "pinprick" (a *punctus*), "and this is the substance of our glory, this is its habitation, here it is that we fill positions of power and covet wealth, and throw mankind into an uproar, and launch even civil wars and slaughter one another to make the land more spacious!"[8] Ironically, to see the entirety of the world in a blink of the eye is also to understand the world's insignificance. To capture the world's vastness is also to contract it. And yet, to hold the world within one's gaze is a singular, hard-won achievement—once accorded to only a chosen few in the visionary space of the dream, now legible and reproducible in print.

These juxtapositions, held in tight allegorical suspension, also describe the historical and imaginative trajectories traced in this book. The writers and texts discussed in these pages speak to a deep, lingering nostalgia for a time of coherence and order, even as they celebrate the possibility that we ourselves make up the world's form and meaning. They acknowledge the role of external factors in the shaping of a global environment—institutions, states, economic flows—but also emphasize the metaphysical power of the idea "world" and the centrality of self-conscious human fashioning to its historical evolution.

Such reflections force us to think beyond macro-theoretical categories such as globalization and universalism and instead pursue a finer-grained analysis that exposes the struggle to balance the demands of the particular with a yearning for the whole—to trace, in other words, cultural processes by which the world is unmade and made anew. After over two decades of critique and debate over the new imperialisms of the late twentieth century, born through the globalization of capital and the shrinking of the world in the age of the internet, we are yet to grapple with the lure of the term "world"—its appeal as a goal, a prize, a pinnacle of achievement—in terms that go beyond discourses of power and oppression. *The Worldmakers* tries to suggest ways in which the desire for universal harmony can be understood through local, particular interventions, a move that aims to break down the hegemonic valence of large-scale views, to zoom in from the planispheric level, following Milton's lead, and investigate the topography of the household. Why are ordinary people (and not just barons of finance or would-be imperial conquerors) fascinated by Google Earth? What motivates the excitement of zooming in and out of the global whole? Why does it make us feel cosmopolitan, at once at home *and* in the world?

Writing in the 1920s, on the eve of the aerial surveys that would offer a new perspective on the world, Aby Warburg would look back to the sixteenth-century atlas as a formal means of making meaning in the face of loss and fragmentation. His *Atlas Mnemosyne*, not a cartographic project in any conventional sense, interrogated the structure of knowledge by juxtaposing images of Renaissance art and writing in provocative combinations, raising anew the question of parts and wholes. Writing of Warburg's *Bilderbuch*, Davide Stimilli observes, "We would be wrong . . . in considering *Mnemosyne* just an atlas in the now common sense or just another collection of images: the globe unbinds itself in its plates, and Dante's expression is not out of place if one considers that the ultimate ambition of Warburg's atlas is no less than to gather "into one volume," literally, "what, in the universe, is unbound" ("ciò che per l'universo si squaderna")."[9] Warburg's aesthetic project becomes an epistemological one; his meditation on art, a meditation on the world.

The compositional flexibility of the atlas, with which this book began, harnesses the power of montage as investigative technique, a means to both unmake and make the whole. As Georges Didi-Hubermans notes, in a recent attempt to recreate Warburg's *Atlas* project, "When we arrange different images or different objects . . . on a table, we are free to modify constantly their configuration. We can make piles or constellations. We can discover new analogies, new trajectories of thought. . . . A table is not made for definitively classifying, for exhaustively making an inventory, or for cataloguing once and for all . . . but instead for gathering segments, or parceling out the world, while respecting its multiplicity and its heterogeneity— and for giving a legibility to the underlying relations."[10] The image of the table here reminds us that maps in the sixteenth century—like anatomical engravings—were often called *tabulae,* a term that combined enumeration and cataloguing with both written and visual representation. These carefully framed images simultaneously dissected and gathered together far-flung parts of the world into unity.

A similar desire to bind up the world's heterogeneity underlies such contemporary worldmaking projects as Google Earth. Brian McClendon, one of the founders of the technology, has reflected that the project's ultimate goal is to reach "the end of resolution," creating a virtual planet at a scale of one pixel for each square centimeter of the Earth's surface.[11] This drive mirrors the early modern cosmographic, cosmopoetic drive to fix the world on the page. But for all its technological sophistication and claim to perfect correspondence between world and image, the Google Earth project, like the Renaissance world atlas, depends on a mediation between part and whole as it seeks to capture a mimetic expression of the world conceived as a totality. The epistemic dilemma that confronts both atlas and digital mapping software is entwined with the historiography of modernity itself, for both attract theoretical and aesthetic interest for their ability to portray and meditate simultaneously on the problem of order.

The union of science and the aesthetic imagination, however, is fraught with paradoxes, as the Fool's Cap Map shows. For the speculative desire of *poiesis* makes an uncomfortable partner for normative claims to truth, particularly when the subject in question is the world. In contrast to the history of the scientific atlas, which has evolved as the exemplar of objective, authorized, and normative knowledge, Didi-Hubermans and other contemporary theorists valorize the deconstructive power of the atlas as an aesthetic form. For the atlas's ability to dismantle old assumptions and restructure them in a different order invites us to imagine alternate models for understanding ourselves and the world. Through its commitment to the part, and its impetus to rethink the structure of the whole, the atlas can become a form that resists the closures of hegemonic universalisms even as it seeks to gather up disparate parts into a coherent vision of wholeness.[12]

To see, then, in such contemporary manifestations of the atlas as Google Earth not only an endlessly perfectible "scientific" resource but an aesthetic form with great deconstructive power is to see how the world itself becomes an aesthetic object that can be made and unmade by individuals with a tap of a finger. It is to experience the appeal of conceptualizing the world as whole from the particular position of a thinking self. Worldmaking, at its most exciting, thus pioneers a supple, expansive approach to the problem of expressing a vision of totality—without succumbing to totalizing rigidity. The image of the individual at her computer screen playing with a digital globe that she makes and unmakes with virtual push-pins is finally not so far from the World-Fool depicted on the 1590 engraving: she too makes and is made by a world in perpetual flux.

Acknowledgments

"Le monde n'est qu'une branloire perenne," writes Montaigne, noting that "je ne puis asseurer mon object." What is an elegant trope in the *Essais* has been plainly true for this book, which has groped toward its protean subject, and has moved slowly with a halting, staggering pace. Were it not for the generosity and good humor of the many teachers, colleagues, and friends who have nudged me on—and kept me from falling down, and bumping into things along the way—this project would never have reached completion. It is with humble gratitude and immense pleasure that I thank them here.

My greatest debt is to my teachers, William Oram, David Quint, and Annabel Patterson, who have shaped me (and not just this project) with their enormous learning, cheerful disagreements, and warm, patient generosity over many years. It is because of their example and faith that I dared to pursue this book's elusive subject.

I have been blessed with spirited and giving colleagues at a range of institutions, who have offered counsel, camaraderie, and countless conversations. From my time at Smith College and Yale University, I'd like to thank Bob Babcock, Paul Bushkovitch, Patrick Coby, Elizabeth Dillon, Edwin Duval, Maija Jansson, Traugott Lawler, Larry Manley, Steven Pincus, Joe Roach, John Rogers, Jean-Frédéric Schaub, Harold Skulsky, the late Frank Turner, and Chris Wood. From my time at the Harvard Society of Fellows, I thank Ann Blair, Peter Galison, Peter Gordon, James Hankins, and Amartya Sen, who showed me new ways of thinking and seeing the world.

My colleagues at Stony Brook University deserve special thanks for their enthusiasm, support, and friendship; it is with deep gratitude that I salute Jennifer Anderson, Eugene Hammond, Ann Kaplan, Peter Manning, Gary Marker, Celia Marshik, Adrienne Munich, Benedict Robinson, Susan Scheckel, Steven Spector, and Kathleen Wilson; particular thanks are due to our "reading group" of Andrew Newman, Douglas Pfeiffer, Jeffrey Santa Ana, and Helen Choi, whose energy and collective impetus kept the book in motion. My new colleagues at Yale have been a strong source of motivation and wise counsel in the final stages of writing; I am eager to thank Julia Adams, David Gabriel, Martin Hägglund, David Jackson, Carol Jacobs, and Katie Trumpener.

Beyond the institutions that have supported me, I have had the good fortune to draw on the generosity of colleagues elsewhere. Many of the insights and details that have filled these pages would have been impossible without the kindness of scholars who have stopped to talk and nourish my own very partial knowledge; for timely interventions and ongoing generosity, I would like to thank Ken Borris, Tom Conley, Surekha Davies, Jordana Dym, Timothy Hampton, Jonathan Gil Harris, Elizabeth Harvey, Susannah Hollister, Daniel Javitch, Carol Kaske, Roger Kuin, Jacques Lezra, David Lee Miller, Evelyn Lincoln, James Muldoon, Ricardo Padrón, Anne Prescott, Jon Quitslund, Timothy Reiss, Robert Stillman, and Susanne Wofford. Early on in the book's genesis, David Porter invited me to present at the conference on Comparative Early Modernities, which changed the trajectory of my thinking: thanks are due to colleagues and audiences there, as well as at Harvard University, the University of California at Berkeley, Brown University, Barnard College, New York University, the John Carter Brown Library, and at various venues at Yale and Stony Brook.

Much of the research for this book was done at the Beinecke Rare Book and Manuscript Library and the John Carter Brown (JCB) Library, whose ever-enthusiastic staff saved this book from hitting obstacles. At the Beinecke, special thanks are due to Bob Babcock, Ray Clemens, George Miles, and Ingrid Lennon-Pressey. At the JCB, particular thanks are due to Valerie Andrews, Susan Danforth, Dennis Landis, Margot Nishimura, Kimberly Nusco, and Ken Ward. For financial support of my research I would like to acknowledge the Yale University Graduate School, the Frederick W. Hilles Publication Fund of Yale University, the Beinecke Rare Book and Manuscript Library, the Mrs. Giles Whiting Foundation, the Society of Fellows at Harvard University, Stony Brook University, and the John Carter Brown Library.

This book is also very much a product of the wonderful people at the

University of Chicago Press who have nurtured its proliferating raw material into something more shapely and elegant. My deepest thanks are to Randolph Petilos: no author can ask for a more generous, warm, and invigorating editor. Enormous gratitude is also due to Mara Naselli, who wrought magic with my unwieldy prose at a crucial moment; to Richard Allen, who copyedited the manuscript with a keen and thoughtful eye; and to June Sawyers, who prepared the index. I would also like to thank the people who nudged the book through the production process: Renaldo Migaldi, Lauren Smith, Joan Davies, and Tadd Adcox.

For intellectual fellowship, conviviality, and friendship in the best Renaissance sense, I salute Emily Anderson, Emily Setina, Sarah van der Laan, Anthony Welch, Brett Foster, Tatiana Seijas, and Jane Grogan. My research assistant extraordinaire, Christina Jones, deserves more thanks than I can give in words. For unstinting, affectionate support and bracing sustenance of various kinds I warmly thank Kester Warlow-Harry, Micala Sidore, Diana Kurkovsky, Diana Cooper, and Zosia and Hannes Baumann. And finally, to my family, who have supported me in innumerable ways against sometimes heavy odds, my gratitude knows no bounds. I am privileged to thank Krishnan and Anita Ramachandran, Carlo and Ambra Gazzola, Zazie and Viola, who keep things real, and my partner of fine distinctions and steadfast love, Giuseppe Gazzola.

Notes

INTRODUCTION

1. Pieter Heyns, *Le miroir du monde* (Antwerp, 1579), a prose translation of Heyn's *Spieghel der werelt* (Antwerp, 1577).

2. Gerhard Mercator, *Atlas; ou, représentation du monde universel, et des parties d'icelui, faicte en tables et descriptions tres amples et exactes: divisé en deux tomes* (Amsterdam, 1633).

3. On the importance of Ptolemy's *Geography*, see the introduction to *Ptolemy's Geography: An Annotated Translation of the Theoretical Chapters*, trans. J. Lennart Berggren and Alexander Jones (Princeton: Princeton University Press, 2000); and Patrick Gautier Dalché, "The Reception of Ptolemy's Geography (End of the Fourteenth to Beginning of the Sixteenth Century)," in *The History of Cartography*, vol. 3: *Cartography in the European Renaissance*, ed. David Woodward (Chicago: University of Chicago Press, 2007).

4. Useful overviews are Jerry Brotton, *Trading Territories: Mapping the Early Modern World* (London: Reaktion Books, 1997), and Denis Cosgrove, *Apollo's Eye: A Cartographic Genealogy of The Earth in the Western Imagination* (Baltimore: Johns Hopkins University Press, 2001).

5. Eugenio Garin, *Science and Civic Life in the Italian Renaissance*, trans. Peter Munz (Gloucester: Peter Smith, 1978). The "epistemological crisis" of the seventeenth century has been explained in three different, highly influential ways by Ernst Cassirer, *The Individual and the Cosmos in Renaissance Philosophy*, trans. Mario Domandi (New York: Harper and Row, 1964); Michel Foucault, *Les mots et les choses: une archéologie des sciences humaines* (Paris: Gallimard, 1966); and Timothy J. Reiss, *The Discourse of Modernism* (Ithaca: Cornell University Press, 1982) and *Knowledge, Discovery, and Imagination in Early Modern Europe: The Rise of Aesthetic Rationalism* (New York: Cambridge University Press, 1997).

6. See Lisa Jardine, *Worldly Goods* (London: Macmillan, 1996).

7. For a complementary definition, see Mary Baine Campbell, *Wonder and Science: Imagining Worlds in Early Modern Europe* (Ithaca: Cornell University Press, 1999).

8. Christopher Marlowe, *Tamburlaine the Great, Parts 1 and 2* (New York: Oxford University Press, 1998), Part 2, act 5, 3.

9. Nelson Goodman, *Ways of Worldmaking* (Indianapolis: Hackett, 1978). The debate on worldmaking in contemporary philosophy, associated with Kantian idealism, is summed up and critiqued by Michael Devitt, "Worldmaking Made Hard," *Croatian Journal of Philosophy* 6, no. 1 (2006): 3–25.

10. Nathaniel Fairfax, *A Treatise of the Bulk and Selvedge of the World. Wherein the Greatness,*

Littleness and Lastingness of Bodies Are Freely Handled (London, 1674), 138. On Fairfax's pamphlet, see Michael Cyril William Hunter, *Science and the Shape of Orthodoxy: Intellectual Change in Late Seventeenth-Century Britain* (London: Boydell & Brewer, 1995), 116. Fairfax's use is one of the oldest; the only earlier instance of the word "worldemaker" noted by the OED is from the dedication to Thomas Blenerhasset's *A Revelation of the True Minerva* (1582): "a man may esteeme that his which the great worldemaker . . . conuaieth into him" (sig. *3).

11. Matthew Prior, *The Poetical Works of Matthew Prior* (London: George Bell, 1907), 2.267. On Prior's attitude toward the new natural philosophy, see Monroe K. Spears, "Matthew Prior's Attitude toward Natural Science," *PMLA* 63, no. 2 (1948): 485–507.

12. David Hume, *Dialogues Concerning Natural Religion, the Posthumous Essays, Of the Immortality of the Soul, and of Suicide, from An Enquiry Concerning Human Understanding of Miracles* (New York: Hackett, 1986), 35–36.

13. I draw on Hannah Arendt's analysis of *homo faber* in modernity: see *The Human Condition* (Chicago: University of Chicago Press, 1998), chap. 42.

14. For detailed discussion of the meaning of "world" in English, see C. S. Lewis, *Studies in Words* (Cambridge: Cambridge University Press, 1990), 214–68. See also the discussions in Leo Spitzer, *Classical and Christian Ideas of World Harmony: Prolegomena to an Interpretation of the Word "Stimmung"* (Baltimore: Johns Hopkins University Press, 1963) and Roland Greene, *Five Words: Critical Semantics in the Age of Shakespeare and Cervantes* (Chicago: University of Chicago Press, 2013).

15. I cite from the eighteenth-century English edition, which contains a selection of emblems: see Cesare Ripa, *Iconology*, trans. George Richardson (New York: Garland, 1979), 1.3. Yassu Okayama, *The Ripa Index: Personifications and Their Attributes in Five Editions of the Iconologia* (Doornspijk: Davaco, 1992) collates all the emblems.

16. Giambattista Vico, *New Science: Principles of the New Science Concerning the Common Nature of Nations*, trans. Dave Marsh (London: Penguin Books, 1999), sections 722, 725.

17. Isidore of Seville, *The Etymologies of Isidore of Seville*, trans. Stephen A. Barney (Cambridge: Cambridge University Press, 2006).

18. Jean-Marc Besse, *Les grandeurs de la terre: aspects du savoir géographique à la Renaissance* (Lyon: ENS, 2003), 2.

19. Giambattista Vico, *On the Most Ancient Wisdom of the Italians*, trans. L. M. Palmer (Ithaca: Cornell University Press, 1988), 40–64.

20. See Vico's outline of the "Idea of the Work" in the *New Science*, section 2.3. On Vico's poetic epistemology, see Giuseppe Mazzotta, *The New Map of the World: The Poetic Philosophy of Giambattista Vico* (Princeton: Princeton University Press, 1999), and James Robert Goetsch, *Vico's Axioms: The Geometry and the Human World* (New Haven: Yale University Press, 1995).

21. I follow recent work that connects modernity to enchantment; for contemporary philosophical meditations on the topic, see Jane Bennett, *The Enchantment of Modern Life: Attachments, Crossings, and Ethics* (Princeton: Princeton University Press, 2001) and David L. Martin, *Curious Visions of Modernity: Enchantment, Magic, and the Sacred* (Cambridge, Mass.: MIT Press, 2011); Michael T. Saler, *As If: Modern Enchantment and the Literary Prehistory of Virtual Reality* (New York: Oxford University Press, 2012) and Eric Hayot, *On Literary Worlds* (Oxford: Oxford University Press, 2012) discuss the literary significance of modernity's enchantments.

22. Martin Heidegger, "The Age of the World-Picture," in *The Question Concerning Technology and Other Essays*, trans. William Lovitt (New York: Harper Torchbooks, 1977), 127.

23. On the rhetoric of "lord of the world" and its theological resonance, see Anthony Pagden, *Lords of All the Worlds: Ideologies of Empire in Spain, Britain and France, 1500–1850* (New Haven: Yale University Press, 1995), especially 23ff.

24. Charles Taylor, *A Secular Age* (Cambridge, Mass.: Harvard University Press, 2007).

25. Laura Doyle, "Notes Toward a Dialectical Method: Modernities, Modernisms, and the Crossings of Empire," *Literature Compass* 7, no. 3 (2010): 197.

26. Walter Mignolo, *The Darker Side of the Renaissance: Literacy, Territoriality, and Colonization* (Ann Arbor: University of Michigan Press, 1995).

27. See Doyle's analysis in "Dialectical Method."

28. Sanjay Subrahmanyam, "Holding the World in Balance: The Connected Histories of the Iberian Overseas Empires, 1500–1640," *The American Historical Review* 112, no. 5 (December 2007): 1359–85.

29. See especially Martha C. Nussbaum, "Compassion & Terror," *Daedalus* 132, no. 1 (January 1, 2003): 10–26; Anthony Appiah, *Cosmopolitanism: Ethics in a World of Strangers* (New York: W. W. Norton, 2007); and Seyla Benhabib, *Another Cosmopolitanism* (Oxford: Oxford University Press, 2006).

30. On its earliest usages, see Paul Hazard, "Cosmopolite," in *Mélanges d'histoire littéraire générale et comparée offerts à Fernand Baldensperger* (Paris: Champion, 1930), 1:354–64.

31. Martin W. Lewis, *The Myth of Continents: A Critique of Metageography* (Berkeley and Los Angeles: University of California Press, 1997), 16.

32. I refer to Dipesh Chakrabarty, *Provincializing Europe: Postcolonial Thought and Historical Difference* (Princeton: Princeton University Press, 2000). See also Chakrabarty's critique in "The Muddle of Modernity," *The American Historical Review* 116, no. 3 (June 2011): 663–75. I explore what such an analysis in a non-European context might look like in "A War of Worlds: Becoming 'Early Modern' and the Challenge of Comparison," in *Comparative Early Modernities: 1100–1800*, ed. David Porter (New York: Palgrave, 2012), 15–46.

33. *De Constantia* 1.9; cited from Justus Lipsius, "The First Book on Constancy," trans. John Stradling, *Philosophical Forum* 37, no. 4 (2006): 389–426.

34. Archibald MacLeish, "A Reflection: Riders on Earth Together, Brothers in Eternal Cold," *New York Times*, December 25, 1968.

35. Robert K. Poole, *Earthrise: How Man First Saw the Earth* (New Haven: Yale University Press, 2008), 3.

CHAPTER ONE

1. Mercator's title is notoriously difficult to translate because of the Latin wordplay ("fabrica mundi et fabricati figura"). Most English translations follow the pirated 1635 London edition by rendering the title as "cosmographical description of the fabric and figure of the world." Even the multimedia CD based on the Library of Congress copy renders the title as "Atlas or Cosmographic Meditations on the Fabric of the World and the Figure of the Fabrick'd": Gerhard Mercator, *Atlas Sive Cosmographicae Meditationes de Fabrica Mundi et Fabricati Figura Duisburg, 1595*, ed. Robert W. Karrow, trans. David Sullivan (Oakland, Calif.: Octavo, 2000); for the LC digital facsimile, see http://hdl.loc.gov/loc.rbc/rosenwald.0730.2. All subsequent citations and translations from the *Atlas* will be from this edition and noted as "Mercator, *Atlas*." I give citations to both a physical copy at the John Carter Brown Library, indicated as "JCB," and the digital copy published by Octavo and based on the Library of Congress copy, indicated as "LC." Citations of the JCB copy refer to signature, page, and verso or recto, as in 2r; citations of the LC digital edition refer to image number and verso or recto. Different copies contain different paratextual matter and some (such as JCB) are bound with other texts.

2. On the influence of emblem books on the visual rhetoric of maps, see Lucia Nuti, "The World Map as an Emblem: Abraham Ortelius and the Stoic Contemplation," *Imago Mundi* 55, no. 1 (2003): 38–55. Also useful is Francesca Fiorani, *The Marvel of Maps: Art, Cartography and Politics in Renaissance Italy* (New Haven: Yale University Press, 2005).

3. Muki Haklay, Alex Singleton, and Chris Parker, "Web Mapping 2.0: The Neogeography of the GeoWeb," *Geography Compass* 2, no. 6 (2008): 2011–39.

4. See transcript of Scott Simon segment on NPR at: http://www.npr.org/2012/04/28/

151583081/a-way-to-see-the-world-and-ourselves#commentBlock. On critiques of the Mercator projection and the "map wars," see Mark S. Monmonier, *Rhumb Lines and Map Wars* (Chicago: University of Chicago Press, 2004) and Sumathi Ramaswamy, "Conceit of the Globe in Mughal Visual Practice," *Comparative Studies in Society and History* 49, no. 4 (2007): 751-82. On maps and ideology, see J. B. Harley, "Silences and Secrecy: The Hidden Agenda of Cartography in Early Modern Europe," *Imago Mundi* 40, (1988) 57-76, and "Maps, Knowledge and Power," in *The Iconography of Landscape: Essays on the Symbolic Representation, Design, and Use of Past Environments* (Cambridge: Cambridge University Press, 1988), 277-312.

5. Steven Van den Broecke, "Dee, Mercator, and Louvain Instrument Making: An Undescribed Astrological Disc by Gerard Mercator (1551)," *Annals of Science* 58, no. 3 (2001): 219-40; and Gerard Le Turner, "The Three Astrolabes of Gerard Mercator," *Annals of Science* 51, no. 4 (July 1994): 329.

6. Bartholomaeus Mercator, *Breves in Sphaeram meditatiunculae, includentes methodum et isagogen in universam cosmographiam* . . . (Colonia, 1563), fol. B7v. See discussion in Steven Van den Broecke, *The Limits of Influence: Pico, Louvain, and the Crisis of Renaissance Astrology* (Amsterdam: Brill, 2003), 211. The term might refer to Mercator's projected treatise on the creation of the world, "De mundi creatione ac fabrica," eventually published in the *Atlas*.

7. This may explain a repeatedly dismissed anecdote in Walter Ghim's biography that Mercator conceived the idea of a world atlas before Ortelius (Mercator, *Atlas*, ✠ v [JCB]; 10v [LC]). The suggestion that Mercator withheld his atlas until Ortelius profited from sales of the *Theatrum* has struck critics as preposterous. The comment is especially strange if we consider that Mercator did publish several maps in 1585, before the completion of the *Atlas*. But the comment and publication history appear more coherent if we believe, along with Jean-Marc Besse (*Les grandeurs de la terre*), that the idea for the *Theatrum* originated in the 1560s during Ortelius and Mercator's travels together in France, and that Ghim's account of their friendship is reasonably accurate.

8. Gerhard Mercator, *Chronologia* (Coloniae Agrippinae: Birckmannus, 1569), p. 3.

9. "Primum ab adolescentia studium mihi fuit Geographia, in quo dum progrederer, adhibita naturali & geometrica speculatione . . . admiranda inveni, non in geographia tantum, sed in universae huius mundanae machinae constitutione, quorum multa hactenus a nemine perspecta sunt." Dedication to Wilhelm, Duke of Julich and Cleve, in Gerhard Mercator, *Tabvlae Geographicae Cl: Ptolomei Ad Mentem Autoris Restitutae & Emendate Per Gerardum Mercatorem* (Cologne: Godfried von Kempen, 1578). On this passage, see Van den Broecke, *Limits of Influence*, 210-11, and Mike A. Zuber, "The Armchair Discovery of the Unknown Southern Continent: Gerardus Mercator, Philosophical Pretensions and a Competitive Trade," *Early Science and Medicine* 16, no. 6 (2011): 516.

10. "exigebat operis distributio & ordo, ut primum de mundi fabrica & distributione partium in universum: deinde coelestium corporum ordine & motu: tertio de eorundem natura, radiatione & operantium conflexu, ad veriorem Astrologiam inquirendam: Quarto de Elementis: Quinto de regnorum, & totius terrae descriptione: Sexto de Principium à [sic] condito mundo genealogiis, ad emigrationes gentium, & primas terrarum habitationes, rerumque iuventarum tempora, & antiquitates indagandas tractarem. Hic enim rerum naturalis est ordo, qui causas & origines eurundem facile commonstrat & ad verum scientiam sapientiamque optimus dux est &c." Mercator, *Atlas*, ✠ 2r (JCB), 10r (LC); translation, 15 (slightly modified).

11. "Geographia designatrix imitatio e[st] totius cogniti orbis cum his quae fere universaliter sibi iunguntur." *In Hoc Opere Haec Continentur Geographiae Cl. Ptolemaei a Plurimis Uiris Utriusq[ue] Linguae Doctiss. Eme[n]data* (Rome: Bernardinus Venetus de Vitalibus, 1508), [E6r].

12. Christian Jacob, *The Sovereign Map: Theoretical Approaches in Cartography Throughout History*, ed. Edward H. Dahl, trans. Tom Conley (Chicago: University of Chicago Press, 2006), 29-30.

13. Erich Auerbach, "Figura," in *Scenes from the Drama of European Literature* (New York: Meridian Books, 1959), 11-76.

14. Lorraine J. Daston and Peter Galison, *Objectivity* (Brooklyn: Zone Books, 2007), 49.

15. Daston and Galison define a "working object" in science as "any manageable, communal representative of the sector of nature under investigation" (*Objectivity*, 19), and understand atlases as "systematic compilations of working objects" (22). At its inception, the world atlas sought to capture *the* paradigmatic working object as whole—the world itself.

16. A digital search of books published between 1470 and 1600, currently held in European libraries, yields over one thousand items with many duplicates; there is a striking increase of books with this title between 1540 and 1560. The main popular exceptions to anatomical texts with the word 'fabrica' in the title are Francesco Alunno's compendium of quotations from well-known Italian authors, *La fabrica del mondo* (first published in Vicenza, 1548) and Giovanni Lorenzo d'Anania's cosmographic text, *Le universal fabrica del mondo* (first published in Naples, 1573). Other texts that use 'fabrica' in their titles are primarily technical manuals ranging from architectural design to the making of astrolabes and medicines.

17. Andreas Vesalius, *De humani corporis fabrica*, ed. Jackie Pigeaud (Paris: Belles lettres, 2001), xviii–xxi.

18. For an overview of the trope's history, see Leonard Barkan, *Nature's Work of Art: The Human Body as Image of the World* (New Haven: Yale University Press, 1975).

19. *Ptolemy's Geography*, 57.

20. Peter Apian, *Cosmographicus liber Petri Apiani Mathematici studiose collectus* (Landshut, 1524), fols. 3–4. Subsequent editions throughout the sixteenth century retain these images.

21. Christian Jacob, "La mimesis géographique en Grèce antique," in *Espace et Représentation: Sémiotique de l'architecture: actes du colloque international tenu à Albi du 20 au 24 Juillet 1981*, ed. Alain Renier (Paris: Editions de la Villette, 1982), 57.

22. There is considerable recent work on the theory and practice of early modern anatomy from a variety of angles. See especially Andrea Carlino, *Books of the Body: Anatomical Ritual and Renaissance Learning* (Chicago: University of Chicago Press, 1999); Devon L. Hodges, *Renaissance Fictions of Anatomy* (Amherst: University of Massachusetts Press, 1985); and Jonathan Sawday, *The Body Emblazoned: Dissection and the Human Body in Renaissance Culture* (London: Routledge, 1995). On the overlap between medicine and history, pertinent to the discursive and conceptual overlaps between medicine and geography, see Nancy G. Siraisi, *History, Medicine, and the Traditions of Renaissance Learning* (Ann Arbor: University of Michigan Press, 2007).

23. See Valerie Traub, "The Nature of Norms in Early Modern England: Anatomy, Cartography, King Lear," *South Central Review* 26, no. 1–2 (2009): 42–81, and "Mapping the Global Body," in *Early Modern Visual Culture: Representation, Race and Empire in Renaissance England*, ed. Peter Erickson and Clark Hulse (Philadelphia: University of Pennsylvania Press, 2000), 44–97. Traub's essays discuss the discursive relations between anatomy and cartography in the sixteenth century, though she does not argue for particular historical connections between the two disciplines.

24. These shared relations are noted by Robert Karrow, "Intellectual Foundations of the Cartographic Revolution" (Ph.D. diss., Loyola University, 1999), 64–65. For the use of the earth-body analogy, see Matthew H. Edney, "Mapping Empires, Mapping Bodies: Reflections on the Use and Abuse of Cartography," *Treballs de la Societat Catalana de Geografia* 63 (2007): 83–104.

25. On this topic see Jean-François Staszak, *La géographie d'avant la géographie: le climat chez Aristote et Hippocrate* (Paris: Ed. L'Harmattan, 1995) and Jorge Cañizares-Esguerra, "New World, New Stars: Patriotic Astrology and the Invention of Indian and Creole Bodies in Colonial Spanish America, 1600–1650," *American Historical Review* 104, no. 1 (February 1, 1999): 33–68.

26. The use of the term "hemisphere" for the lobes of the brain first occurs in Thomas Willis, *Cerebri anatome* (London, 1664), 67.

27. Despite the development of new interdisciplinary fields in medical humanism and medical geography, the still operative disciplinary divisions between histories of medicine and cartography obscure seemingly important connections between the development of medical and geographic thought and their expression in printed books in the sixteenth century. On the early modern conjunction between medicine and geography, see Hiro Hirai, ed., *Cornelius Gemma: Cosmology, Medi-*

cine, and Natural Philosophy in Renaissance Louvain (Pisa: Serra, 2008) and C. B. Valencius, "Histories of Medical Geography," *Medical History Supplement*, no. 20 (2000): 3–28. Studies on scientific illustration and prints rarely consider map books alongside books on instrument making, botany, or anatomy, though they were frequently printed at the same presses and involved overlapping networks of artisans, authors, and publishers; for books that do explore these connections, see Volker Remmert, *Picturing the Scientific Revolution* (Philadelphia: Saint Joseph's University Press, 2011), an abridged English translation of *Widmung, Welterklärung und Wissenschaftslegitimierung: Titelbilder Und Ihre Funktionen in Der Wissenschaftlichen Revolution* (Wiesbaden: Harrassowitz in Kommission, 2005); and Sachiko Kusukawa and Ian Maclean, eds., *Transmitting Knowledge: Words, Images, and Instruments in Early Modern Europe* (Oxford: Oxford University Press, 2006).

28. On tropes of anatomical discovery, see Florence Bourbon and Roberto Lo Presti, "Introduction," *Renaissance and Reformation/Renaissance et Réforme* 33, no. 3 (July 28, 2011): 5.

29. On the importance of autopsy as a mode of knowing for Vesalius, see Sachiko Kusukawa, *Picturing the Book of Nature: Image, Text, and Argument in Sixteenth-Century Human Anatomy and Medical Botany* (Chicago: University of Chicago Press, 2012). On autopsy in cosmography and travel narratives, see Frank Lestringant, *Mapping the Renaissance World: The Geographical Imagination in the Age of Discovery* (Berkeley and Los Angeles: University of California Press, 1994) and Andrea Frisch, *The Invention of the Eyewitness: Witnessing and Testimony in Early Modern France*, (Chapel Hill: University of North Carolina Press, 2004).

30. For an analysis of the epistemological basis of anatomy, see Roberto Lo Presti, "Anatomy as Epistemology: The Body of Man and the Body of Medicine in Vesalius and His Ancient Sources (Celsus, Galen)," *Renaissance and Reformation/Renaissance et Réforme* 33, no. 3 (July 28, 2011): 27–60. For a geographic analysis, see Besse, *Les grandeurs de la terre*.

31. For Vesalius's reference to Gemma, see Andreas Vesalius, *De Humani Corporis Fabrica: Basel, 1543*, facsimile on CD-ROM (Palo Alto, Calif: Octavo, 2003), 161. On Gemma Frisius's students and his interest in medicine/anatomy, see Van den Broecke, *Limits of Influence*; Fernand Hallyn, *Gemma Frisius, arpenteur de la terre et du ciel* (Paris: Champion, 2008); and Hirai, *Cornelius Gemma*.

32. Mercator's "Orbis imago," as an imitation of Finé's cordiform map has recently attracted scholarly attention. Key essays that discuss cordiform maps are Ruth Watson, "Cordiform Maps Since the Sixteenth Century: The Legacy of Nineteenth-Century Classificatory Systems," *Imago Mundi* 60, no. 2 (2008): 182–94, and "The Decorated Hearts of Oronce Fine: The 1531 Double Cordiform Map of the World," *The Portolan: Journal of the Washington Map Society* 65 (2006): 13–27; George Kish, "The Cosmographic Heart: Cordiform Maps of the 16th Century," *Imago Mundi* 19 (1965): 13–21; and Monique Pelletier, "Le monde dans un coeur: Les deux mappemondes d'Oronce Finé," *Cartographica Helvetica* 12 (1995): 27–37.

33. Watson, "Decorated Hearts," and Tom Conley, *The Self-made Map: Cartographic Writing in Early Modern France* (Minneapolis: University of Minnesota Press, 1996), 88–134.

34. William Harvey, *Exercitatio anatomica de motu cordis et sanguinis in animalibus* (Frankfurt, 1628), 3. On the astrological significance of the sun and heart, see Eugenio Garin, *Astrology in the Renaissance: The Zodiac of Life* (London: Routledge and Kegan Paul, 1983), 11.

35. Antoine Mizauld, *Aesculapii et uraniae medicum simul & astronomicum ex colloquio conjugium harmoniam microcosmi cum macrocosmo: sive humani corporis cum caelo, paucis figurans, & perspicue demonstrans* (Lyons, 1550).

36. On medical astrology in the Renaissance see especially, Anthony Grafton and Nancy G. Siraisi, "Between the Election and My Hopes: Girolamo Cardano and Medical Astrology," in *Secrets of Nature: Astrology and Alchemy in Early Modern Europe*, ed. William R. Newman and Anthony Grafton (Cambridge, Mass.: MIT Press, 2001), 69–132; and Nancy G. Siraisi, *The Clock and the Mirror: Girolamo Cardano and Renaissance Medicine* (Princeton: Princeton University Press, 1997).

37. On astrology in Louvain, and Mercator's involvement in the field, see Van den Broecke, *Limits of Influence*; and Annelies van Gijsen, "L'astrologie," in *Gérard Mercator Cosmographe: Le temps et l'espace* (Antwerp: Fonds Mercator Paribas, 1994), 220–33.

38. On Mercator's correspondence with Dee, and Dee's dedication of his *Propaedeumata apho-ristica* (1558) to Mercator, see Van den Broecke, *Limits of Influence*, 168–71. On Mizauld's friend-ship with Gemma, see Hiro Hirai, "'Prisca Theologia' and Neoplatonic Reading of Hippocrates in Fernel, Cardano and Gemma," in *Cornelius Gemma: Cosmology, Medicine and Natural Philosophy in Renaissance Louvain*, ed. Hiro Hirai (Pisa: Serra, 2008), 91–104.

39. On the geographic particularities of the map vis-à-vis its contemporaries, see Gilbert A. Cam, "Gerard Mercator: His 'Orbis Imago' of 1538," *Bulletin of the New York Public Library* 41, no. 5 (May 1937): 376–77.

40. The full text reads: "Terras hic esse certum est, sed quantas quibusque limitibus finitas incertum."

41. For a discussion of Mercator's view on the distribution of continents see Zuber, "Armchair Discovery."

42. "Proposuimus aut partitionem orbis in gene tantum quam deinceps in particularibus ali-quot regionibus. . . ." Crane, *Mercator,* 94, discusses the map's ambition.

43. The point is noted by Cosgrove, *Geographical Imagination*, and Crane, *Mercator,* 96. Both gesture toward the symbolism of interiority associated with the heart but do not develop its impli-cations for the philosophical underpinnings of world-mapping.

44. On the iconography of the heart see Louis Gougaud, *Dévotions et pratiques ascétiques du Moyen Âge* (Paris: Desclée de Brouwer, 1925); Anne Sauvy, *Le miroir du coeur: quatre siècles d'images savantes et populaires* (Paris: Editions du Cerf, 1989); and Eric Jager, *The Book of the Heart* (Chicago: University of Chicago Press, 2000).

45. For possible Lutheran aspects of the heart image, see Crane, *Mercator,* 96–98, who notes the importance of the heart image to Melanchthon and the map's dedication to J. Drosius, who may have also been implicated in the Louvain heresy trials of 1543–44. Jerry Brotton reiterates these points in "A 'Devious Course': Projecting Toleration on Mercator's 'Map of the World', 1569," *The Cartographic Journal* 49, no. 2 (May 2012): 101–6.

46. Robert Karrow, "Commentary" in LC facsimile of Mercator, *Atlas,* 7–8.

47. Hodges, *Renaissance Fictions,* 5.

48. On this challenge in Vesalius's anatomy, see Nancy G. Siraisi, "Vesalius and Human Diver-sity in *De Humani Corporis Fabrica*," *Journal of the Warburg and Courtauld Institutes* 57 (January 1, 1994): 60–88.

49. This is how Nicholas Crane describes Finé's map in *The Man Who Mapped the Planet* (Lon-don: Weidenfeld & Nicolson, 2002), 96.

50. For Mercator's innovations in calligraphy, see A. S. Osley, *Mercator: A Monograph on the Lettering of Maps, Etc. in the 16th Century Netherlands with a Facsimile and Translation of His Trea-tise on the Italic Hand and a Translation of Ghim's Vita Mercatoris* (London: Faber, 1969); and Ton Croiset van Uchelen, "L'écriture et la calligraphie," in *Gérard Mercator Cosmographe: Le temps et l'espace*, ed. Marcel Watelet (Antwerp: Fonds Mercator Paribas, 1994), 150–61.

51. Pamela H. Smith, *The Body of the Artisan: Art and Experience in the Scientific Revolution* (Chi-cago: University of Chicago Press, 2004). For discussion of Mercator's attempt to define himself outside the class-status of "artisan," see Zuber, "Armchair Discovery."

52. On the iconography of Atlas, see James R. Akerman, "Atlas, la genèse d'un titre," in *Gerardi Mercatoris Atlas Europae: facsimilé des cartes de Gérard Mercator contenues dans l'Atlas de l'Europe vers 1570–1572*, ed. Marcel Watelet (Antwerp: Fonds Mercator, 1994), 15–29, and Volker R. Rem-mert, "Visual Legitimisation of Astronomy in the Sixteenth and Seventeeth Centuries: Atlas, Her-cules and Tycho's Nose," *Studies in History and Philosophy of Science, Part A* 38, no. 2 (June 2007): 327–62.

53. See Peter van der Krogt, "Mercators Atlas: Geschichte, Editionen, Inhalt," in *Gerhard Mer-cator und die geistigen Strömungen des 16. und 17. Jahrhunderts*, ed. Hans H. Blotevogel and Rienk Vermij (Bochum: Brockmeyer, 1995), 49–64, at 55. For a contrary view, see Akerman, "La genèse d'un titre."

54. Much has been written on the mapbook as a theater: see especially Besse, *Les grandeurs*

de la terre; Peter van der Krogt, "The *Theatrum Orbis Terrarum*: The First Atlas?" in *Abraham Ortelius and the First Atlas: Essays Commemorating the Quadricentennial of His Death, 1598–1998*, ed. Martin van der Broecke, Peter van der Krogt, and Peter Meurer ('t Goy-Houten, the Netherlands: HES Publishers, 1998), 55–78; W. Waterschoot, "The Title Page of Ortelius's *Theatrum Orbis Terrarum*," *Quaerendo* 9 (1979): 43–68; John Gillies, *Shakespeare and the Geography of Difference* (Cambridge: Cambridge University Press, 1994); and Ann Blair, *The Theater of Nature: Jean Bodin and Renaissance Science* (Princeton: Princeton University Press, 1997).

55. One strain of the visual iconography of Atlas/astronomy emphasizes the contemplative mode: see Dürer's print of the astronomer and Jacques de Gheyn's print of Saturn.

56. For an analysis of the Renaissance significance of the divine artifex figure, see Alexandre Koyré, *From the Closed World to the Infinite Universe* (Baltimore: Johns Hopkins University Press, 1968), 274–76.

57. Zuber suggests that the impressive beard that distinguishes both Atlas and Mercator (in his portrait) further offers a visual connection: see "Armchair Discovery," 526–27.

58. "Et quia summorum virtus imitanda virorum est, / Hunc avus exemplo credidit esse sibi" from "In Atlantem / Gerardi Mercatoris Avi sui." Mercator, *Atlas*, ✠ 4v (JCB), 13v (LC); translation, 34.

59. Daston and Galison, *Objectivity*, 37.

60. On the complex relationship between science and art in the making of knowledge, see Lorraine Daston, "Fear and Loathing of the Imagination in Science," *Daedalus* 127, no. 1 (January 1, 1998): 73–95; Daston and Galison, *Objectivity*, specifically chap. 2; Alexander Marr, *Between Raphael and Galileo: Mutio Oddi and the Mathematical Culture of Late Renaissance Italy* (Chicago: University of Chicago Press, 2011), 168–74; and Smith, *Body of the Artisan*. Bernardus Furmerius notes Mercator's evident pride in his own hand ("manuque sua") in a poem prefacing the *Atlas*.

61. Akerman, "La genèse d'un titre," 15.

62. Mercator's apparent rejection of the Titanic genealogy puzzles scholars. Keuning suggests that Mercator's recreation of a human genealogy for Atlas suggests his desire to come up with a new name for his project ("The History of an Atlas: Mercator-Hondius," *Imago Mundi* 4 [1947] 37–62, at 38). Akerman argues there is no confusion ("La genèse d'un titre," 23).

63. Diodorus Siculus, *Bibliotheca Historica* (Cambridge, Mass.: Harvard University Press, 1961), 4.27.5.

64. Akerman, "La genèse d'un titre," 24, who credits an unpublished manuscript by Robert Karrow (see note 33).

65. James Akerman, "From Books with Maps to Books as Maps: The Editor in the Creation of the Atlas Idea," in *Editing Early and Historical Atlases*, ed. Joan Winnearls (Toronto: University of Toronto Press, 1995), 4–48, at 34.

66. For a historical overview of atlas structure before and after Mercator, see James Akerman, "On the Shoulders of a Titan: Viewing the World of the Past in Atlas Structure" (Ph.D. diss., Pennsylvania State University, 1991).

67. Daston and Galison, *Objectivity*, 17.

68. Ibid., 25–26. Also see Mi Gyung Kim's critique, "A Historical Atlas of Objectvity," *Modern Intellectual History* 6, no. 3 (2009) 569–96, at 571.

69. Mapping still remains a powerfully aesthetic endeavor: see Jacob, *Sovereign Map*; David Woodward, *Art and Cartography: Six Historical Essays* (Chicago: University of Chicago Press, 1987); and Denis E. Cosgrove, *Geography and Vision: Seeing, Imagining and Representing the World* (London and New York: Palgrave Macmillan, 2008).

70. The OED notes that "cosmos" derives from the Greek κόσμος, meaning "order, ornament, world or universe (so called by Pythagoras or his disciples 'from its perfect order and arrangement')." Several Renaissance cosmographers allude to this etymology, including Jacopo d'Angelo, Simon Grynaeus, Ortelius, Hondius and Blaeu; even Alexander von Humboldt names his geographic synthesis *Kosmos* in a nod toward the aesthetic dimensions of the term (Cosgrove, *Geography and Vision*, 34–39).

71. Indeed, while defending his choice to translate the title of Ptolemy's *Geography* as *Cosmographia*, Jacopo d'Angelo, the first translator of the text into Latin, celebrated the term as an equivalent to the Latin *mundus*. See the dedicatory epistle to Pope Alexander V, in *Cosmographia*, ed. Filippo Beroaldo, trans. Jacopo D'Angelo (Bologna, 1477), Ar.

72. The interest classical rhetorical and poetic theory takes in the harmonious integration of parts into wholes offers cosmographers a well-developed vocabulary for discussing their own integrative challenge. Particularly relevant is Aristotle's *Poetics* 7–8; Horace's dictum that the beautiful whole was "simplex . . . et unum" (*Ars Poetica* 24) is also relevant.

73. On *eusunopton*, see Jacob, "La mimesis géographique," 53–79; and Christian Jacob, "L'oeil et la mémoire: sur la périégèse de la terre habitée de Denys," in *Arts et légendes d'espaces: figures du voyage et rhétoriques du monde,* eds. Christian Jacob and Frank Lestringant (Paris: Presses de l'Ecole normale supérieure, 1981), 21–97, at 34–37.

74. On the convergence of art and science in pre-1800 scientific atlases, see Daston and Galison, *Objectivity,* 79.

75. On the importance of Mercator's Ptolemy for the development of the historical atlas as a genre, see W. A. Goffart, *Historical Atlases: The First Three Hundred Years, 1570–1870* (Chicago: University of Chicago Press, 2003). Mercator's historicized Ptolemy may also have influenced Ortelius's *Parergon,* which is often identified as the first historical atlas. Significantly, Mercator's Ptolemy is combined into a single volume with Ortelius's *Parergon* by Petrus Bertius entitled *Theatri geographiae veteris* in 1618.

76. Henry Newton Stevens and Edward Everett Ayer, *Ptolemy's Geography: A Brief Account of All the Printed Editions down to 1730* (London: H. Stevens, Son and Stiles, 1908), 22.

77. There were two different Greek versions of "Ptolemy" circulating in the thirteenth and fourteenth centuries, the A-recension, having a corpus of twenty-seven maps, and the B-recension, covering the same territory in sixty-five or more smaller maps; 38 of the 51 recorded Greek manuscripts had no maps at all. However, 41 of 58 Latin manuscripts included maps, all of which were based on the A-recension. Of the 222 maps printed in the fifteenth century recorded by Campbell, 117 were versions of Ptolemy's twenty-seven maps; 32 editions of Ptolemy with maps were published between 1477 and 1600 (Akerman, "From Books with Maps," 8–9). These figures are based on O.A.W. Dilke, J. B. Harley, and David Woodward, "Cartography in the Byzantine Empire," in *The History of Cartography,* vol. 1, *Cartography in Prehistoric, Ancient, and Medieval Europe and the Mediterranean,* ed. J. B. Harley and David Woodward (Chicago: University of Chicago Press, 1987), 267–74; and Tony Campbell, *The Earliest Printed Maps, 1472–1500* (London: British Library, 1987).

78. Girolamo Ruscelli, ed., *La geografia di Claudio Tolomeo Alessandrino* (Venice: Vincenzo Valgrisi, 1561).

79. The printed notes to the binder at the end of this edition include specific directions to bind the "Tavola universale nuova" (the new map of the world) before the "Tavola prima universale antica di tutta la Terra fin' a' tempi di Tolomeo" ("the first universal, ancient map of the entire world known since Ptolemaic times"), suggesting a self-consciousness about the book's organization and its effect. See pages ⚹ r, A1r.

80. Bruno Latour, "Visualization and Cognition: Thinking with Eyes and Hands," *Knowledge and Society* 6 (1986): 1–40. Latour notes how the gathering of old and new in one place makes visible contradictions that in turn produce new knowledge (12).

81. See text accompanying the map which includes a list of sources.

82. Jean Paul Richter, ed., *The Notebooks of Leonardo da Vinci* (New York: Dover, 1970), 2:111.

83. See the useful 'map index' of text and graphics in Akerman, "From Books with Maps," 21.

84. Nicolas Sanson, *Atlas nouveau: contenant toutes les parties du monde* (Paris, 1696); and Gilles Robert de Vaugondy, *Atlas universel* (Paris, 1757). Sanson's frontispiece even retains the Atlas image: this figure is a hybrid between Mercator's Atlas with his athletic, classical body and older versions showing an old man holding the world on his shoulders. Sanson's enormous atlas is notable for its detailed "tables geographiques" which preface each section of maps, while Vau-

gondy's, in a throwback to older Ptolemaic editions, contains a long prefatorial "essai" on the development of geography followed by the maps. The "Orbis vetus" in this edition corresponds not to the Ptolemaic world but to the world as it was known in the seventeenth century.

85. A characteristic feature of Dutch mapbooks, including Mercator's, is an emphasis on maps of the Low Countries in utter disproportion to other parts of even Europe.

86. "Cum . . . necessitate ordo semper requirat generalia particularibus anteponere, totumque parti . . . ego devinctus lege huic primo . . . tomo, universalem orbis terrae typum, euisque quatuor partes . . . & singulis quoque deinceps tomis consequentibus, ut is etiam perfectum semper habeat opus, ac totius universi descriptionem, nec hac utili speculatione privetur, qui vel suae tantum patriae delineationem sibi comparaverit. Iucunda etenim est & maximopere necessaria generalium contemplatio, ei qui vel minimam mundi . . . cognitionem habere cupit." Mercator, *Atlas*, "Orbis Terrae Typus," [A3r] JCB), 32r (LC); translation, 152–53.

87. Akerman, citing Dirk de Vries, notes that much of Mercator's coverage of Europe was compiled as larger, wall-sized maps, which had apparently been broken into constituent atlas-sized sheets. Some of Mercator's sheets could thus be taken out of the atlas and assembled into single wall-sized maps: de Vries identifies nine such multi-sheet maps for various parts of Europe. See Akerman, "From Books with Maps," 32; and Dirk de Vries, "The 'Helvetia' Wall-map by Gerhard Mercator," *Cartographica Helvetica* 5 (1992): 3–10.

88. For discussion of the 'narrative' aspects of atlases, see Denis Wood, "Pleasure in the Idea: The Atlas as Narrative Form," *Cartographica: The International Journal for Geographic Information and Geovisualization* 24, no. 1 (1987): 24–46.

89. Blaeu, *Atlas Maior, siue, Cosmographia Blauiana*, 6 (following "Orbis Terrarum" section). Sanson's detailed "tables geographiques" for the world and each of its regions in the *Atlas nouveau* may derive from this kind of schematic diagram in Blaeu's atlas. The publication success of the plates of Mercator's *Atlas* in the seventeenth century, when the principal Dutch publishing houses of Hondius, Jansson, and Blaeu purchased and used them, also ensured that the formal structure of the 1595 *Atlas* set the standard for understanding the order of the world.

90. Abraham Ortelius, *Theatre de l'univers: contenant les cartes de tout le monde: avec une brieve declaration d'icelles* (Antwerp: Plantin, 1572), epigraph.

91. Cicero, *De natura deorum*, 2.37: ". . . ipse autem homo ortus est ad mundum contemplandum et imitandum—nullo modo perfectus, sed est quaedam particula perfecti. Sed mundus quoniam omnia conplexus est neque est quicquam, quod non insit in eo, perfectus undique est" (Marcus Tullius Cicero, *De natura deorum; Academica*, trans. H. Rackham [Cambridge, Mass.: Harvard University Press, 1956]). Besse, *Les grandeurs de la terre*, 335, discusses the passage's significance for Ortelius and Neo-Stoicism.

92. See Nuti, "World Map"; Giorgio Mangani, *Il "mondo" di Abramo Ortelio: misticismo, geografia e collezionismo nel Rinascimento dei Paesi Bassi* (Modena: Franco Cosimo Panini, 1998); Denis Cosgrove, "Globalism and Tolerance in Early Modern Geography," *Annals of the Association of American Geographers* 93, no. 4 (2003): 852–70; and Brotton, "A 'Devious Course.'"

93. On the genre of the cosmographic meditation, see the essays collected in Frank Lestringant, ed., *Les méditations cosmographiques à la renaissance*, Cahiers V. L. Saulnier 26 (Paris: Presses de l'Université Paris-Sorbonne, 2009). On the relationship between meditations and scientific practice, see Pierre Hadot, *Philosophy as a Way of Life: Spiritual Exercises from Socrates to Foucault* (Oxford and New York: Blackwell, 1995).

94. For reflections on the relationship between geography and "sense of place," see Yi-fu Tuan, *Cosmos and Hearth: A Cosmopolite's Viewpoint* (Minneapolis: University of Minnesota Press, 1996).

95. Lestringant, *Mapping the Renaissance World*, 5.

96. On the literary and philosophical traditions of *kataskopos*, see Patrick Gautier Dalché, "Les antécédents médiévaux de la méditation géo-cartographique," in *Les méditations cosmographiques*, ed. Frank Lestringant, Cahiers V. L. Saulnier 26 (Paris: Presses de l'Université Paris-Sorbonne, 2009), 19–40.

97. "Huic operi titulum imposuit: Atlas sive cosmographicae speculationis libri quinque" [He titled this work: Atlas or Cosmographical speculations in five books]. Mercator, *Atlas*, ⚹ 2r (JCB); 10r (LC); translation, 16.

98. Christian Jacob, "Quand les cartes reflechissent," *Espaces temps* 62–63 (1996): 36–49, at 46.

99. Besse, *Les grandeurs de la terre*, 336; Cosgrove, "Globalism and Tolerance," 862.

100. Seneca *Naturales Quaestiones* 1, pref. 7–11.

101. "Tu interim Lector vale & fruere, ac huius tuae habitationis ac tibi ad tempus tantum concessae gloriam cum Poeta Buchanano diligenter considera, qui sic eam caelestibus comparat, ut animos terrestribus ac rebus hisce caducis immersos extrahat, & ad altiora ac aeterna viam ostendat." Mercator, *Atlas*, "Orbis Terrae Typus," 32r (LC); translation, 154–55.

102. "Percipias rerum sit quantula portio, verbis / Quam nos magnificis in regna superba secamus / Partimur ferro, mercamur sanguine fuso / Ducimus exiguae glebae de parte triumphos. / . . . si cum stellati tegamine caeli / Componas puncti instar erit, vel seminis, unde / Condidit innumeros senior Gargetius orbes. / [. . .] Quantula pars rerum est, in qua se gloria tollit, / Ira fremit, metus exanimeat, dolor urit, egestas / Cogit opes ferro, insidus flamma atque veneno? / Scilicet & trepido ferment humana tumultu." Mercator, *Atlas*, "Orbis Terrae Typus," 32r (LC); translation, 154–55.

103. Noting the geographical root of "ethos," Denis Cosgrove argues that the principle purpose of world-mapping in a pre-secular age was ethical, for it concerned the place of human life in an ordered creation (*Geography and Vision Seeing*, 38). He connects this idea to Plato *Timaeus* 3.38 and its legacies in Aristotle's *Metaphysics* and *De anima*.

104. "Qui quantum patet universus Orbis, / & quicquid fuit universe in Orbe . . ." in Mercator, *Atlas*, [no sign, leaf 2v] (JCB); 8v (LC), translation, 5.

105. "Sidera cum terris coniunxit, sacra prophanis / Addidit, at rectem fecit utrumque tamen. / Astra Mathematicus radio descripsit acuto, / Et dedit in parvo conspicienda globo. / In tabulas terrae spaciosum contulit orbem, / [. . .] Sacraque detexit vatum mysteria, Christem / Praecones iussit quattuor ire simul. / Atque ea sic fecit superos ut vinceret omnes / Artifices, proprio marte, manuque sua" in Mercator, *Atlas*, [no sign, leaf 4v] (JCB); 7v (LC); translation, 2.

106. "Hunc Atlantem tam insignem eruditione, humanitate ac sapientia virum mihi imitandum proposui, quo ad ingenium & vires suppetunt, Cosmographiam veluti ex alta animi specula contemplaturus . . . à [*sic*] creatione incipiens, partes eius omnes, quatenus istud methodica ratio postulat, iuxta creationis ordinem enumerabo, & physicem contemplabor, quò causae rerum innotescant . . . sic totum mundum tanquam in speculo proponam, ut ad inveniendas rerum causas, sapientiam & prudentiam assequendam, sint aliqualia rudimenta, & lectorem ad altiores speculationes ducere possint." Mercator, *Atlas*, A1r (JCB); 14r (LC); translation, 37.

107. For studies of the relationship between cartography and theology, see Pauline Moffitt Watts, "The European Religious Worldview and Its Influence on Mapping," in *The History of Cartography*, vol. 3, *Cartography in the European Renaissance*, ed. David Woodward, (Chicago: University of Chicago Press, 2007), 382–400, at 397–98.

108. See Catherine Delano-Smith and Elizabeth Morley Ingram, *Maps in Bibles, 1500–1600* (Geneva: Droz, 1991); Fiorani, *Marvel of Maps*; and Kenneth Nebenzahl, *Maps of the Holy Land* (New York: Abbeville Press, 1986). Most mapmakers in the sixteenth and seventeenth centuries appear to have had Reformist leanings: it is a striking fact that the production and use of maps seems more firmly associated with Protestant regions than with Catholic ones. Delano Smith and Ingram's study decisively shows that this practice was a Reformist one, as maps began to appear in Bibles in the 1520s, coinciding with Luther's break from the Catholic Church. The great cycle of maps in the Vatican Galleria may in fact be a post-Tridentine response to the widespread Protestant use of maps in the service of religious contemplation (Fiorani, *Marvel of Maps*).

109. David Woodward, "Cartography in the Renaissance: Continuity and Change," in *The History of Cartography*, vol. 3.1, *Cartography in the European Renaissance*, ed. David Woodward (Chicago: University of Chicago Press, 2007), 3–24, at 10. See also figure 6.2 in Robert Karrow, "Intellectual Foundations," 241.

110. ". . . toutesfois il n'est rien au pris de l'aucteur, qui a les mains si grandes, qu'en une il contient tout le monde, & entre deux ou trois doigts il tourne toute la terre." André Thevet, *La cosmographie universelle* (Paris, 1575), ā5r (copy in Beinecke library).

111. "Ceste discipline Cosmographique donques sert pour descouvrir la vanité de ce en quoy nous nous arrestons, puis abaissant nostre orgeuil, elle adresse nostre esprit à ce qui est grand, & ne le permect plus s'arrester à ce qui n'est rien. Et pour ceste cause ie pense qu'il n'y a science, après la Theologie, qui ayt plus grande vertu de nous faire cognoistre la grandeur & puissance divine, & l'avoir en admiration que celle la." Thevet, *La cosmographie*, a4v-5r (copy in Beinecke library).

112. See for instance fig. 16: God as geometer in the Codex Vindobonensis 2554.

113. See the studies by Crane, *Mercator*; Brotton, "A 'Devious Course'"; Marcel Watelet, ed., *Gérard Mercator Cosmographe: le temps et l'espace* (Antwerp: Fonds Mercator Paribas, 1994); Rienk Vermij, "Mercator and the Reformation," in *Mercator und Wandlungen der Wissenschaften im 16. und 17. Jahrhundert* (Bochum: Brockmeyer, 1993), 77–90; and Alastair Hamilton, *The Family of Love* (Cambridge: J. Clarke, 1981), 65–82.

114. Some suggest that Mercator's arrest may have in fact been linked to the Reformation symbolism in his first two maps: see Crane, *Mercator* and Brotton, "A 'Devious Course'."

115. Brotton, "A 'Devious Course,'" 104.

116. Mercator, *Atlas*, ✠ v-2r (JCB); 10 (LC); translation, 13–14.

117. Letter of March 24, 1583 to Heresbach, in Maurice Van Durme, *Correspondence Mercatorienne* (Antwerp: Nederlandsche Boekhandel, 1959).

118. "Axioma est omni menti opificium huius mundane machinae aliquomodo inspicienti, Deum ipsius autorem immensae potentiae, sapientiae & bonitatis esse. . . . Id enim molimur, dum Cosmographiam tradimus, ut ex mirabili omnium rerum in unum Dei finem Concordia, & ex inperscrutabili in compositione providentia, infinita Dei sapientia, & inexhausta eius bonitas conspiciantur . . ." Mercator, *Atlas*, A2r (JCB), 14r (LC); translation, 38–39.

119. The *Atlas*'s English translator (1635) adds frequent quotations from Josuah Sylvester's *The Divine Weeks*, a translation of *La Sepmaine*, to amplify Mercator's commentary.

120. Bartholomaeus Mercator, *Breves in Sphaeram meditatiunculae*.

CHAPTER TWO

1. "Nostre monde vient d'en trouver un autre (et qui nous respond si c'est le dernier de ses freres, puis que les Daemons, les Sibylles et nous, avons ignoré cettuy-cy jusqu'asture?) . . ." All citations from Montaigne's *Essais* refer to Michel de Montaigne, *Les essais*, ed. Pierre Villey, Verdun L. Saulnier, and Marcel Conche (Paris: Quadrige/PUF, 2004). Translations are from *The Complete Essays of Michel de Montaigne*, trans. Donald Frame (Stanford: Stanford University Press, 1958). Text and translations will be cited as "Montaigne" followed by volume and essay number, and page numbers from the Villey and Frame editions: Montaigne, 3.6; 908/693.

2. "Et de cette mesme image du monde qui coule pendant que nous y sommes, combien chetive et racourcie est la cognoissance des plus curieux!" Montaigne, 3.6; 908/693.

3. Montaigne, 3.6; 908/693: "Iamque adeo affecta est aetas, affectaque tellus." The reference is to *De rerum natura* 2.1.136; all translations of Lucretius are from Titus Lucretius Carus, *De rerum natura*, trans. W. H. D. Rouse (Cambridge, Mass.: Harvard University Press, 1967).

4. "Monde" appears 380 times in the *Essais*, while "univers" appears 16 times.

5. Montaigne, "Au lecteur," 3/2: "je suis moy-mesmes la matiere de mon livre."

6. Montaigne, 2.17; 657/499.

7. See Philip Ford, *The Montaigne Library of Gilbert de Botton at Cambridge University Library* (Cambridge: Cambridge University Library, 2008). Montaigne's copies of the 1562 edition of Strabo's geography and the 1551 edition of Apian's *Cosmographia* both survive, as does his heavily annotated copy of Lucretius's *De rerum natura*.

8. On connections between the three essays, see Tom Conley, "Mapping Montaigne: The 'Apologie' as Diagram and Discourse," *Montaigne Studies* 6 (1994): 79; Michel de Certeau, *Heterologies: Discourse on the Other* (Minneapolis: University of Minnesota Press, 1986), 68; Giuliano Gliozzi, *Adamo e il nuovo mondo: la nascita dell'antropologia come ideologia coloniale: dalle genealogie bibliche alle teorie razziali (1500–1700)* (Firenze: La nuova Italia, 1977); and Géralde Nakam, *Les Essais de Montaigne, miroir et procès de leur temps: témoignage historique et création littéraire* (Paris: Nizet, 1984)—though none discuss them from the perspective of the *Essais*'s overall structure.

9. Montaigne, "Au lecteur," 3/2.

10. Montaigne, 1.26; 157/116; this passage is famously echoed by Descartes in the *Discours de la méthode.*

11. On the book of nature as a trope, see Ernst Robert Curtius, *European Literature and the Latin Middle Ages*, trans. Willard R. Trask (Princeton: Princeton University Press, 1990), 319–25; William Eamon, *Science and the Secrets of Nature: Books of Secrets in Medieval and Early Modern Culture* (Princeton: Princeton University Press, 1994); and James J. Bono, *The Word of God and the Languages of Man: Interpreting Nature in Early Modern Science and Medicine* (Madison: University of Wisconsin Press, 1995). On the *speculum mundi* and its relevance for self-portraiture, see Michel Beaujour, *Miroirs d'encre: rhétorique de l'autoportrait* (Paris: Seuil, 1980), 29–41.

12. Noting this shift of metaphors, Timothy Hampton claims that the world's function as book is subverted by its function as mirror, because in the latter analogy individuals learn how to judge healthily only by recognizing their own infirmity (*Writing from History*, 146). His observation suggests how the function of the book itself was changing: for writers such as Montaigne, the book was no longer a final and finite epistemic authority but itself a shifting, partial, and inevitably flawed representation; for Montaigne, as for Mercator, the conception of the book-as-object had to be reimagined in order to reflect and contain a shifting world.

13. Maurice Merleau-Ponty, *Signs* (Evanston: Northwestern University Press, 1964), 199.

14. Montaigne, 1.26; 146/107.

15. For related discussion of this trope see Colin Dickson, "Geographic Imagination in the *Essais* and Geomorphism in Montaigne Criticism," in *Geo/Graphies: Mapping the Imagination in French and Francophone Literature and Film* (Amsterdam: Rodopi, 2003), 29–58.

16. Montaigne, 3.2; 804–5/610–11.

17. Montaigne, 1.31; 204/151.

18. Montaigne, 2.12; 572/430.

19. On the vision of the world as a body, see George Tucker, "Déchets, déchéance et recyclage-corps, corps du monde et corps-texte chez Joachim Du Bellay et Michel de Montaigne," in *Strategic Rewriting*, ed. David Lee Rubin, (Charlottesville, Va.: Rookwood, 2002).

20. Montaigne, 1.21; 99–100/70.

21. On the Ebstorf map, see Evelyn Edson, *Mapping Time and Space: How Medieval Mapmakers Viewed Their World* (London: British Library, 1997).

22. Charles Taylor, *Sources of the Self: The Making of Modern Identity* (Cambridge: Cambridge University Press, 1992), 182.

23. Montaigne, 3.13; 1072/821.

24. The literature on this essay is vast; influential essays in this vein include: Certeau, *Heterologies*; Gérard Defaux, "A propos 'Des coches' de Montaigne (III, 6): de l'écriture de l'histoire à la représentation du moi," *Montaigne Studies* 6 (1994), and "Un cannibale en haut de chausses: Montaigne, la différence et la logique de l'identité," *MLN* 97, no. 4 (1982); Frank Lestringant, "Le Cannibalisme des 'Cannibales,' I," *Bulletin de la Société des Amis de Montaigne* 9–10 (1982), and "Le Cannibalisme des 'Cannibales,' II: De Montaigne à Malthus," *Bulletin de la Société des Amis de Montaigne* 11–12 (1982); Daniel Martin, "Cannibals and Kings: Montaigne and the Valladolid Hearings of 1550–1551," *History of European Ideas* 20, no. 1–3 (1995); Peter Eubanks, "Montaigne's Cannibalistic Communion," *Montaigne Studies* 22 (2010): 53–60; and Edward Tilson, "Le discours sauvage: cannibalisme et autres rites barbares dans les *Essais* de Montaigne," *Montaigne Studies* 22, *Montaigne et le Nouveau Monde* (2010): 137–58.

25. See Frank Lestringant, ed., *Le Brésil de Montaigne: Le Nouveau Monde des Essais (1580-1592)* (Paris: Editions Chandeigne, 2005), and Michel de Certeau, *The Writing of History* (New York: Columbia University Press, 1988), 209–43. Also useful is Andrea Frisch, "Montaigne and the Poetry of Antarctic France," *Montaigne Studies* 22, *Montaigne et le Nouveau Monde* (2010): 39–52, and *Invention of the Eyewitness*; and Lison Baselis-Bitoun, "Jean de Léry précursor de Montaigne," *Montaigne Studies* 22, *Montaigne et le Nouveau Monde* (2010): 61–70.

26. The discovery of the New World had prompted, among other things, a careful reexamination of classical texts for any mention of its existence. The best candidates were Plato's Atlantis and the Carthaginian islands of the Pseudo-Aristotle, both of which Montaigne invokes only to reject as possible sources. The most thorough analysis of the Atlantis story in Montaigne remains Gliozzi, *Adamo e il nuovo mondo*. On the Atlantis debate, see also Harold John Cook, "Ancient Wisdom, the Golden Age, and Atlantis: The New World in Sixteenth-century Cosmography," *Terrae incognitae* 10 (1978) 25–43; and Dickson, "Geographic Imagination in the *Essais*." Apart from the Platonic dialogues, there is no other mention of Atlantis in classical antiquity, though several post-classical and Byzantine accounts mention islands in the Atlantic.

27. See Timothy Hampton, "The Subject of America: History and Alterity in Montaigne's 'Des Coches,'" in *The Project of Prose in Early Modern Europe and the New World*, ed. Elizabeth Fowler and Roland Greene (Cambridge: Cambridge University Press, 1997), 89.

28. On the importance of oral accounts and hearsay in the making of cosmographies, see Lestringant, *Mapping the Renaissance World*; Frisch, *Invention of the Eyewitness*; and María M. Portuondo, *Secret Science: Spanish Cosmography and the New World* (Chicago: University of Chicago Press, 2009).

29. "We need a man either very honest, or so simple that he has not the stuff to build up false inventions and give them plausibility. . . . [E]ach man calls barbarism whatever is not his own practice; for indeed it seems we have no other test of truth and reason than the example and pattern of the opinions and customs of the country we live in." Montaigne, 1.31; 205/152.

30. On the philosophical status of the Atlantis story, see Luc Brisson, *Plato the Myth Maker*, trans. Gerard Naddaf (Chicago: University of Chicago Press, 1999); Gerard Naddaf, "Plato and the *peri phuseos* Tradition," in *Interpreting the Timaeus-Critias: Proceedings of the IV Symposium Platonicum, Selected Papers*, ed. Luc Brisson and Tomás Calvo (Sankt Augustin: Academia Verlag, 1997), 27–36; and Enrico Berti, "L'oggetto dell'eikos logos nel *Timeo* di Platone," in *Interpreting the Timaeus-Critias* (Sankt Augustin: Academia Verlag, 1997), 119–32.

31. Though most of the text is given to Timaeus's description of the cosmos (27c–92c), the central monologue is framed in relation to the political and ethical issues raised in the *Republic*. This elaborate framing device has led scholars to connect four Platonic dialogues—*Republic-Timaeus-Critias-Laws*—in terms of their dramatic and philosophic situations. Each of these deals with ideal cities and worlds—the best city of the *Republic* prompts the tale of Atlantis and the *eikos logos* of Timaeus's cosmos, both of which can be considered a preamble to the constitution of Magnesia, the best possible earthly city, in the *Laws*. See Brisson, *Plato the Myth Maker*.

32. Montaigne, 1.31; 206–7/152–53.

33. The existence of Montaigne's servant is a matter of debate: see Defaux, "Un Cannibale en haut de chausses," who suggests that it is an elaborate fictional device. Scholars have also debated whether the traveler to the New World is an oblique reference to Jean de Léry, whose account of his voyage to Brazil is a source for "Des cannibales."

34. Montaigne, 1.31; 205/152.

35. See Anthony Pagden, *The Fall of Natural Man: The American Indian and the Origins of Comparative Ethnology* (Cambridge: Cambridge University Press, 1986).

36. In "Mapping Montaigne," Tom Conley notes that this passage "sets the reader on the razor's edge of a paradox. On one hand, the cosmographic 'style' of abundant, fluvial, and amplified description of an invisible world of force and flow shows us all that the essayist has at his behest to write of the topic undertaken. He must reject the ground on which his observations are placed"

(70). However, he ultimately argues that Montaigne invokes figures of cosmographical authority in order to impugn cosmography itself; in his view, Montaigne valorizes topographic detail above fluvial cosmographic whole (69).

37. See David L. Schaefer, *The Political Philosophy of Montaigne* (Ithaca: Cornell University Press, 1990), 179.

38. On this punning use of "histoires," see André Pérouse, "Le mot 'histoire' dans les *Essais* de Montaigne," in *Montaigne et l'histoire: actes du colloque international de Bordeaux, 29 septembre-1er octobre 1988*, ed. Claude-Gilbert Dubois (Paris: Klincksieck, 1988), 15-17.

39. It is worth noting that the figure of a traveler who brings news of an unknown land or new knowledge goes back to Plato's *Symposium* and *Phaedo*. In the context of Montaigne, the *Phaedo*, with its extensive discussion of epistemology, particularly the use of hypothesis, is a relevant subtext.

40. For a summary and overview, see Frank Lestringant, *Cannibals: The Discovery and Representation of the Cannibal from Columbus to Jules Verne* (Cambridge: Polity Press, 1997).

41. See David Quint, *Montaigne and the Quality of Mercy: Ethical and Political Themes in the Essais* (Princeton: Princeton University Press), 75-101; Claude Rawson, "'Indians' and Irish: Montaigne, Swift, and the Cannibal Question," *Modern Language Quarterly* 53, no. 3 (1992): 299-363; and George Hoffmann, "Anatomy of the Mass: Montaigne's 'Cannibals,'" *PMLA* 117, no. 2 (2002): 207-21.

42. Montaigne, 1.31; 214/159.

43. Merleau-Ponty, *Signs*, 198.

44. The most recent and extensive study of Montaigne's translation of Sebond, including discussions of manuscripts, editions, and a comparison with Jean Martin's 1566 translation, is Mireille Habert, *Montaigne traducteur de la Théologie Naturelle: Plaisantes et sainctes imaginations* (Paris: Editions Classiques Garnier, 2010), which updates the classic study of Joseph Coppin (*Montaigne: Traducteur de Raymond Sebon* (Lille: Imprimerie H. Morel, 1925)). Unlike Coppin and other critics who emphasize Montaigne's fidelity to Sebond, however, Habert argues that the many modifications that Montaigne makes in his translation suggests that he is engaged in an ongoing and complex dialogue with his predecessor. See also Philip Hendrick, *Montaigne et Sebond: L'art de la traduction* (Paris: Champion, 1996). On the biography and influence of Sebond, see Eusebi Colomer, "Raimond Sebond: un humaniste avant la lettre," in *Montaigne, "Apologie de Raimond Sebond": De la Theologia à la Théologie*, ed. Claude Blum (Paris: Champion, 1990); Comparot, *Amour et vérité* and *Augustinisme et aristotélisme: de Sebon à Montaigne* (Lille: Atelier national de reproduction des thèses, 1985); Alain Guy, "La *Theologia Naturalis* en son temps: structure, portée, origines," in *Montaigne, "Apologie de Raimond Sebond": De la Theologia à la Théologie*, ed. Claude Blum (Paris: Champion, 1990); and Jesus Martinez de Bujanda, "L'influence de Sebond en Espagne au XVIe siècle," *Renaissance and Reformation/Renaissance et Reforme* 10 (1974), 78-84.

45. See the essays in *Montaigne Studies* 5 (1993); recent scholarship on Montaigne has focused more on his activity as a translator.

46. For reconstructions and analyses of the circumstances in which Montaigne may have undertaken his translation, see Michel Simonin, "La préhistoire de l'Apologie de Raimond Sebond," in *Montaigne, "Apologie de Raimond Sebond": De la Theologia à la Théologie*, ed. Claude Blum (Paris: Champion, 1990), 85-116; Jean Balsamo, "Un gentilhomme et sa théologie," in *Dieu à nostre commerce et société: Montaigne et la théologie*, ed. Philippe Desan (Geneva: Droz, 2008), 105-26; and Philippe Desan, "Apologie de Sebond ou justification de Montaigne?" in *Dieu à nostre commerce et société: Montaigne et la théologie*, ed. Philippe Desan (Geneva: Droz, 2008), 175-200.

47. On the editions and translations of Sebond's *Theologia*, see Alain Guy, "Bibliographie: La Theologia Naturalis," in *Montaigne, "Apologie de Raimond Sebond": De la Theologia à la Théologie*, ed. Claude Blum (Paris: Champion, 1990), 301-19; Bujanda, "L'influence de Sebond"; and Balsamo, "Un gentilhomme et sa théologie," 124-26. The critical edition of the *Theologia* is also useful:

Raimundus Sabundus, *Theologia Naturalis Seu Liber Creaturarum. Mit Literargeschichtlicher Einfuhrung und Kritischer Edition des Prologs und des Titulus*, ed. Friedrich Stegmüller (Stuttgart-Bad Cannstatt: Frommann, 1966). The contemporary importance of Sebond is also evident from the entries he merited in Bayle's dictionary ("Sebonde, Raymond," in *Dictionnaire historique et critique* [Rotterdam: 1697], and "Sebonde, Raymond," in *Dictionnaire historique et critique* [Amsterdam: 1740]), and Louis Ellies du Pin's polemical *Histoire des controverses et des matières ecclésiastiques traitées dans le XVe siècle* (Paris: 1698), 310.

48. Guy, "La *Theologia Naturalis* en son temps," 19.

49. Sebond was writing in what might be described as a period of early humanist resurgence: by 1431, the earliest suggested date for the manuscript of Sebond's work, Lorenzo Valla had already written and circulated the Epicurean *De voluptate*, Ficino was engaged in translating Plato, Nicholas of Cusa was writing *De docta ignorantia*, and the Spanish humanist Juan Fernández de Heredia was translating Plutarch and Valerius Maximus. Sebond's anthropocentrism and Montaigne's possible discomfort with it is a question that is extensively discussed in much criticism on the *Théologie naturelle*, especially with regard to the changes that Montaigne makes in translating the Prologue: see for instance, Mireille Habert, "L'inscription du sujet dans la traduction de la *Theolgia Naturalis*," *Montaigne Studies* 5, no. 1-2 (1993): 27-48.

50. On Montaigne's knowledge (or lack thereof) of Sebond's condemnation by the Roman censors, see Desan, "Apologie de Sebond."

51. I am not suggesting that Sebond and Montaigne shared the same basic theological beliefs, but rather that their analyses may have a shared epistemological agenda.

52. Michel de Montaigne, *Oeuvres complètes de Michel de Montaigne*, ed. A. Armaingaud (Paris: Conard, 1935), 9:iii. My translation.

53. For discussion of the dedicatory letter to the *Theologie naturelle*, see Habert, *Montaigne traducteur*, 10-15. Habert, however, connects the use of the word "barbaresque" to "De la praesumption" (2.17) and the geographic regions associated with "Barbarie" (i.e. North Africa west of Egypt), an association that allows her to link Sebond to the Moors and the conditions in Spain before the Reconquista. While "barbaresque" could refer specifically to North Africans in the sixteenth century, it is a rarely used adjective: Nicot's *Thresor de la langue françoyse* (1606) does not record it (though it does list *barbare, barbarie, barbares, barbarisme*, and *barbarique*), while Cotgrave's *Dictionarie of the French and English Tongues* (1611) lists "barbare" as a synonym for "Barbaresque."

54. On Sebond's use of the book of nature metaphor, see Habert, "L'inscription du sujet," 27-28.

55. See Frank Lestringant, "Crisis of Cosmography," 153-79, and Ann M. Blair, *Too Much to Know: Managing Scholarly Information Before the Modern Age* (New Haven: Yale University Press, 2010).

56. It is likely that Sebond's work haunts the essay on education. Villey suggests that "De l'institution des enfans" and the "Apologie" were written about the same time: see Villey, *Les sources et l'évolution*, 1:304, 11. Hampton suggests that "De l'institution" (1.26) is the secular counterpart of the "Apologie" (Hampton, *Writing from History*, 142, note 10).

57. Conley, "Mapping Montaigne," 76.

58. Montaigne, 2.12; 530/395.

59. Montaigne, 2.12; 528/393.

60. Montaigne, 2.12; 531, 533/396, 397.

61. Montaigne, 2.12; 537/401.

62. Montaigne, 2.12; 511/379.

63. Montaigne explores the usefulness of conjecture in a related passage: see 2.12; 507/375.

64. Hans Blumenberg's discussion of young Augustine's skepticism suggests one way in which skepticism and faith might be closely related: see *The Legitimacy of the Modern Age*, trans. Robert M. Wallace (Cambridge, Mass.: MIT Press, 1983), 270-71. My argument seeks to show that the

apparent cleavage between Montaigne as translator and Montaigne as essayist may not be so deep after all; the theologian and the skeptic may have more in common than hitherto acknowledged.

65. Montaigne, 2.12; 571–72/430; my emphases.

66. Thus, Conley suggests that the "Apologie" is a cosmography among the "topographical atlas of the *Essais*" ("Mapping Montaigne," 73).

67. Significantly, this section, added in the Bordeaux copy, is framed by a double reference to the *Timaeus*: it opens with an allusion to Timaeus's account of the cosmos and closes with an allusion to Critias's account of Solon's visit to Saïs, before resuming the 1588 (B) text.

68. Montaigne, 2.12; 572–73/430–31.

69. On the importance of theories regarding the plurality of worlds in early modern thought, see Blumenberg, *Legitimacy of the Modern Age*, 156–57; Fernand Hallyn, *The Poetic Structure of the World: Copernicus and Kepler* (Cambridge: Zone Books, 1990).

70. I discuss this ongoing juxtaposition of Plato and Lucretius at some length in a pair of essays on the English poet Edmund Spenser's cosmological reflections; see "Mutabilitie's Lucretian Metaphysics: Skepticism and Cosmic Process in Spenser's Cantos," in *Celebrating Mutabilitie: Essays on Edmund Spenser's Mutabilitie Cantos*, ed. Jane Grogan (Manchester: Manchester University Press, 2010), 220–45, and "Edmund Spenser, Lucretian Neoplatonist: Cosmology in the Fowre Hymnes," *Spenser Studies: A Renaissance Poetry Annual* 24, no. 1 (2009): 373–411.

71. "Il n'est chose en quoy le monde soit si divers qu'en coustumes et loix." Montaigne, 2.12; 580/437.

72. Montaigne, 2.12; 601–3/455–57.

73. Lucretius *De rerum natura* 5.827–30.

74. The impression this section (5.330ff) made on Montaigne is clearly evident from his annotations in his copy of Lucretius: see M. A. Screech, *Montaigne's Annotated Copy of Lucretius: A Transcription and Study of the Manuscript, Notes and Pen-Marks* (Geneva: Droz, 1998). See discussion in Eric MacPhail, "Montaigne's New Epicureanism," *Montaigne Studies* 12 (2000): 99–100.

75. On Montaigne's use of the *translatio imperii*, see Hampton, "Subject of America."

76. "Notre monde vient d'en trouver un autre . . . non moins grand, plain et membru que luy, toutesfois si neuf et si enfant qu'on luy aprend encore son a, b, c: il n'y a pas cinquante ans qu'il ne sçavoit ny lettres, ny pois, ny mesure, ny vestements, ny bleds, ny vignes. . . . Si nous concluons bien de nostre fin, et ce poëte de la jeunesse de son siecle, cet autre monde ne faira qu'entrer en lumiere quand le nostre en sortira. L'univers tombera en paralisie; l'un membre sera perclus, l'autre en vigueur." Montaigne, 3.6; 908–9/693.

77. Lucretius *De rerum natura* 5.831–35.

78. "Aussi jugeoient-ils, ainsi que nous, que l'univers fut proche de sa fin, et en prindrent pour signe la desolation que nous y apportames." Montaigne, 3.6; 913–14/698.

79. Most commonly associated with the Nahuatl-speaking peoples of Mexico, the myth is recorded in some twenty different variants in Central Mexico within about a hundred years of the conquest. It appears in the works of all the main Spanish chroniclers—Duran, Andres de Olmos, Sahagún, Motolinia, Mendieta, and López de Gómara, among others—and enters into the extended European imagination of the New World through a spate of translations.

80. "Le premier perit avec toutes les autres creatures par universelle inondation d'eaux; le second, par la cheute du ciel sur nous, qui estouffa toute chose vivante, auquel aage ils assignent les geants . . . le troisiesme, par feu qui embrasa et consuma tout; le quatriesme, par une émotion d'air et de vent qui abbatit jusques à plusieurs montaignes . . . apres la mort de ce quatriesme Soleil, le monde fut vingt-cinq ans en perpetuelles tenebres, au quinziesme desquels fut creé un homme et une femme qui refeirent l'humaine race; dix ans apres, à certain de leurs jours, le Soleil parut nouvellement creé; et commence, depuis, le compte de leurs années par ce jour là. Le troisiesme jour de sa creation, moururent les Dieux anciens; les nouveaux sont nays depuis, du jour à la journée . . ." Montaigne, 3.6; 914/698.

81. On allusions to the Bible in "Des coches," see Conley, *Self-Made Map*. In the "Apologie

de Raymond Sebond," Montaigne also invokes New World customs as precursors or parallels to Christian belief and practice (see Montaigne, 2.12; 573-74/432-33).

82. On the struggle between European and Amerindian cosmological systems for ideological dominance, see Mignolo, *Darker Side*.

83. On the Requirimiento and its uses, see Patricia Seed, "Taking Possession and Reading Texts: Establishing the Authority of Overseas Empires," *The William and Mary Quarterly* 3d ser., 49, no. 2 (1992). Claude Lévi-Strauss offers an interpretation of these myths in *The Story of Lynx*, trans. Catherine Tihanyi (Chicago: University of Chicago Press, 1995).

84. Edwin Duval, "Lessons of the New World," 95-112.

85. Hampton, "Subject of America," 97-98.

86. Hampton, "Subject of America," 97.

87. Montaigne, 3.6; 907/692.

88. The careful framing with reference to Plato is a C-text revision.

89. "[C] Et la narration de Solon, sur ce qu'il avoit apprins des prestres d'Aegypte . . . ne me semble tesmoignage de refus en cette consideration. <<*Si interminatam in omnes partes magnitudinem regionum videremus et temporum, in quam se injiciens animus et intendens ita late longeque peregrinatur, ut nullam oram ultimi videant in qua possit insistere: in hac immensitate infinita vis innumerabilium appareret formarum.*>>" Montaigne, 3.6; 907/692.

90. "Cuius operam profecto non desideraretis, si inmensam et interminatam in omnis partis magnitudinem regionum videretis, in quam se iniciens animus et intendens ita late longeque peregrinatur, ut nullam tamen oram ultimi videat, in qua possit insistere." From *De natura deorum* 1.20; the entire, extended passage (not quoted here in full) is relevant for Montaigne. Latin citations of Cicero are from Marcus Tullius Cicero, *De natura deorum. Academica*, trans. H. Rackham (Cambridge, Mass.: Harvard University Press, 1967). English translations are by Francis Brooks (translator of the 1896 Loeb edition of *De natura deorum*).

91. For a discussion of Montaigne's use of Lucretius in the context of this passage, see MacPhail, "Montaigne's New Epicureanism."

CHAPTER THREE

1. On the succession crisis and Philip's strategy, see Fernando J. Bouza Alvarez, *Felipe II y el Portugal dos povos: imágenes de esperanza y revuelta* (Valladolid: Universidad de Valladolid, Secretariado de Publicaciones e Intercambio Editorial, 2010). On the connection between Camões and the cultural capital of the Portuguese nation, particularly evident in the afterlife of his poem in Portugal and Spain, see Miguel Martínez, "A Poet of Our Own: The Struggle for *Os Lusíadas* in the Afterlife of Camões," *Journal for Early Modern Cultural Studies* 10, no. 1 (2010): 71-94.

2. For reflections on the link between nation-building and colonial experience, see Tamar Herzog, *Defining Nations: Immigrants and Citizens in Early Modern Spain and Spanish America* (New Haven: Yale University Press, 2003); Barbara Fuchs, "Imperium Studies: Theorizing Early Modern Expansion," in *Postcolonial Moves: Medieval Through Modern*, ed. Patricia Clare Ingham and Michelle R. Warren (New York: Palgrave, 2003), 71-90; and the roundtable essays in *The William and Mary Quarterly* 64, no. 2 (2007).

3. The terms "nation" and "empire" (and their derivatives) have a fiercely contested status in recent scholarship. On one level, these terms are anachronistic (when measured by some historical and sociological definitions) and suggest more monolithic political formations than what historically existed. While looser and less specific terms such as "national sentiment" or "imperial sentiment" would perhaps capture the fluidity of the political and institutional contexts more accurately, my use of the conventional terms is meant to encompass, expose, and reflect upon that fluidity.

4. For this definition of epic and its relation to empire, see David Quint, *Epic and Empire:*

Politics and Generic Form from Virgil to Milton (Princeton: Princeton University Press, 1993), particularly chapter 1.

5. On the conceptual and political uses of "imperium," see Pagden, *Lords of All the World*; and James Muldoon, *Empire and Order: The Concept of Empire, 800–1800* (New York: St. Martin's Press, 1999).

6. "La materia poetica adunque pare amplissima oltre tutte l'altre, però che abbraccia le cose alte e basse . . . le incognite e le conosciute, le nuove e l'antiche, le nostre e le straniere, le sacre e le profane, le civili e le naturali, l'umane e le divine. Laonde i suoi termini non pare che siano i monti o mari che dividono l'Italia o la Spagna, non il Tauro, non l'Atlante, non Battro, non Tile, non il mezzo giorno o 'l settentrione o 'l'oriente o l'occidente, ma il cielo e la terra . . . percioché Dante, innalzandosi dal centro, ascende sovra tutte le stelle fisse a sovra tutti i giri celesti; e Virgilio e Omero ci descrissero non solamente le cose che sono sotto la terra, ma quelle ancora che a pena con l'intelletto possiamo considerare. . . . È dunque grandissima la varietà delle cose trattate da loro e da gli altri che prima o dopo hanno poetato, e grandissima la diversità dell'opinioni, o più tosto la contrarietà de' giudicii, la mutazione delle favelle, de' costume, delle leggi, delle cerimonie, delle republiche, de' regni, de gl'imperatori e quasi del mondo istesso, il quale pare che abbia mutata faccia, e ci si rappresenta quasi in un'altra forma e in un'altra sembianza." Torquato Tasso, *Discorsi dell'arte poetica e del poema eroico*, ed. Luigi Poma (Bari: Laterza, 1964), 79; English translation is from *Discourses on the Heroic Poem*, trans. Mariella Cavalchini and Irene Samuel (Oxford: Clarendon Press, 1973), 21–22.

7. See J. R. Goodman, *Chivalry and Exploration, 1298–1630* (Woodbridge, UK: Boydell Press, 1998); Joan P. Linton, *The Romance of the New World: Gender and the Literary Formations of English Colonialism* (New York: Cambridge University Press, 1998); Mary Baine Campbell, *The Witness and the Other World: Exotic European Travel Writing, 400–1600* (Ithaca: Cornell University Press, 1988).

8. On the tension between epic and voyage in *Os Lusíadas*, see Helder Macedo and Fernando Gil, *The Traveling Eye: Retrospection, Vision, and Prophecy in the Portuguese Renaissance* (Dartmouth: University of Massachusetts Dartmouth, 2009). For readings that discuss the poem's genre and commercial contexts, see Richard Helgerson, *Forms of Nationhood: The Elizabethan Writing of England* (Chicago: University of Chicago Press, 1992), 155–63; and Shankar Raman, *Framing "India": The Colonial Imaginary in Early Modern Culture* (Stanford: Stanford University Press, 2002).

9. On the emergence of the nation in time, and the uses of national history, see Andrew Escobedo, *Nationalism and Historical Loss in Renaissance England: Foxe, Dee, Spenser, Milton* (Ithaca: Cornell University Press, 2004).

10. Thomas M. Greene, *The Descent from Heaven: A Study in Epic Continuity* (New Haven: Yale University Press, 1963), 17–19.

11. Much of the rich scholarship on postcolonial approaches to early modern texts focuses on these trajectories; for a recent synthesis, see Shankar Raman, *Renaissance Literature and Postcolonial Studies* (Edinburgh: Edinburgh University Press, 2011).

12. I discuss the critical tradition associated with each poem in some detail below. Classical scholarship, particularly on Virgil, has grappled with the problem of "other voices" in narrative texts: see especially R.O.A.M. Lyne, *Further Voices in Vergil's Aeneid* (Oxford: Oxford University Press, 1987); Stephen Hinds, *Allusion and Intertext: Dynamics of Appropriation in Roman Poetry* (Cambridge: Cambridge University Press, 1997); and Alessandro Barchiesi, *La traccia del modello: effetti omerici nella narrazione virgiliana* (Pisa: Giardini, 1984).

13. "Quid enim videatur ei magnum in rebus humanis, cui aeternitas omnis totiusque mundi nota sit magnitudo?" Cicero, *Tusculan Disputations* 4.17. Cited from Marcus Tullius Cicero, *Tusculan Disputations*, The Loeb Classical Library (Cambridge, Mass.: Harvard University Press, 1960).

14. Balachandra Rajan, "Milton and Camões: Reinventing the Old Man," *Portuguese Literary and Cultural Studies* 9 (1998): 178.

15. C. M. Bowra, *From Virgil to Milton* (London: Macmillan, 1957), 86. Bowra's characteriza-

tion is often cited in claims for Camões as a "world-poet"—see most recently, George Monteiro, "Camões's *Os Lusíadas*: The First Modern Epic," in *The Cambridge Companion to Epic* (Cambridge: Cambridge University Press, 2010), 119–32.

16. See Hélio J. S. Alves, "Post-Imperial Bacchus: The Politics of Literary Criticism in Camões Studies, 1940–2001," *Portuguese Literary and Cultural Studies* 9 (1998): 95–106; several scholars in the volume raise the issue, and the volume itself is tellingly entitled, "Post-Imperial Camões." For a historiographical perspective, see David Inglis, "Mapping Global Consciousness: Portuguese Imperialism and the Forging of Modern Global Sensibilities," *Globalizations* 8, no. 5 (2011): 687–702. In a now much-cited comment, Thomas Greene pointed to the poem's seeming entrapment by the pace of global historic transformation: "*Os Lusíadas* today seems almost swamped by the twentieth century. Of the two great forces which animate it, imperialism and nationalism, the first is largely discredited in our time, and the second is beginning to be suspect" (*Descent from Heaven*, 220).

17. My translation. All citations from *Os Lusíadas* follow Luís de Camões, *Os Lusíadas*, ed. Emanuel Paulo Ramos (Porto: Porto Editora, 2006); all translations are from Luís Vaz de Camões, *The Lusiads: Luis Vaz de Camões*, trans. Landeg White (Oxford: Oxford University Press, 1997) unless otherwise indicated.

18. Connections between the Ilha dos Amores and Carthage have been repeatedly noted since the celebrated edition of Manuel Faria e Sousa. See more recently, Quint, *Epic and Empire*, 119–20; Anna Klobucka, "Lusotropical Romance: Camões, Gilberto Freyre, and the Isle of Love," *Portuguese Literary and Cultural Studies* 9 (2002): 121–37; and Vítor Manuel de Aguiar e Silva, *Camões: Labirintos e fascínios* (Lisboa: Cotovia, 1994).

19. See Silva, *Camões: Labrinitos e fascínios*; Klobucka, "Lusotropical Romance"; Quint, *Epic and Empire*.

20. Philip R. Hardie, *Virgil's Aeneid: Cosmos and Imperium* (Oxford: Clarendon Press, 1986).

21. On how expanding European empires used the language of universal dominion, see Pagden, *Lords of All the World*; Muldoon, *Empire and Order*; and S. L. Collins, *From Divine Cosmos to Sovereign State: An Intellectual History of the Idea of Order in Renaissance England* (New York: Oxford University Press, 1989).

22. For this context, see Hardie, *Virgil's Aeneid*; D. J. Furley, *Cosmic Problems: Essays on Greek and Roman Philosophy of Nature* (Cambridge: Cambridge University Press, 1989); Monica Gale, *Lucretius and the Didactic Epic* (London: Bristol Classical Press, 2001); Thomas Leinkauf, ed., *Plato's Timaeus and the Foundations of Cosmology in Late Antiquity, the Middle Ages and Renaissance* (Leuven: Leuven University Press, 2005); and Zdravko Planinc, *Plato Through Homer: Poetry and Philosophy in the Cosmological Dialogues* (Columbia: University of Missouri Press, 2003).

23. Strabo *Geography* 1.1.2. For a discussion of Strabo's interest in Homer, see Lawrence Kim, "The Portrait of Homer in Strabo's Geography," *Classical Philology* 102, no. 4 (October 2007): 363–88.

24. Charles R. Boxer, *The Portuguese Seaborne Empire, 1415–1825* (Manchester: Carcanet, 1991).

25. I follow Hardie's analysis of the link between the tales of Iopas, Aeneas, and Anchises: see *Virgil's Aeneid*, 52–84.

26. See my discussion of the Atlas figure in chapter 1.

27. *Aeneid* 1.743–44. All translations from Virgil follow *Aeneid*, trans. Robert Fitzgerald (New York: Random House, 1983).

28. It is worth noting that Camões frames the Nymph's song (instead of Tethys's prophecy) with references to Iopas and Demodocus. However, the narrator's invocation at 10.8 frames both accounts and invites comparison with Homer and Virgil.

29. Quint, *Epic and Empire*, 119; Klobucka, "Lusotropical Romance," esp. 123–25. See also Gilberto Freyre, *The Portuguese and the Tropics: Suggestions Inspired by the Portuguese Methods of Integrating Autochthonous Peoples and Cultures Differing from the European in a New, or Luso-tropical, Complex of Civilisation* (Lisbon: Executive Committee for the Commemoration of the Vth Cen-

tenary of the Death of Prince Henry the Navigator, 1961); and Fernando Gil, *O efeito-Lusíadas* (Lisboa: João Sá da Costa, 1997).

30. Josiah Blackmore, *Moorings: Portuguese Expansion and the Writing of Africa* (Minneapolis: University of Minnesota Press, 2009), 146.

31. Klobucka, "Lusotropical Romance," 130.

32. Luís de Camões, *Lusíadas de Luis de Camoens, Príncipe de los poetas de España*, ed. Manuel de Faria e Sousa (Madrid, 1639), 1.254. Faria e Sousa counts twenty-seven discrete allusions to Lucretius in *Os Lusíadas*, often juxtaposing Virgillian and Lucretian intertexts.

33. On Lucretian influence in sixteenth-century literature and culture, see Stephen Greenblatt, *The Swerve: How the World Became Modern* (New York: W. W. Norton, 2011); Gerard Passannante, *The Lucretian Renaissance: Philology and the Afterlife of Tradition* (Chicago: University of Chicago Press, 2011); and Ada Palmer, *Reading Lucretius in the Renaissance* (Cambridge, Mass.: Harvard University Press, 2014). See also the critical survey in my "Mutabilitie's Lucretian Metaphysics," 220–45.

34. On this trope, see Katharine Park, "Nature in Person: Medieval and Renaissance Allegories and Emblems," in *The Moral Authority of Nature*, ed. Lorraine Daston and Fernando Vidal (Chicago: University of Chicago Press, 2004).

35. On Camões's use of geographic discourse, see Bernhard Klein, "Mapping the Waters: Sea Charts, Navigation, and Camões's *Os Lusíadas*," *Renaissance Studies* 25, no. 2 (2011): 228–47. On anthropomorphic geography, see Jacob, "Quand les cartes reflechissent," 36–49.

36. For discussion of a similar, feminized representation of the Spanish empire on a map made in Manila, see Ricardo Padrón, *The Spacious Word: Cartography, Literature, and Empire in Early Modern Spain* (Chicago: University of Chicago Press, 2004), 232–33 and fig. 35.

37. Livy *Historia* 1.16.

38. The peculiar hesitation on the part of the major imperial powers to use the term "empire" to describe their possessions is discussed in J. H. Elliott, "A Europe of Composite Monarchies," *Past & Present* no. 137 (November 1, 1992): 48–71; Pagden, *Lords of All the World*; and Muldoon, *Empire and Order*.

39. On the competing justifications for Portuguese expansion and their rival factions at court, see Sanjay Subrahmanyam, *The Portuguese Empire in Asia, 1500–1700: A Political and Economic History* (New York: Longman, 1993) and *The Career and Legend of Vasco da Gama* (Cambridge: Cambridge University Press, 1997).

40. See the argument of Helgerson in *Forms of Nationhood*, 149–92.

41. On the importance of imagining the earth as an integrated terraqueous globe, see W.G.L. Randles, *De la terre plate au globe terrestre: une mutation epistémologique rapide, 1480–1520* (Paris: A. Colin, 1980) and Pierre Duhem, *Le système du monde: histoire des doctrines cosmologiques de Platon à Copernic* (Paris: A. Hermann, 1913). See also David Wootton's review essay, "No Words for World," *TLS: The Times Literary Supplement* no. 5715 (October 12, 2012): 8–9, which critiques contemporary map historians for ignoring this shift.

42. On universal empire and the Christian church, see the useful synthesis in Muldoon, *Empire and Order*.

43. Alexander von Humboldt, *Cosmos: A Sketch of a Physical Description of the Universe* (Baltimore: Johns Hopkins Press, 1997), 1:57–59.

44. On the contrast of force and cunning in the Homeric epics, see Marcel Detienne and Jean Pierre Vernant, *Cunning Intelligence in Greek Culture and Society* (Atlantic Highlands, N.J.: Humanities Press, 1978), 11–23. In writing of force and fraud as the two great organizing principles of imaginative literature, Northrop Frye locates the dichotomy in the contrast between the Homeric epics, "the story of *forza* in the *Iliad*, the story of wrath (*menis*) of Achilles, of *froda* in the *Odyssey*, the story of the guile (*dolos*) of Ulysses" (*The Secular Scripture: A Study of the Structure of Romance* [Cambridge, Mass.: Harvard University Press, 1976], 65–66). For reflections on the feminist implications of the myth of Metis, see Lillian E. Doherty, *Siren Songs: Gender, Audiences, and Narrators*

in the Odyssey (Ann Arbor: University of Michigan Press, 1995), 1–8; and Jay Dolmage, "Metis, Mêtis, Mestiza, Medusa: Rhetorical Bodies across Rhetorical Traditions," *Rhetoric Review* 28, no. 1 (January 2009): 1–28.

45. Klein, "Mapping the Waters," 241–42. On the demands of mobile, oceanic space, see Steve Mentz, *At the Bottom of Shakespeare's Ocean* (London: Continuum, 2009).

46. Martín Cortés, *The Arte of Nauigation*, trans. Richard Eden (London, 1561), fol. 56r.

47. As Anthony Pagden notes, however, establishing an empire on the ocean was oxymoronic, since, "as every imperialist knew, 'empire' implied rulership, and that, on the British and Dutch understanding of the law of nations, could not be exercised at sea" (*Lords of All the World*, 4).

48. From the sonnet, "Mudam-se os tempos, mudam-se as vontades." Citations and translations from the sonnets are from Luís de Camões, *Selected Sonnets: A Bilingual Edition*, trans. William Baer (University of Chicago Press, 2008).

49. Helder Macedo, "Conceptual Oppositions in the Poetry of Camões," *Portuguese Literary and Cultural Studies* 9 (1998): 64–66. See also Macedo's readings in "Love as Knowledge: The Lyric Poetry of Camões," *Portuguese Studies* 14 (1998): 51–64; and Gil and Macedo, *Traveling Eye*.

50. On the rhetorical and cartographic status of islands in the early modern period, see Frank Lestringant, *Le livre des îles: atlas et récits insulaires de la genèse à Jules Verne* (Geneva: Librairie Droz, 2002); and Dodds Klaus and Royle A. Stephen, "The Historical Geography of Islands. Introduction: Rethinking Islands," *Journal of Historical Geography* 29, no. 4 (October 2003): 487–98. In the *History of the World*, Walter Raleigh describes the imaginary naming of an island by the Spanish explorer Pedro de Sarmiento, highlighting the questionable nature of insular locations: "he told me merrily that it was to be called the Painter's Wife's Island: saying, that whilst the fellow drew that map, his wife sitting by desired him to put in one country for her, that she, in imagination, might have an island of her own." Walter Raleigh, *The History of the World: In Five Books* (Edinburgh: Archibald Constable and Co., 1820), 3.340–42.

51. See Telmo dos Santos Verdelho, *Indice reverso de "Os lusíadas"* (Coimbra: Biblioteca Geral da Universidade, 1981). "Império" in the sense of authority or control occurs eight additional times.

52. Raman is one of the few critics who discusses the imminent obsolescence of the worldview celebrated in the poem's final cantos: see *Framing "India."* The cosmology of the epic, though, has received some recent critical interest: see J. M. Costa, "Camões e a Cosmogonia," *Bulletin of the Astronomical Society of Brazil* 23 (August 2003): 74–75; and James Nicolopulos, *The Poetics of Empire in the Indies: Prophecy and Imitation in La Araucana and Os Lusíadas* (College Station: Penn State University Press, 2000).

53. See Costa, "Camões e a Cosmogonia"; and Raman, *Framing "India."* On the pedagogical use of Sacrobosco, see Owen Gingerich, "Sacrobosco as a Textbook," *Journal for the History of Astronomy* 19, no. 4 (1988): 269–72.

54. See Pedro Nunes, *Tratado da sphera com a theorica do sol e da lua e ho primeiro livro da geographia de Claudio Ptolomeo: reproduction fac-similé de l'exemplaire appartenant à la Bibliothèque du Duc de Brunswick à Wolfenbüttel, edition 1537, Lisbonne* (Munich: J. B. Obernetter, 1915).

55. Lucretius *De rerum natura* 5.97: "moles et machina mundi." The metaphor of the world-machine is found in a range of twelfth- and thirteenth-century writers including Robert Grosseteste, Alain de Lille, Hugh of St Victor, Bernard Sylvester, and Sacrobosco. On the use of the metaphor in the early modern period, see Giovanni di Pasquale, "Il concetto di machina mundi in Lucrezio," in *Lucrezio, la natura e la scienza*, ed. Marco Beretta and Francesco Citti (Florence: Olschki, 2008), 35–50; and Marcus Popplow, "Setting the World Machine in Motion: The Meaning of Machina Mundi in the Middle Ages and the Early Modern Period," in *Mechanics and Cosmology in the Medieval and Early Modern Period*, ed. Massimo Bucciantini, Michele Camerota, and Sophie Roux (Florence: Olschki, 2007), 45–71.

56. On the international importance of Nunes's mathematics, see the synthesis in Bruno Almeida, "On the Origins of Dee's Mathematical Programme: The John Dee–Pedro Nunes Connection," *Studies in History and Philosophy of Science Part A* 43, no. 3 (September 2012): 460–69.

57. For biographical context, see Andrew Hadfield, *Edmund Spenser: A Life* (New York: Oxford University Press, 2012); Christopher Burlinson and Andrew Zurcher, *Edmund Spenser: Selected Letters and Other Papers* (Oxford: Oxford University Press, 2009).

58. All citations from *The Faerie Queene* are from Edmund Spenser, *The Faerie Queene*, ed. A.C. Hamilton, H. Yamashita, and T. Suzuki (New York: Longman, 2001).

59. Chorographic readings of *The Faerie Queene* were pioneered by Helgerson, *Forms of Nationhood*; more recently, see Bernhard Klein, *Maps and the Writing of Space in Early Modern England and Ireland* (Basingstoke: Palgrave, 2001); Rhonda Lemke Sanford, *Maps and Memory in Early Modern England: A Sense of Place* (New York: Palgrave, 2002); and Joan Fitzpatrick, "Marrying Waterways: Politicizing and Gendering the Landscape in Spenser's Faerie Queene River-Marriage Canto," in *Archipelagic Identities: Literature and Identity in the Atlantic Archipelago, 1550–1800*, ed. Philip Schwyzer and Simon Mealor (Aldershot: Ashgate, 2004), 81–91; The cosmographic scope of Spenser's poetry is beginning to attract critical attention: see Patrick G. Cheney and Lauren Silberman, *Worldmaking Spenser: Explorations in the Early Modern Age* (Lexington: University Press of Kentucky, 2000), and Tamsin Badcoe, "'The Compasse of That Islands Space': Insular Fictions in the Writing of Edmund Spenser," *Renaissance Studies* 25, no. 3 (2011): 415–32.

60. See my "Edmund Spenser, Lucretian Neoplatonist," 373–411, for a discussion of cosmology in the late poems.

61. Stephen Greenblatt first proposed similarities between the Bower of Bliss and the New World in *Renaissance Self-fashioning: From More to Shakespeare* (Chicago: University of Chicago Press, 1980). The Cave of Mammon has also been identified as a possible New World locale: see Thomas H. Cain, *Praise in The Faerie Queene* (Lincoln: University of Nebraska Press, 1978); David Read, *Temperate Conquests: Spenser and the Spanish New World* (Detroit: Wayne State University Press, 2000), 66–78; and David Landreth, *The Face of Mammon: The Matter of Money in English Renaissance Literature* (Oxford: Oxford University Press, 2012).

62. See Torquato Tasso, *Gerusalemme liberata*, ed. Bruno Maier (Milan: Biblioteca Universale Rizzoli, 1982), canto 15, which is closely imitated in *The Faerie Queene* 2.1–37, though Spenser carefully leaves out Tasso's Columbus encomium at 15.31–32. Theodore Cachey points to the existence of an Ambrosiana manuscript of Tasso's epic in which the fifteenth canto is suggestively entitled "La navigazione del mondo nuova del poema di T. Tasso" (The navigation to the New World of the Poem by T. Tasso), most likely based on Antonio Pigafetta's account of the Magellan circumnavigation; see "Tasso's Navigazione del Mondo Nuovo and the Origins of the Columbus Encomium (GL, xv, 31–32)," *Italica* 69, no. 3 (1992), and Sergio Zatti, *L'ombra del Tasso: epica e romanzo nel Cinquecento* (Milano: B. Mondadori, 1996); also relevant is Jane Tylus, "Reasoning Away Colonialism: Tasso and the Production of the *Gerusalemme liberata*," *South Central Review* 10, no. 2 (1993). Spenser's erasure of the remaining Columbus stanzas while incorporating the New World into his epic raises the tantalizing possibility that perhaps he may have known something of this parallel text.

63. The difficulty of drawing a map for the poem emerges in Wayne Erickson, *Mapping the Faerie Queene: Quest Structures and the World of the Poem* (New York: Garland, 1996). A peculiar instance of this difficulty is apparent in the fold-out chart that attempts to illustrate the topography of Faeryland in Waldo F. McNeir and Foster Provost, *Edmund Spenser: An Annotated Bibliography, 1937–1972* (Pittsburgh: Duquesne University Press, 1975). In addition to the cosmographic impulse, we may also identify the influence of what Patrick Cheney has described as "counternational" epic projects, such as that of Ovid, in *The Faerie Queene*; see *Marlowe's Counterfeit Profession: Ovid, Spenser and Counter-Nationhood* (Toronto: University of Toronto Press, 1997).

64. The most influential account of this episode remains Harry Berger, *The Allegorical Temper: Vision and Reality in Book II of Spenser's Faerie Queene* (New Haven: Yale University Press, 1957), 77–85. See also Leonard Barkan, *Nature's Work of Art*; Michael Murrin, *The Veil of Allegory: Some Notes toward a Theory of Allegorical Rhetoric in the English Renaissance* (Chicago: University of Chicago Press, 1969) chap. 4; Robert L. Reid, "Alma's Castle and the Symbolization of Reason in *The*

Faerie Queene," *Journal of English and Germanic Philology* 80 (1981): 512 – 27, and David Lee Miller, *The Poem's Two Bodies: The Poetics of the 1590 Faerie Queene* (Princeton: Princeton University Press, 1988), 183 – 91.

65. This vision of poetic practice sheds new light on the much discussed Letter to Raleigh. Spenser's famous description of "a Poet historicall" as one who "thrusteth into the middest, euen where it most concerneth him, and there recoursing to the thinges forepaste, and divining of thinges to come, maketh a pleasing Analysis of all" (*The Faerie Queene,* 716 –17; my emphases) looks more like a gloss on the allegory of mind and world in book 2.

66. The literature on this topic is vast. I have found particularly useful: Judith H. Anderson, "The Antiquities of Fairyland and Ireland," *Journal of English and Germanic Philology* 86, no. 2 (1987): 199 – 214; Linda Gregerson, "Colonials Write the Nation: Spenser, Milton, and England on the Margins," in *Edmund Spenser: Essays on Culture and Allegory,* ed. Jennifer Klein Morrison and Matthew Greenfield (Aldershot: Ashgate, 2000), 89 – 105; Hadfield, *Shakespeare, Spenser, and the Matter of Britain*; Rebecca Helfer, "Falling into History: Trials of Empire in Spenser's *Faerie Queene,*" in *Fantasies of Troy: Classical Tales and the Social Imaginary in Medieval and Early Modern Europe,* ed. Alan Shepard and Stephen D. Powell (Toronto: Centre for Reformation and Renaissance Studies, 2004), 237 – 52; and Bart Van Es, *Spenser's Forms of History* (New York: Oxford University Press, 2002).

67. Miller, *Poem's Two Bodies,* 206. A comparison to Camões seems almost inevitable: the unbroken line of succession in Spenser's Faery chronicle is clearly what Camões seeks for King Sebastião and the endangered Avis succession.

68. For a theoretical and ethno-historical analysis of this point, see Anthony D. Smith, *Myths and Memories of the Nation* (Oxford: Oxford University Press, 1999); P. J. Geary, *The Myth of Nations: The Medieval Origins of Europe* (Princeton: Princeton University Press, 2002); and Caspar Hirschi, *The Origins of Nationalism: An Alternative History from Ancient Rome to Early Modern Germany* (Cambridge: Cambridge University Press, 2011).

69. On the epistemological challenge of world-description for early modern historians, see Serge Gruzinski, *Les quatre parties du monde: histoire d'une mondialisation* (Paris: Martinière, 2004), and Sanjay Subrahmanyam, "Connected Histories: Notes Toward a Reconfiguration of Early Modern Eurasia," *Modern Asian Studies* 31, no. 3 (1997): 735 – 62, and "On World Historians in the Sixteenth Century" *Representations* 91 (2005): 26 – 57.

70. Harry Berger has argued that the two chronicles present different versions of the same history, such that Guyon's book offers an elegant fictional revision of the same historical pattern offered to Arthur—an identification that is borne out by the correspondence in stanzas 2.10.75 – 76 of Elferon, Oberon, and Tanaquill with Henry VII, Henry VIII, and Elizabeth (see *Allegorical Temper*). However, my argument suggests that this culminating correspondence may also be read oppositionally, contrasting national and world histories, which necessarily overlap at crucial points.

71. The point was first made by Helgerson, who stresses the link between the British chronicle and chorographic surveys of the land as central to a developing sense of the nation in *The Faerie Queene* (see *Forms of Nationhood,* 105 – 48).

72. For theoretical and historiographical debates on the association of the nation with territorial specificity, see Eric J. Hobsbawm, *Nations and Nationalism since 1780: Programme, Myth, Reality* (New York: Cambridge University Press, 1990); Adrian Hastings, *The Construction of Nationhood: Ethnicity, Religion, and Nationalism* (Cambridge: Cambridge University Press, 1997); and Anthony D. Smith, *The Nation in History: Historiographical Debates about Ethnicity and Nationalism* (Hanover: University Press of New England, 2000).

73. See Smith, *Myths and Memories and Nationalism and Modernism: A Critical Survey of Recent Theories of Nations and Nationalism* (London: Routledge, 1998).

74. Subrahmanyam, "On World Historians," 36.

75. Ibid.

76. The most extensive description of Britain by a Roman historian is Tacitus's *Agricola;* chap-

ter 10 offers a particularly vivid account of the landscape. The analogies to both India and the New World are suggestively analyzed by Elizabeth Bellamy from a postcolonial perspective; see "Spenser's Faeryland and 'the Curious Genealogy of India,'" in *Worldmaking Spenser: Explorations in the Early Modern Age,* ed. Patrick Cheney and Lauren Silberman (Lexington: University of Kentucky Press, 2000), 177–92.

77. The giants are identified as cannibalistic in 3.9.49; moreover, the extermination-and-settlement method had already been used by the Spanish in America (vividly recorded, for instance, in Montaigne's contemporary essay "Des coches") and it has been suggested that the English drew precisely on these Spanish methods to subdue Irish rebellions. On the politics of the giants, see Susanne Wofford, *The Choice of Achilles: The Ideology of Figure in the Epic* (Stanford: Stanford University Press, 1992), 334–53, and Walter Stephens, *Giants in Those Days: Folklore, Ancient History, and Nationalism* (Lincoln: University of Nebraska Press, 1989); on analogies between Brutus's conquest and the New English in Ireland, see Christopher Ivic, "Spenser and the Bounds of Race," *Genre* 32 (1999): 153–55; on the Spanish connection see Barbara Fuchs, "Spanish Lessons: Spenser and the Irish Moriscos," *SEL: Studies in English Literature 1500–1800* 42, no. 1 (2002).

78. For analysis of the *translatio* trope in the chronicles and its relationship to both India and the New World, see Bellamy, "Spenser's Faeryland."

79. Gruzinski, *Les quatre parties du monde,* 22. My translation.

80. Similarly, see Charlotte Artese, "King Arthur in America: Making Space in History for *The Faerie Queene* and John Dee's *Brytanici Imperii Limites,*" *Journal of Medieval and Early Modern Studies* 33, no. 1 (2003): 125–41, who discusses why, in a historically imaginative revision, John Dee argued that Arthur had originally discovered America.

81. The Lucretianism of the Gardens of Adonis was first raised by Edwin Greenlaw ("Spenser and Lucretius," *Studies in Philology* 17 [1920]: 455–84) and much debated in the Variorum edition. The issue has been raised more recently in a pair of essays by Anthony Esolen: "Spenser's 'Alma Venus': Energy and Economics in The Bower of Bliss," *English Literary Renaissance* 23, no. 2 (Spring 1993): 267–86, and "Spenserian Chaos: Lucretius in The Faerie Queene," *Spenser Studies* 11 (1994): 31–51; see my extended discussion of Spenser's Lucretianism and Platonism in "Edmund Spenser, Lucretian Neoplatonist."

82. Hannah Arendt, *Between Past and Future,* trans. Jerome Kohn (New York: Penguin Classics, 2006), 91–141.

83. Ibid., 91.

84. Hardie, *Virgil's Aeneid,* 135. See the response, and claim for a more complex understanding of gigantomachic imagery in James J. O'Hara, *Inconsistency in Roman Epic: Studies in Catullus, Lucretius, Vergil, Ovid and Lucan* (Cambridge: Cambridge University Press, 2007), esp. 96ff.

85. On Lucretius's gigantomachy, see Diskin Clay, "Lucretius's Gigantomachy," in *Lucretius and His Intellectual Background,* ed. Keimpe Algra, Mieke H. Koenen, and Pieter H. Schrijvers (Amsterdam: Kluwer Academic Publishers, 1997), 187–92; and Monica Gale, *Myth and Poetry in Lucretius* (Cambridge: Cambridge University Press, 1994). Clay notes the connections to the *Sophist* and Aristotle's *On Philosophy*: see also his more extended discussion of Epicureanism in *Paradosis and Survival: Three Chapters in the History of Epicurean Philosophy* (Ann Arbor: University of Michigan Press, 1998). The Giants are explicitly mentioned twice in *De rerum natura* (at 4.138ff. and 5.117ff.), and gigantomachic imagery also appears at the opening proem of book 1.

86. Quint discusses the gigantomachic implications of the episode; see *Epic and Empire,* 122–23.

87. See Quint, *Epic and Empire,* 117–25; and Lawrence Lipking, "The Genius of the Shore: Lycidas, Adamastor, and the Poetics of Nationalism," *PMLA* 111, no. 2 (March 1996): 205–21.

88. Lucretius *De rerum natura* 6.424ff. Classical sources for the description include Lucan and Homer; Camões's allusion to Lucretius seems to have been first noticed by Faria e Sousa and repeated by successive commentators in the eighteenth and nineteenth centuries.

89. Lucretius *De rerum natura* 6.424, 6.435.

90. Quint, *Epic and Empire*, 123.

91. See the detailed discussion in Gale, *Myth and Poetry*.

92. See especially, Silva, *Camões: Labrinitos e fascínios*; and Klobucka, "Lusotropical Romance."

93. See Lucretius *De rerum natura* 2.655–60 for the demystification of Neptune and Ceres. For discussion of the contrast between Camões and Lucretius on the mythological treatment of gods, see Milton Torres, "Lucrécio, Camões e os deuses," *Aletria: Revista de estudos de literatura* 19 (2009): 191–204. Torres however does not note the possible allusion; indeed, it seems difficult to reconcile Camões's emphasis on a militant Christianity with his use of the atheist Lucretius, though he is hardly unique in this in the sixteenth century: see extensive recent work on Lucretian influence by Valentina Prosperi, *Di soavi licor gli orli del vaso: la fortuna di Lucrezio dall'umanesimo alla controriforma* (Torino: N. Aragno, 2004); Simone Fraisse, *L'influence de Lucrèce en France au seizième siècle: une conquête du rationalisme* (Paris: A. G. Nizet, 1962); Alison Brown, *The Return of Lucretius to Renaissance Florence* (Cambridge, Mass.: Harvard University Press, 2010) and Palmer, *Reading Lucretius in the Renaissance*.

94. According to myth, Thetis was prophesied to give birth to a son greater than his father: hence her marriage to the hero Peleus; her son would be the hero Achilles. She is also linked to a tale of gigantomachy in the *Iliad*.

95. Spenser uses "world" (or forms of it) fifteen times in the two *Cantos of Mutabilitie*; the frequency of use in the other complete twelve-canto books of *The Faerie Queene* are as follows: book 1 (40); book 2 (41); book 3 (54); book 4 (23); book 5 (23); book 6 (24).

96. While several scholars have suggested that Mutabilitie represents original sin because she "death for life exchanged foolishlie: / Since which, all liuing wights haue learn'd to die," closer inspection raises questions about such logic. For sustained analysis of the association of Mutabilitie with the Fall in Genesis 3, see Harold L. Weatherby, *Mirrors of Celestial Grace: Patristic Theology in Spenser's Allegory* (Toronto: University of Toronto Press, 1994); Sherman Hawkins, "Mutabilitie and the Cycle of the Months," in *Form and Convention in the Poetry of Edmund Spenser: Selected Papers from the English Institute*, ed. William Nelson (New York: Columbia University Press, 1961), 76–102; and Judith Anderson, "Mutability and Mortality: Reading Spenser's Poetry," in *Celebrating Mutabilitie: Essays on Edmund Spenser's Mutabilitie Cantos*, ed. Jane Grogan (Manchester: Manchester University Press, 2010), 246–74. Mutability may represent the *effects* of the Fall, but it is hard to imagine how she *causes* it. It is more likely that Mutabilitie's questioning of the basis of divine order itself represents a fall from grace. For Spenser, such a move toward the demystification of all forms of authority, including the transcendental order of a providential deity, must certainly have been a matter of great anxiety—and considerable fascination.

97. Berger notes that this is a "piece of Lucretian materialism" that aims at "demythologizing the elements" (*Revisionary Play*, 261).

98. Spenser may have been aided in his rationalizing efforts by Natale Conti's influential *Mythologiae*, whose discussion of Titan, from whom Mutabilitie traces her descent, offers a naturalistic explanation, drawn from Lucretius, that supports a materialistic metaphysics: see Natale Conti, *Natale Conti's Mythologiae*, trans. John Mulryan and Steven Brown (Tempe: Arizona Center for Medieval and Renaissance Studies, 2006), 542. See my discussion of this allusion in "Mutabilitie's Lucretian Metaphysics."

99. On the politics of the Egalitarian Giant, see Annabel Patterson, "The Egalitarian Giant: Representations of Justice in History/Literature," *Journal of British Studies* 31 (1992): 97–132; Richard A. McCabe, *Spenser's Monstrous Regiment: Elizabethan Ireland and the Poetics of Difference* (Oxford: Oxford University Press, 2002); Elizabeth Fowler, "The Failure of Moral Philosophy in the Work of Edmund Spenser," *Representations* 51 (Summer 1995): 47–76; and James Holstun, "The Giant's Faction: Spenser, Heywood, and the Mid-Tudor Crisis," *Journal of Medieval and Early Modern Studies* 37, no. 2 (March 20, 2007): 335–71. Spenser explicitly links the Giant, through a series of biblical allusions, to radical, anti-creationist positions. The claim to "weigh" the winds, the waters, and the earth derives from rhetorical challenges posed by God (or on God's

behalf) to mortals who dare question divine order and control of the universe. This theological dimension resonates with Mutabilitie's challenge against eternal stability toward the seeming fallenness of change.

100. Holstun, "Giant's Faction," 340, notes that the Giant is sympathetic and recalls both Astraea and Artegall (5.2.30; 5.1.7, 11).

101. Both Patterson and Holstun argue in favor of the Giant and point to the subversive power of his claims.

102. At the beginning of the second book of the *Naturalis historia*, Pliny explicitly connects world and nature: see the extended discussion at 2.1.

103. Anne Prescott, "Giants," in *The Spenser Encyclopedia*, ed. A. C. Hamilton (Toronto: University of Toronto Press, 1997), 332–33.

104. I discuss the philosophical implications of Nature's verdict at length in "Mutabilitie's Lucretian Metaphysics."

105. Behind the concision of Spenser's stanzas lies an army of sources and interpretations of the classical myth ranging from Christianizing efforts by the early Church fathers to symbolic analyses by Renaissance mythographers. Classical sources for the myth of Prometheus include Ovid *Metamorphoses* 1; Horace *Odes* 2.13; Hesiod *Theogony* 507–616 and *Works and Days* 42–89. The main Renaissance mythographers who discuss Prometheus are Natale Conti in his *Mythologiae* and Vincenzo Cartari in the *Imagini degli dei degl'antichi*. On the interpretation and reception of the myth, see Susanna Barsella, "The Myth of Prometheus in Giovanni Boccaccio's *Decameron*," *MLN* 119 Supplement (2004): S120–41; Dora and Erwin Panofsky, *Pandora's Box: The Changing Aspects of a Mythical Symbol* (New York: Pantheon Books, 1962); and Olga Raggio, "The Myth of Prometheus: Its Survival and Metamorphoses Upto the Eighteenth Century," *Journal of the Warburg and Courtauld Institutes* 21, no. 1/2 (1958): 44–62.

106. See Arthur Golding, trans., *Ovid's Metamorphoses*, Epistle, 439–44; I cite from *Ovid's Metamorphoses*, ed. Madeleine Forey, trans. Arthur Golding (London: Penguin, 2002).

CHAPTER FOUR

1. René Descartes, *Oeuvres de Descartes*, eds. Charles Adam and Paul Tannery, 11 vols. (Paris: J. Vrin, 1973), 1:270–71. Henceforth, "AT" followed by volume and page number. All translations are mine unless otherwise noted.

2. Descartes and Galileo are rarely discussed in relation to each other; for a recent exception that carefully traces the significance of Galileo for Descartes, see Michael Friedman, "Descartes and Galileo: Copernicanism and the Metaphysical Foundations of Physics," in *A Companion to Descartes*, ed. Janet Broughton and John Carriero (Oxford: Blackwell, 2008), 69–83. Essays that consider the question in useful ways include: Katharine Park, Lorraine J. Daston, and Peter L. Galison, "Bacon, Galileo, and Descartes on Imagination and Analogy," *Isis* 75, no. 2 (June 1984): 287–89; Roger Ariew, "Introduction: Galileo in Paris," *Perspectives on Science* 12, no. 2 (2004): 131–34 and "Descartes as Critic of Galileo's Scientific Methodology," *Synthese* 67, no. 1 (April 1986): 77–90; Blake D. Dutton, "Physics and Metaphysics in Descartes and Galileo," *Journal of the History of Philosophy* 37, no. 1 (1999): 49–71; and William R. Shea, "Descartes as Critic of Galileo," in *New Perspectives on Galileo*, ed. Robert E. Butts and Joseph C. Pitt (Dordrecht: D. Reidel, 1978), 139–59.

3. The importance of *Le monde* as the first major work of systematic mechanistic natural philosophy is discussed by Friedman, "Descartes and Galileo," 80–100; see also Stephen Gaukroger, *Descartes: An Intellectual Biography* (Oxford: Oxford University Press, 1995); and William R. Shea, *The Magic of Numbers and Motion: The Scientific Career of René Descartes* (Canton: Science History Publications, U.S.A., 1991).

4. Descartes to Mersenne, November 13, 1629; AT 1:70. All translations are my own unless otherwise stated.

5. On the evolution and expression of Descartes's skepticism and the anti-skeptical project of the *Meditationes*, see Richard Popkin, *The History of Scepticism: From Savonarola to Bayle* (New York: Oxford University Press, 2003), 143–73.

6. Stephen Menn, "The Intellectual Setting," in *The Cambridge History of Seventeenth-Century Philosophy*, ed. Daniel Garber and Michael Ayers (Cambridge: Cambridge University Press, 1998), 34.

7. "Frustra hic Senecam invocat Galilaeus, frustra hic luget nostril temporis calamitem, quod vera ac certa mundanarum partium dispositio non teneatur . . ." From *The Assayer*, 6.8; cited from Galileo Galilei, *Il Saggiatore*, ed. Ottavio Besomi and Mario Helbing (Roma: Antenore, 2005), 115–16; translation from Galileo Galilei, *The Essential Galileo*, ed. and trans. Maurice Finocchiaro (Indianapolis: Hackett, 2008), 179.

8. John Donne, *The First Anniversarie. An Anatomy of the World* (London, 1621); I cite from *The Complete Poetry and Selected Prose of John Donne*, ed. Charles M. Coffin and Denis Donoghue (New York: Modern Library, 2001).

9. "Essendo, dunque, sicuramente falsi li due sistemi, e nullo quello di Ticone, non dovrebbe il Sarsi riprendermi se con Seneca desidero la vera constituzion dell'universo." From *The Assayer*, 6.41–42; Galilei, *Il Saggiatore*, 120; Galilei, *Essential Galileo*, 184.

10. Galileo too had used "world" (*mondo*) in the sense of universe in his *Dialogo sopra i due massimi sistemi del mondo* (1632).

11. The French tradition of scholarship on Descartes primarily regards him as a metaphysician, which is a perspective that goes at least as far back as the work of Nicolas Malebranche; classic studies include Ferdinand Alquié, *La découverte métaphysique de l'homme chez Descartes* (Paris: Presses universitaires de France, 1950); Etienne Gilson, *La liberté chez Descartes et la théologie* (Paris: Alcan, 1913); Henri Gouhier, *La pensée métaphysique de Descartes* (Paris: J. Vrin, 1962) and *La pensée religieuse de Descartes* (Paris: J. Vrin, 1972); Martial Guéroult, *Descartes' Philosophy Interpreted According to the Order of Reasons*, trans. Roger Ariew, 2 vols. (Minneapolis: University of Minnesota Press, 1983); and Jean-Luc Marion, *Sur la théologie blanche de Descartes: Analogie, création des vérités éternelles et fondement* (Paris: Presses universitaires de France, 1981). The Anglo-American tradition of Descartes scholarship, particularly in the context of the history of science in the last twenty years, has sought to reclaim the fundamentally "scientific" bases of Descartes's philosophy; major studies in this area include Desmond M. Clarke, *Descartes's Philosophy of Science* (Manchester: Manchester University Press, 1982); Daniel Garber, *Descartes' Metaphysical Physics* (Chicago: University of Chicago Press, 1992); and Stephen Gaukroger, *Descartes' System of Natural Philosophy* (Cambridge: Cambridge University Press, 2002).

12. These include five large volumes of (mostly) scientific correspondence with almost all major contemporary mathematicians and natural philosophers in Europe and several tentative essays where he applied his methods to specific scientific problems. On the discrepancy between the unpublished scientific writings and the published philosophy, see Clarke, *Descartes's Philosophy of Science*, 1–5, and Jean Laporte, *Le rationalisme de Descartes* (Paris: Presses universitares de France, 1945), 477.

13. I draw extensively on two recent biographies of Descartes for the details of his career. See Gaukroger, *Descartes*, and Geneviève Rodis-Lewis, *Descartes: Biographie* (Paris: Calmann-Lévy, 1995).

14. On March 20, 1629 Scheiner had observed a particularly striking appearance of "multiple suns" at Frascati. The phenomena clearly excited Descartes, who wrote to Mersenne in October 1629 seeking fuller information. See AT 1:22–29.

15. Descartes to Mersenne, November 25, 1630, AT 1:179.

16. Descartes to Constantijn Huygens, January 1642, AT 3:523.

17. "Fatetur autem se illas cogitationes paucas, quas de Mundo habuit, summa cum voluptate reminisci, maximeque aestimare, nec cum ullis aliis materiae commutare velle." From René Descartes and Frans Burman, *Entretien Avec Burman, Manuscrit De Göttingen*, ed. Charles Adam

(Paris: Boivin et cie, 1937), 96-98; translation from John Cottingham, ed., *Descartes's Conversation with Burman* (Oxford: Clarendon Press, 1976), 39. The *Conversation* remains a contested source in Cartesian scholarship: see Roger Ariew's objections to Cottingham's treatment of the *Conversation* in "The Infinite in Descartes' Conversation with Burman," *Archiv Für Geschichte Der Philosophie* 69, no. 2 (1987): 140-63, which considers the scholarly debates. In his magisterial edition of Descartes's work, Adam too cautions against reading too much into the *Conversation*: see AT 12:483-84. For a critical view of its utility, see Cottingham, *Descartes's Conversation with Burman*, xviii. My own use of the *Conversation* in this chapter is limited to instances where further corroboration is available in other Cartesian works.

18. The first edition of *Le monde* appeared in 1664, published in Paris by Jacques Le Gras. Clerselier reiussed *Le monde* along with its companion *L'Homme* in 1677.

19. There are competing accounts of the trial of 1632, but see Maurice A. Finocchiaro, ed., *The Galileo Affair: A Documentary History* (Berkeley and Los Angeles: University of California, 1989). For a recent account of the trial and its stakes see J. L. Heilbron, *Galileo* (New York: Oxford University Press, 2010); its legacy is discussed in Maurice A. Finocchiaro, *Retrying Galileo, 1633-1992* (Berkeley and Los Angeles: University of California Press, 2005).

20. For recent discussion of these intersections, see Frédérique Aït-Touati, *Fictions of the Cosmos: Science and Literature in the Seventeenth Century*, trans. Susan Emanuel (Chicago: University of Chicago Press, 2011).

21. Ibid., 10-11, and 37-40.

22. Fernand Hallyn, *Les structures rhétoriques de la science: de Kepler à Maxwell* (Paris: Seuil, 2004), 130-35.

23. Alexandre Koyré, *Closed World*, 2.

24. My thinking on Descartes builds on Ernst Cassirer's classic account of sixteenth-century philosophic thought in *Individual and the Cosmos*, extending its themes into the seventeenth century.

25. For a reading of Cartesian philosophy as theodicy, see Zbigniew Janowski, *Cartesian Theodicy: Descartes' Quest for Certitude* (Boston: Kluwer Academic, 1999). See also Cottingham's recent comments on the place of God in Cartesian philosophy in "The Role of God in Descartes's Philosophy," in *A Companion to Descartes*, ed. Janet Broughton and John Carriero (Oxford: Blackwell, 2008), 288-301.

26. Descartes, *Oeuvres Philosophiques*, 1:343; Henceforth, "Alquié" followed by volume and page numbers. AT 11:31. I cite both the Alquié and the standard Adam-Tannery edition for all of Descartes's major works; correspondence and minor works are cited from Adam-Tannery only. All translations are my own unless otherwise noted.

27. The most striking example in the period of a physical theory that contradicted sense impressions was Copernicanism itself: Galileo praises Copernicus for rejecting the evidence of the senses and arriving at truth through mathematical demonstrations.

28. René Descartes, *The World and Other Writings*, ed. Stephen Gaukroger (Cambridge: Cambridge University Press, 1998), xxix.

29. The idea of the world is not an innate idea for Descartes, but one that is arrived at through cognition as he argues in the Sixth Meditation; see especially John Carriero's interpretation in *Between Two Worlds: A Reading of Descartes's Meditations* (Princeton: Princeton University Press, 2009).

30. In his analysis of the *Meditationes*, Carriero suggests that Descartes's break with Augustine brings a new understanding of the relation between intellectual and sensory cognition as well as a "new, fundamental characterization of the mind according to which what is fundamental is not the ability to cognize universally (to abstract universals) but to see that something is true (to make judgments)" (*Between Two Worlds*, 16-17).

31. ". . . car depuis deux ou trois mois, je me suis engagé fort avant dans le Ciel . . . Car encore qu'elles paroissent fort irregulierement éparses . . . je ne doute point toutefois qu'il n'y ait un ordre

naturel entr'elles, lequel est regulier & determiné; & la connaissance de cét ordre est la clef & le fondement de la plus haute & plus parfaite science, que les hommes puissent avoir, touchant les choses materielles; d'autant que par son moyen on pourroit connoistre à priori toutes les diverses formes & essences des cors terrestres, au lieu que, sans elle, il nous faut contenter de les deviner à posteriori, & par leurs effets." Descartes to Mersenne, [May 10, 1632], AT 1:250–51.

32. Stanley Rosen, *The Quarrel Between Philosophy and Poetry: Studies in Ancient Thought* (New York: Routledge, 1988), ix.

33. See Dalia Judovitz, *Subjectivity and Representation in Descartes: The Origins of Modernity* (Cambridge: Cambridge University Press, 1988); Hassan Melehy, *Writing Cogito: Montaigne, Descartes, and the Institution of the Modern Subject* (Albany: State University of New York Press, 1997); Jean-Luc Nancy, *Ego Sum* (Paris: Flammarion, 1979); and Timothy J. Reiss, *Mirages of the Selfe: Patterns of Personhood in Ancient and Early Modern Europe* (Stanford: Stanford University Press, 2003).

34. Nancy, *Ego Sum*, 98.

35. This emphasis on personal consciousness rather than on an independent external reality underlies a strong strand of scholarship on Descartes and derives from a phenomenological approach that perhaps originates most notably in Husserl's *Cartesian Meditations*.

36. The importance of fiction, particularly for Renaissance philosophies, is discussed by Harry Berger, *Second World and Green World: Studies in Renaissance Fiction-Making* (Berkeley and Los Angeles: University of California Press, 1988).

37. Alquié 1:343; AT 11:31–32.

38. For insightful histories of the term, including Descartes's use of it, see John D. Lyons, *Before Imagination: Embodied Thought from Montaigne to Rousseau* (Stanford: Stanford University Press, 2005); and Matthew William Maguire, *The Conversion of Imagination: From Pascal Through Rousseau to Tocqueville* (Cambridge, Mass.: Harvard University Press, 2006).

39. Descartes to Guez de Balzac, [April 1631], AT 1:198–99.

40. On this use of the meditation, see Peter Dear, "Mersenne's Suggestion: Cartesian Meditation and the Mathematical Model of Knowledge in the Seventeenth Century," in *Descartes and His Contemporaries: Meditations, Objections, and Replies*, ed. Roger Ariew and Marjorie Greene (Chicago: University of Chicago Press, 1995), 44–62.

41. On Descartes's dreams, particularly the much discussed third dream of 10 Nov. 1619, where Descartes describes having seen a table with an encyclopedia and a collection of poems, see John R. Cole, *The Olympian Dreams and Youthful Rebellion of René Descartes* (Urbana: University of Illinois Press, 1992); Alan Gabbey and Robert E. Hall, "The Melon and the Dictionary: Reflections on Descartes's Dreams," *Journal of the History of Ideas* 59, no. 4 (1998); Gaukroger, *Descartes: An Intellectual Biography*, 106–11; and Geneviève Rodis-Lewis, *L'Oeuvre de Descartes* (Paris: J. Vrin, 1971), 1:50–55. Descartes's earliest biographer, Adrien Baillet, connects the dreams to Descartes's desire to unite poetry and philosophy; see AT 10:184. These concerns return later, in the *Regulae*, particularly the much discussed Regula 3 where Descartes distinguishes between the actions of the imagination and the intellect. On this point see especially, Dennis Sepper, *Descartes's Imagination: Proportion, Images, and the Activity of Thinking* (Berkeley and Los Angeles: University of California Press, 1996); and Jones, *Good Life*.

42. Descartes's 1616 *Law Theses* open with a citation of Lucretius *De rerum natura* 4.2–10; see transcript in Jean-Robert Armogathe, Vincent Carraud, and Robert Fernstra, "La licence en droit de Descartes: un placard inédit de 1616," *Nouvelles de la republique des lettres* Anno 8, no. 1988-II (1988): 125.

43. "Mirum videri possit, quare graves sententiae in scriptis poetarum, magis quam philosophorum. Ratio est quod poetae per entusiasmum et vim imaginationis scripsere: sunt in nobis semina scientiae, ut in silice, quae per rationem a philosophis educuntur, per imaginationem a poetis excutiuntur magisque elucent" (AT 10:217). Adam and Tannery note that the language of the dream visions is almost identical to this note in the *Cogitationes privatae* (AT 10:184); on this connection, see also Sepper, *Descartes's Imagination*, 46–49.

44. On Galileo's use of the poetic argument, see the most recent account in Heilbron, *Galileo*. See also the classic essays by Erwin Panofsky, "Galileo as a Critic of the Arts: Aesthetic Attitude and Scientific Thought," *Isis* 47, no. 1 (March 1, 1956): 3–15; and Stillman Drake, "Galileo's Language: Mathematics and Poetry in a New Science," *Yale French Studies* no. 49 (January 1973): 13–27.

45. Pierre Alain Cahné, *Un autre Descartes: Le philosophe et son langage* (Paris: J. Vrin, 1980), 101.

46. Macrobius lays the groundwork for philosophical argument, following Neoplatonists such as Porphyry and Proclus in distinguishing between the *fabula* (fable) and the *narratio fabulosa* (fabulous narrative). The former is not admissible in philosophy; the latter is admissible when "a decent, dignified conception of holy truths, with respectable events and characters, is presented beneath a modest veil of allegory" (Hoc totum fabularum genus, quod solas aurium delicias profitetur . . . aut sacrarum rerum notio sub pio figmentorum velamine honestis et tecta rebus et vestita nominibus enuntiatur) (1.2.8, 11). Latin citation is from *Commentaire au Songe de Scipion*, trans. Mireille Armisen-Marchetti (Paris: Les Belles Lettres, 2001); English translation is from *Commentary on the Dream of Scipio*, trans. William Harris Stahl (New York: Columbia University Press, 1952), 84–85)).

47. Cahné, *Un autre Descartes*, 117. My translation.

48. Alquié, 1.343; AT 11:31.

49. This point is also discussed in Nancy, *Ego Sum*, 98, who notes the connection with the portrait but argues that Descartes's world is an entirely subjective construct rooted in a new idea of the autonomous self.

50. For a discussion of the illusion/reality trope in baroque literature, particularly in the context of theatrical illusion, and the possible relationship of Descartes's fable to it, see Jean-Pierre Cavaillé, *Descartes: La fable du monde* (Paris: J. Vrin, 1991).

51. The "old quarrel between poetry and philosophy," is invoked in Plato's *Republic* 607b.

52. For a survey of the reception history and interpretive tradition surrounding the *Timaeus*, see Étienne, "La réception du Timée", and more generally the essays in, Neschke-Hentschke, *Le Timée de Platon*, and Gretchen J. Reydams-Schils, *Plato's Timaeus as Cultural Icon* (Notre Dame, Ind.: University of Notre Dame Press, 2003). On Renaissance translations and commentaries on the *Timaeus*, in the context of the Platonic revival, see Allen, *Synoptic Art*; Étienne, "Visages d'un interprète"; Hankins, *Plato in the Italian Renaissance*.

53. For discussion of the tricky concept of *eikos logos* and the cosmological account in the Timaeus, see Anne Freire Ashbaugh, *Plato's Theory of Explanation: A Study of the Cosmological Account in the Timaeus* (Albany: State University of New York Press, 1988), 7–47. Other useful studies include Harold Cherniss, "The Relation of the Timaeus to Plato's Later Dialogues," in *Selected Papers*, ed. Leonardo Tarán (Leiden: E. J. Brill, 1977); E. Howald, "Eikos Logos," *Hermes* 57 (1922); David Runia, "The Literary and Philosophical Status of Timaeus' Proemium," in *Interpreting the Timaeus-Critias*, ed. Tomás Calvo and Luc Brisson (Sankt Augustin: Academia Verlag, 1997); and G. Turrini, "Contributo all'analisi del termine eikos II. Linguagio, verosimiglianza e immagini in Platone," *Acme* 32 (1979).

54. Wilson, "Descartes and Augustine," 46, citing the classic study by Wilhelm Max Wundt, *Sinnliche Und Übersinnliche Welt* (Leipzig: Kröner, 1923), 161. See also Menn, *Descartes and Augustine*, who shows how indebted Descartes's Platonism is to Augustine.

55. On the question of "imaginary spaces" in Scholastic philosophy and its relation to problems of infinitude, see Edward Grant, "Medieval and Seventeenth-Century Conceptions of an Infinite Void Space Beyond the Cosmos," *Isis* 60, no. 1 (April 1, 1969): 39–60; Pierre Duhem, *Medieval Cosmology: Theories of Infinity, Place, Time, Void, and the Plurality of Worlds*, ed. and trans. Roger Ariew (Chicago: University of Chicago Press, 1985), especially vol. 7, chaps. 1 and 2; Cees Leijenhorst, "Jesuit Concepts of Spatium Imaginarium and Thomas Hobbes's Doctrine of Space," *Early Science and Medicine* 1, no. 3 (October 1, 1996): 355–80; and Roger Ariew, *Descartes and the Last Scholastics* (Ithaca: Cornell University Press, 1999).

56. Collegium Conimbricense, *Commentariorum Collegii Conimbricensis Societatis Iesu, in Octo*

Libros Physicorum Aristotelis Stagiritae (Venice, 1602), book 8, chap. 10, question 2, article 4: "Quidnam sit imaginarium spatium?" col. 518. This passage and its antecedents are discussed in detail in Grant, "Beyond the Cosmos," 51ff.

57. On the distinctions associated with the problem of infinitude (of the World and of God), see Duhem, *Medieval Cosmology*; Edward Grant, *Much Ado about Nothing: Theories of Space and Vacuum from the Middle Ages to the Scientific Revolution* (Cambridge: Cambridge University Press, 1981); and Koyré, *Closed World*. On Descartes's conception of infinitude, see the thorough recent account (including an analysis of the scholarly debates thereof) in Ariew, "Descartes' Conversation with Burman," who argues vigorously against Cottingham's claim that "even within the Cartesian corpus proper, the distinction between the indefinite and the infinite is on the verge of collapsing" (John Cottingham, "Reply to Roger Ariew," *Studia Cartesiana* 1 [1979]: 187).

58. See Ariew on the Third Meditation's argument for the existence of God (an infinite substance) and its dependence on the impossibility of arriving at the infinite from the finite through its indefinite increase. For Ariew, then, the indefinite is classed with the finite (Ariew, "Descartes' Conversation with Burman," 149).

59. See Koyré, *Closed World*, 106-24; Henri Gouhier, *La pensée métaphysique* (Paris: J. Vrin, 1962).

60. Koyré, *Closed World*, 109.

61. AT 5:51.

62. Cusa argues for both a positive (absolute God) and privative infinite (i.e. without limits which is the universe) (see De Docta Ignorantia, 2, chap. 1). See Ariew, "Descartes' Conversation with Burman," 155, for an assessment of the difference between Descartes and Cusa.

63. On the matter of infinitude, mostly expressing prudential and pious concerns, see Letter to Mersenne, May 27, 1638 (AT 2:138); Letter to More, February 5, 1649 (AT 5:274-75); Letter to More, April 15, 1649 (AT 5:345); and Letter to Mersenne, 28 January 1641(AT 3:293).

64. Descartes may have also found his technique of comparing human and divine makers in the *Timaeus*, where Timaeus's own creatively constructed *eikos logos* narrates and implicitly parallels the Platonic demiurge who creates the universe in an extraordinary act of creative construction; see *Timaeus* 29c.

65. The form of cosmogony itself instantly locates *Le monde* within the immensely popular genre of Genesis interpretations. Such works ranged from poetic elaborations of the biblical account and the constantly reprinted copies of the patristic hexamera, especially those of St. Augustine, to complex natural philosophical interpretations of Genesis, such as Marin Mersenne's enormous *Quaestiones celeberrimae in Genesim*.

66. As Paolo Rossi notes, "What else had Descartes done . . . if not present an alternative account to that of Genesis?" (*The Dark Abyss of Time: The History of the Earth and the History of Nations from Hooke to Vico* (Chicago: University of Chicago Press, 1984), 45).

67. AT 1:194.

68. See Jacques Roger, "The Cartesian Model and Its Role in the Eighteenth-Century 'Theory of the Earth,'" in *Problems of Cartesianism*, ed. Thomas M. Lennon, John M. Nicholas, and John W. Davis (Montreal: McGill-Queen's University Press, 1982), 95-102. Paolo Rossi, on the other hand, suggests that Descartes was very aware of the problem (*Dark Abyss*, 41-52). Alquié acknowledges the problem but asserts that Descartes effected an unproblematic resolution of religion and science on metaphysical grounds; since the truths of science and religion were not of the same categorical order, they could not be contradictory; see Alquié, *La découverte métaphysique*, 122-23.

69. "sum jam in describenda nativitate mundi, in qua spero me comprehensurum maximam Physicæ partem. Dicam autem me, relegendo primum caput Geneseos non sine miraculo deprehendisse, posse secundum cogitationes meas totum explicari multo melius, uti quidem mihi videtur, quam omnibus modis quibus illud interpretes explicuerunt, quod antehac nunquam speraveram: nunc vero, post novæ meæ Philosophiæ explicationem, mihi propositum est clare ostendere illam cum omnibus fidei veritatibus multo melius consentire, quam Aristotelicam."

René Descartes, *Correspondance*, ed. Charles Adam and Gaston Milhaud (Paris: F. Alcan, 1936), 1:421.

70. See Thomas Aquinas, *Summa theologiae* 1.1.1.

71. See my discussion of this trope in Montaigne's *Essais* in chapter 2.

72. Alquié, 1:348–49; AT 11:36.

73. For analysis of *Le monde* through the lens of possible-worlds theory, see Hallyn, *Kepler à Maxwell*, 123–70.

74. On the question of Descartes's understanding of the divine creation, see Thomas M. Lennon, "The Cartesian Dialectic of Creation," in *The Cambridge History of Seventeenth-Century Philosophy*, ed. Daniel Garber and Michael Ayers (Cambridge: Cambridge University Press, 1998).

75. "Quantum ad Theologiam, cùm una veritas alteri adversari nunquam possit, esset impietas timere, ne veritates in Philosophiâ inventae iis quae sunt de fide adversentur. Atque omnino profiteor nihil ad religionem pertinere, quod non aeque ac etiam magis facilè explicetur per mea principia, quàm per ea quae vulgo recepta sunt" (AT 7:581).

76. This argument is frequently asserted in the *Meditationes*, where God's creative power provides the metaphysical foundation for Cartesian science. See the classic works by Henri Gouhier, *La pensée religieuse*; Gilson, *La liberté chez Descartes*; and Marion, *Sur la théologie blanche*. See also the recent argument by Michael Della Rocca in "Descartes, the Cartesian Circle, and Epistemology Without God," *Philosophy and Phenomenological Research* 70, no. 1 (2005): 1–33.

77. Alquié, 1:615; AT 6:42.

78. For extended consideration of these issues in Descartes's physics and metaphysics, see Clarke, *Descartes's Philosophy of Science*; Garber, *Descartes' Metaphysical Physics*; Gaukroger, *System of Natural Philosophy*; Gaukroger, Schuster, and Sutton, *Descartes's Natural Philosophy*; Marion, *Sur la théologie blanche*; and Shea, *Magic of Numbers*.

79. There is a long tradition of Genesis commentaries that attempt to "explain" the events of the creation in quasi-scientific terms. For an overview of some of these accounts, see Frank Egleston Robbins, *The Hexaemeral Literature: A Study of the Greek and Latin Commentaries on Genesis* (Chicago: University of Chicago Press, 1912), and Arnold Williams, *The Common Expositor: An Account of the Commentaries on Genesis, 1527-1633* (Chapel Hill: University of North Carolina Press, 1948). On the *Le monde* as such an explanation, see Nancy, *Ego Sum*, 102–3.

80. Cottingham, "Role of God," 299.

81. See the contrasting analyses of Jean-Luc Marion and Stephen Gaukroger.

82. John Paul II, *Memory and Identity: Conversations at the Dawn of a Millennium* (New York: Rizzoli, 2005), 9.

83. Cottingham argues that the Pope here confuses Descartes's epistemic priority with a metaphysical priority: he suggests that for Descartes, as for St. Augustine, the priority of the cogito over esse/God is purely an epistemic priority and not a metaphysical one; see Cottingham, "Role of God," 292.

84. The cited phrase is from Michael Della Rocca, "Causation Without Intelligibility and Causation Without God in Descartes," in *A Companion to Descartes*, ed. Janet Broughton and John Carriero (Oxford: Blackwell, 2008), 245. Della Rocca suggests that Descartes's position on causation resembles Hume's break with divine causation; see also Rocca, "Epistemology Without God."

85. "Et sane non dubium est quin ea omnia quae doceor a naturâ aliquid habeant veritatis: per naturam enim, generaliter spectatam, nihil nunc aliud quàm vel Deum, ipsum, vel rerum creatarum coordinationem a Deo institutam intelligo" (AT 7:80); translation from René Descartes, *The Philosophical Writings of Descartes*, ed. John Cottingham (Cambridge: Cambridge University Press, 1984), 2:56. The Cottingham edition and translation is hereafter cited as "CSM" followed by volume and page number.

86. On the Voetius affair, see J. A. Van Ruler, *The Crisis of Causality: Voetius and Descartes on God, Nature, and Change* (Leiden: Brill, 1995); and Theo Verbeek, *Descartes and the Dutch: Early Reactions to Cartesian Philosophy, 1637-1650* (Carbondale: Southern Illinois University Press, 1992).

87. Jean-Robert Armogathe and Vincent Carraud, "The First Condemnation of Descartes's *Oeuvres*: Some Unpublished Documents from the Vatican Archives," in *Oxford Studies in Early Modern Philosophy*, vol. 1, ed. Daniel Garber and Steven Nadler (Oxford: Clarendon Press, 2003), 74–76.

88. Armogathe and Carraud note that Spinula refers constantly to *Principia* 3.45–47; see notes to "Descartes's *Oeuvres*," 75–76.

89. I cite the French text (i.e. the *Principes de philosophie*) because it contains additional phrases, probably overseen by Descartes himself, that are not present in the original Latin version. Added text is indicated by "<>." AT 9-2:125–26; CSM 1:257.

90. Alquié, 1:346; AT 11:34.

91. Isaac Newton, *Opticks, or, a Treatise of the Reflections, Refractions, Inflections and Colours of Light* (London: 1721), 377–78.

92. Isaac Newton, *Sir Isaac Newton's Mathematical Principles of Natural Philosophy and His System of the World*, ed. Florian Cajori, trans. Andrew Motte and Florian Cajori (Berkeley and Los Angeles: University of California Press, 1947), 544.

93. Rossi, *Dark Abyss*, 43.

94. "Cartesian" has become synonymous with rigorous logic, rational clarity, or more pejoratively, a lack of imagination. On the breakdown of the world of correspondences and the epistemic rupture of the seventeenth-century epitomized by Descartes, see Foucault, *Les mots et les choses*. On Descartes's role in the seventeenth-century segregation of poetry and philosophy, see Basil Willey, *The Seventeenth Century Background: Studies in the Thought of the Age in Relation to Poetry and Religion* (New York: Columbia University Press, 1952). On Descartes and the rise of aesthetic rationalism, see Reiss, *Knowledge, Discovery, and Imagination*.

95. I use the word "paradigm" here in the specific sense described by Thomas Kuhn, *The Structure of Scientific Revolutions* (Chicago: University of Chicago Press, 1996), 10.

96. Heilbron describes Galileo in quixotic terms: see *Galileo*, chapter 7. The physician Antoine Menjot claims that Pascal "appelait la philosophie cartésienne le roman de la nature, semblable à peu près à la histoire de Don Quichotte." In Blaise Pascal, *Oeuvres complètes*, ed. Henri Gouhier and Louis Lafuma (Paris: Editions du Seuil, 1963), 641.

97. Christiaan Huygens, *Oeuvres complètes* (The Hague: M. Nijhoff Société hollandaise des sciences, 1888), 10:403, 405–6.

98. "I should also add a word of advice touching on the best way to read this book, which is that I would like the reader first to read through the entire work, as one would a novel" (AT 9-2:11).

99. Robert Boyle, "The Excellency of Theology, Compar'd with Natural Philosophy," in *The Works of Robert Boyle*, vol. 8: *Publications of 1674–76*, ed. Michael Hunter and Edward B. Davis (London: Pickering and Chatto, 1999), 57. For Boyle, however, the romance-like structure of the natural world also evoked the fundamental impossibility of attaining complete knowledge of it; Descartes uses the same analogy to reach the opposite conclusion.

100. See James Daniel Collins, *Descartes' Philosophy of Nature* (Oxford: Blackwell, 1971), 7–8. For a contrasting view and analysis of *Le monde* as a didactic work, see John Schuster, "Descartes and the Scientific Revolution, 1618–1634: An Interpretation" (Ph.D. diss., Princeton University, 1977). The tactical argument is most often the explanation for Descartes's use of a fiction in *Le monde*. But such a view is to take Descartes's claims to avoid controversy at face value, or to read back into the actual conceptualization of *Le monde* the worries over persecution in the *Discours* (and later works). As William Shea points out incisively, Descartes started writing *Le monde* long before the condemnation of Galileo, and thus, though the condemnation was certainly responsible for Descartes's suppression of the treatise, it is very unlikely that worries over persecution led Descartes to use the form of fiction; see "*Le monde*," 73–74 and Marjorie Grene, *Descartes* (Minneapolis: University of Minnesota Press, 1985).

101. See Dear, *Revolutionizing the Sciences*, 149–67. The importance of this literary logic is discussed by Fernand Hallyn, but primarily in the context of Kepler and Copernicus; see *Poetic*

Structure. For a related argument, see also Hans Blumenberg, *The Genesis of the Copernican World* (Cambridge, Mass.: MIT Press, 1987).

102. AT 9-2:125; CSM 1:257. For Newton's comment, "I frame no hypotheses," see Newton, *Mathematical Principles*, 547.

103. The same worry lies at the core of chief counselor Melchior Inchofer's report to the Inquisition after Galileo's first hearing at the trial of 1633: "the author [Galileo] claims to discuss a mathematical hypothesis, but he gives it physical reality, which mathematicians never do." See summary in Giorgio de Santillana, *The Crime of Galileo* (Chicago: University of Chicago Press, 1955), 246–47. Timothy Reiss describes this problem in terms of "the dispute as to the relationship between mathematical hypothesis (system) and the description of reality (as existent); see "The 'Concevoir' Motif in Descartes," in *La cohérence intérieure: études sur la littérature française du XVIIe siècle presentées en homage à Judd D. Hubert,* ed. Jacqueline van Baelen and David Rubin (Paris: Chez Jean-Michel Place, 1977), 205.

104. Though Descartes and Hume are traditionally opposed, there are some interesting continuities: see Rocca, "Causation Without Intelligibility." It is Hume, too, who uses the term "worldmaking" in the *Dialogues* to contrast various versions of the world's creation (*Dialogues*, 35–36).

105. Matthew Prior, *Dialogues of the Dead, and Other Works in Prose and Verse*, ed. A. R. Waller (Cambridge: Cambridge University Press, 1907), 267.

106. The OED cites Nathaniel Fairfax's use of "world-maker" in precisely this sense in *A Treatise of the Bulk and Selvedge of the World* (1674). Newtonian scholar and mathematician John Keill also uses the term in his first book, *An Examination of Dr Burnet's Theory of the Earth* (1698), an attack on Burnet's cosmogony, which is clearly based on Descartes's. The ultimate goal of Keill's attack was Descartes himself "whom he convicted of having made fashionable his foolish ambition to make a world of his own"; see Roger, "Cartesian Model," 109.

CHAPTER FIVE

1. *Paradise Lost* 12.1–6. All references are from John Milton, *Paradise Lost,* ed. Alastair Fowler (New York: Longman, 1998).

2. *Paradise Lost* 5.268, 3.565–66, 3.673–74.

3. "Adam" appears 119 times, "Eve" 118 times, "God" 271 times, and "Earth" 213 times. "Universe" appears only seven times, while "cosmos" is never used. These searches were conducted electronically using the Chadwyck-Healey Literature Online database and cross-checked against Celia Florén, *John Milton: A Concordance of Paradise Lost* (New York: Olms-Weidmann, 1992).

4. Despite several uses of the phrase "new world" or "new created world," direct allusion to America occurs only once in *Paradise Lost,* in the first simile after the Fall (9.1115–18). For detailed analysis of the poem through the lens of American expansion and imperial conquest, see J. Martin Evans, *Milton's Imperial Epic: Paradise Lost and the Discourse of Colonialism* (Ithaca: Cornell University Press, 1996).

5. On the poem's reception history, see Nicholas von Maltzahn, "The First Reception of Paradise Lost," *Review of English Studies* 47, no. 188 (1996): 479–99; and William Poole, "Two Early Readers of Milton: John Beale and Abraham Hill," *Milton Quarterly* 38, no. 2 (2004): 76–99.

6. "On *Paradise Lost*," ll. 9–10; cited from Milton, *Paradise Lost.* On Marvell's poem see especially: Judith Herz, "Milton and Marvell: The Poet as Fit Reader," *Modern Language Quarterly* 39, no. 3 (1978): 239–63; Sharon Achinstein, "Milton's Spectre in the Restoration: Marvell, Dryden, and Literary Enthusiasm," *Huntington Library Quarterly: Studies in English and American History and Literature* 59 (1997): 1–29; and Kenneth Gross, "'Pardon Me, Mighty Poet': Visions of the Bard in Marvell's 'On Mr. Milton's Paradise Lost," *Milton Studies* 16 (1982): 77–96.

7. *Letter to the Grand Duchess Christina* (1615) cited from Galilei, *Essential Galileo*, 114. Galileo cites Augustine's *De genesi ad litteram,* book 2 (110), a crucial text in debates over science and scripture.

8. The tension between divine and human creators is discussed provocatively in Gordon Teskey, *Delirious Milton: The Fate of the Poet in Modernity* (Cambridge, Mass.: Harvard University Press, 2006).

9. For related analysis see Catherine Gimelli Martin, *The Ruins of Allegory: Paradise Lost and the Metamorphosis of Epic Convention* (Durham: Duke University Press, 1998).

10. Garin, *Science and Civic Life*, 4–5. The context of Garin's comment is Leon Battista Alberti: "Just because Alberti is always a poet, and that means a creator, he is well aware of the risk involved in all creativity, in every construction which amounts to bringing about a fundamental change in what is given to us and indeed in the whole world."

11. Blaise Pascal, *Pensées*, ed. Michel Le Guern, 2 vols. (Paris: Gallimard, 1977), 1:15; my translation.

12. Yi-fu Tuan, *Cosmos and Hearth*, 1, 8.

13. The recent scholarship on Milton and the New Science is expanding. I have found particularly useful: Harinder Singh Marjara, *Contemplation of Created Things: Science in Paradise Lost* (Toronto: University of Toronto Press, 1992); John Rogers, *The Matter of Revolution: Science, Poetry, and Politics in the Age of Milton* (Ithaca: Cornell University Press, 1996); William Poole, "Milton and Science: A Caveat," *Milton Quarterly* 38, no. 1 (2004): 18–34; Catherine Gimelli Martin, "'What If the Sun Be Centre to the World?': Milton's Epistemology, Cosmology, and Paradise of Fools Reconsidered," *Modern Philology* 99, no. 2 (November 2001): 231–65, and "'Boundless the Deep': Milton, Pascal and the Theology of Relative Space," *ELH* 63, no. 1 (1996): 45–78.

14. "At a Vacation Exercise," ll. 35–46. All citations of Milton's shorter poems are from John Milton, *Complete Poems and Major Prose*, ed. Merritt Y. Hughes (New York: Odyssey Press, 1957). On the challenge of dating the Prolusions, see *A Variorum Commentary on the Poems of John Milton*, ed. Merritt Y. Hughes (New York: Columbia University Press, 1970), 2.1:16–18, note 2; see also 136–37. For recent appraisals of the Prolusions, see Gordon Campbell and Thomas N. Corns, *John Milton: Life, Work, and Thought* (Oxford: Oxford University Press, 2008), 58–60; and N. K. Sugimura, *"Matter of Glorious Trial": Spiritual and Material Substance in Paradise Lost* (New Haven: Yale University Press, 2009), 11–29.

15. For recent discussion of the poem and the prolusion in which it appears, see Sugimura, *"Matter of Glorious Trial."* The poem also looks ahead to "De Idea Platonica" and "Il Penseroso"; Sugimura discerns an allusion to Sylvester (p. 22, note 79).

16. The scholarship on Milton and Galileo is substantial. I have drawn on: George F. Butler, "Milton's Meeting with Galileo: A Reconsideration," *Milton Quarterly* 39, no. 3 (2005): 132–39; Maura Brady, "Galileo in Action: The 'Telescope' in Paradise Lost," *Milton Studies* 44 (2005): 129–52; Amy Boesky, "Milton, Galileo, and Sunspots: Optics and Certainty in Paradise Lost," *Milton Studies* 34 (1997): 23–43; Neil Harris, "Galileo as Symbol: The Tuscan Artist in Paradise Lost," *Annali dell'istituto e museo di storia della scienza di Firenze* 10, no. 2 (1985): 3–29; and Judith Herz, "'For Whom This Glorious Sight?': Dante, Milton, and the Galileo Question," in *Milton in Italy: Contexts, Images, Contradictions*, ed. Mario A. Di Cesare (Binghamton, N.Y.: Medieval and Renaissance Texts and Studies, 1991), 147–57.

17. For discussion of the problematic analogy between the temporal and the spiritual realms in *Paradise Lost*, particularly in the context of the new science, see David Quint, "'Things Invisible to Mortal Sight': Light, Vision, and the Unity of book 3 of *Paradise Lost*," *Modern Language Quarterly* 71, no. 3 (September 1, 2010): 229–69, and *Inside Paradise Lost: Reading the Designs of Milton's Epic* (Princeton: Princeton University Press, 2014).

18. See Jean Dietz Moss, "Galileo's Letter to Christina: Some Rhetorical Considerations," *Renaissance Quarterly* 36, no. 4 (December 1, 1983): 547–76. Giorgio de Santillana compares Galileo's letter to Milton's *Areopagitica* (*Crime of Galileo*, 96–98).

19. All three similes are linked by "spot," later linked to the earth itself as a point, a spot.

20. John Guillory, *Poetic Authority: Spenser, Milton, and Literary History* (New York: Columbia University Press, 1983), 161.

21. Brady, "Galileo in Action," 131–32.

22. See Harris, "Galileo as Symbol," 11–12.

23. From *Descriptio Globi Intellectualis* in Francis Bacon, *The Works of Francis Bacon, Baron of Verulam, Viscount St. Albans, and Lord High Chancellor of England*, ed. James Spedding, Robert Leslie Ellis, and Douglas Denon Heath (St. Clair Shores, Mich.: Scholarly Press, 1969), 4:512–13, 536; see discussion of Bacon and Milton in Martin, ""What If the Sun,'" 252.

24. On the problem of error, speculation and the new science, see Quint, *Inside Paradise Lost*; Martin, "What If the Sun."

25. The first simile introduces a series of comparisons that alter the relative shape and size of objects in succession—Satan's shield is like the moon, his spear seems like a ship's mast only to be revealed as far larger, the giant devils suddenly seem like leaves. See Brady, "Galileo in Action," on the moral ambiguities presented by these problems of scale.

26. For a useful perspective on the relationship between geography and science in this context, see David N. Livingstone, "Geography, Tradition and the Scientific Revolution: An Interpretative Essay," *Transactions of the Institute of British Geographers* 15, no. 3, 2 (1990): 359–73.

27. On this topic, see Aït-Touati, *Fictions of the Cosmos*.

28. Anna Friedman Herlihy, "Renaissance Star Charts," in *The History of Cartography*, vol. 3.1: *Cartography in the European Renaissance*, ed. David Woodward (Chicago: University of Chicago Press, 2007), 99–122.

29. On the mapping of the southern sky, see E. Dekker, "Early Explorations of the Southern Celestial Sky," *Annals of Science* 44, no. 5 (1987): 439–70; Frank Verbunt and Robert H. van Gent, "Early Star Catalogues of the Southern Sky: De Houtman, Kepler (Second and Third Classes), and Halley," *Astronomy & Astrophysics* 530 (2011), doi: http://dx.doi.org/10.1051/0004-6361/201116795; and E. B. Knobel, "On Frederick de Houtman's Catalogue of Southern Stars, and the Origin of the Southern Constellations," *Monthly Notices of the Royal Astronomical Society* 77 (1917): 414–32.

30. For the centrality of Plancius in naming the southern constellations see Herlihy, "Renaissance Star Charts"; Elly Dekker, "On the Dispersal of Knowledge of the Southern Celestial Sky," *Der Globusfreund* 35/37: 211–30; and Deborah Jean Warner, *The Sky Explored: Celestial Cartography, 1500–1800* (New York: A. R. Liss, 1979).

31. Figures from the New Testament populate the north celestial sphere, while figures from the Old Testament populate the south: Julius Schiller, *Coelum Stellatum Christianum* (Augsburg: 1627).

32. Milton's interest in geography is well documented: see Elbert N. S. Thompson, "Milton's Knowledge of Geography," *Studies in Philology* 16, no. 2 (April 1, 1919): 148–71; Robert Ralston Cawley, *Milton and the Literature of Travel* (Princeton: Princeton University Press, 1951); Amy Lee Turner, "Milton and Jansson's Sea Atlas," *Milton Quarterly* 4, no. 3 (1970): 36–39; and more recently, Blair Hoxby, *Mammon's Music: Literature and Economics in the Age of Milton* (New Haven: Yale University Press, 2002).

33. See the discussion of the 1592 and 1594 maps in Rodney W. Shirley, *The Mapping of the World: Early Printed World Maps, 1472–1700* (London: Holland Press, 1984).

34. John Milton, *Complete Prose Works of John Milton*, ed. Don M. Wolfe (New Haven: Yale University Press, 1953), 7:494–95.

35. Sanson, *Atlas nouveau*.

36. On Milton's use of Sylvester, see George Coffin Taylor, *Milton's Use of Du Bartas* (Cambridge, Mass.: Harvard University Press, 1934); and recently, Stephen M. Fallon, "Milton's Sin and Death: The Ontology of Allegory in Paradise Lost," *English Literary Renaissance* 17, no. 3 (1987): 329–50; John Rumrich, "Milton's God and the Matter of Chaos," *PMLA* 110, no. 5 (1995): 1035–46; Martin, "What If the Sun"; and Sugimura, "*Matter of Glorious Trial.*"

37. See especially John Gillies, "Space and Place in Paradise Lost," *ELH* 74, no. 1 (2007): 27–57; and Martin, "'Boundless the Deep.'"

38. Robert Hollander, "Milton's Elusive Response to Dante's *Comedy* in *Paradise Lost*," *Milton Quarterly* 45, no. 1 (2011): 1–24.

39. See Galileo Galilei, *Le opere di Galileo Galilei*, ed. Antonio Favaro, 20 vols. (Firenze: Giunti-

Barbera, 1890), vol. 9. For an account of the lectures, see J. L. Heilbron, *Galileo*, 28-33; and John Kleiner, *Mismapping the Underworld: Daring and Error in Dante's Comedy* (Stanford: Stanford University Press, 1994). Milton's knowledge of Galileo's literary interests is unclear.

40. On creation in the Nativity Ode, see Barbara K. Lewalski, *The Life of John Milton: A Critical Biography* (Oxford: Blackwell, 2000), 38; Stella Revard, *Milton and the Tangles of Neaera's Hair: The Making of the 1645 Poems* (Columbia: University of Missouri Press, 1997), 64; and David Quint, "Expectation and Prematurity in Milton's 'Nativity Ode,'" *Modern Philology* 97, no. 2 (1999): 195-219.

41. "On the Morning of Christ's Nativity," ll. 117-24.

42. "At a Vacation Exercise," ll. 43-44.

43. For discussion of the dynamic relationship between chaos and order in *Paradise Lost*, see Regina M. Schwartz, *Remembering and Repeating: On Milton's Theology and Poetics* (Chicago: University of Chicago Press, 1993), 8-39.

44. "Naturam non pati senium," ll. 19-21, 33-36; cited from Milton, *Complete Poems and Major Prose*, 34.

45. See the overview of this tradition in Ramachandran, "Edmund Spenser, Lucretian Neoplatonist," 373-411.

46. The literature on the shifting modes of biblical interpretation in the seventeenth and eighteenth centuries is vast. I have found the following works especially useful: Ana M. Acosta, *Reading Genesis in the Long Eighteenth Century: From Milton to Mary Shelley* (Aldershot: Ashgate, 2006); Klaus Scholder, *The Birth of Modern Critical Theology: Origins and Problems of Biblical Criticism in the Seventeenth Century* (Philadelphia: Trinity Press International, 1990); Peter T. van Rooden, *Theology, Biblical Scholarship and Rabbinical Studies in the Seventeenth Century: Constantijn L'Empereur, 1591-1648* (Leiden: E. J. Brill, 1989); Henning Graf Reventlow, *The Authority of the Bible and the Rise of the Modern World* (Philadelphia: Fortress Press, 1985); Arnold Williams, *The Common Expositor,* and "Renaissance Commentaries on Genesis and Some Elements of the Theology of Paradise Lost," *PMLA* 56, no. 1 (1941): 151-64; C. B. Kaiser, *Creation and the History of Science*, vol. 3 (Grand Rapids: W. B. Eerdmans Publishing, 1991), and *Creational Theology and the History of Physical Science: The Creationist Tradition from Basil to Bohr* (Leiden: E. J. Brill, 1997); and Joseph Gebara, *Le dieu créateur: histoire de la doctrine de la création, de Philon d'Alexandrie à Théophile d'Antioche* (Lille: Atelier national de reproduction des thèses, 2003). On atheism, deism, and other challenges to theological orthodoxy, see Michael J. Buckley, *At the Origins of Modern Atheism* (New Haven: Yale University Press, 1987); Margaret C. Jacob, *The Radical Enlightenment: Pantheists, Freemasons, and Republicans* (London: Allen & Unwin, 1981); Roger D. Lund, ed., *The Margins of Orthodoxy: Heterodox Writing and Cultural Response, 1660-1750* (Cambridge: Cambridge University Press, 1995); and Mariangela Priarolo and Maria Emanuela Scribano, eds., *Fausto Sozzini e la filosofia in Europa: atti del convegno, Siena, 25-27 Novembre 2004* (Siena: Accademia senese degli Intronati, 2005).

47. See Scholder, *The Rise of Critical Theology.*

48. Early examples include: Jakob Boehm's *Mysterium Magnum: An Exposition of the First Book of Genesis* (1654), Alexander Ross's *An Exposition on the Fourteen First Chapters of Genesis* (London, 1626), and Thomas White's *De mundo dialogi tres* (Paris, 1642). Toward the end of the century, in the wake of Newton's discoveries, a flurry of works, such as Thomas Burnet's *Telluris theoria sacra* (1681), John Ray's *Miscellaneous Discourses concerning the Dissolution and Changes of the World* (1692), Thomas Robinson's *New Observations on the Natural History of this World of Matter, and this World of Life* (1696), and the Socinian Jean Le Clerc's *Twelve Dissertations out of Monsieur Le Clerk's Genesis* (1696), make their appearance.

49. Milton's radical theology is the subject of much recent scholarship. I have found especially useful: John Rogers, *Matter of Revolution*; Michael Lieb, *Theological Milton: Deity, Discourse and Heresy in the Miltonic Canon* (Pittsburgh: Duquesne University Press, 2006); Stephen B. Dobranski and John Peter Rumrich, eds., *Milton and Heresy* (Cambridge: Cambridge University Press,

1998); Mark R. Kelley, Michael Lieb, and John T. Shawcross, eds., *Milton and the Grounds of Contention* (Pittsburgh: Duquesne University Press, 2003); David Loewenstein and John Marshall, eds., *Heresy, Literature and Politics in Early Modern English Culture* (Cambridge: Cambridge University Press, 2006).

50. For useful overviews, see Acosta, *Reading Genesis*, and Reventlow, *Authority of the Bible*.

51. For discussion of genre and interpretive purpose, see Rooden, *Biblical Scholarship and Rabbinical Studies*, 132–58.

52. See H. F. Fletcher, *The Use of the Bible in Milton's Prose* (Urbana: University of Illinois, 1929). Milton's authorship of *De doctrina christiana* is now contested; see William Bridges Hunter, *Visitation Unimplor'd: Milton and the Authorship of De doctrina christiana* (Pittsburgh: Duquesne University Press, 1998) and the response in Gordon Campbell et al., "The Provenance of *De doctrina christiana*," *Milton Quarterly* 31, no. 3 (1997): 67–117. I follow the suggestion in Campbell et al. that Milton was involved in the *De doctrina* manuscript.

53. Milton's quite frequent identification as a Socinian and Deist suggests that his commitment to rational explanation, evident in the expositions of cosmic order in *Paradise Lost*, was recognized as part of a wider movement away from traditional biblical exegesis and toward a new, dangerous direction. Milton's Socinian leanings have been the subject of much recent scholarship; see Lieb, *Theological Milton*; Bryson, *Tyranny of Heaven*; and John Rogers, "Milton and the Heretical Priesthood of Christ," in *Heresy, Literature and Politics in Early Modern English Culture*, ed. David Loewenstein and John Marshall (Cambridge: Cambridge University Press, 2006), 203–20. Abraham Hill thought Milton was a Socinian, pointing to a passage in Milton's *Artis Logicae* (see William Poole, "The Early Reception of Paradise Lost," *Literature Compass* 1, no. 1 [2004]). In *Life of Milton*, John Toland cites Evelyn on Milton's dubious orthodoxy (see discussions in Abraham Stoll, "Discontinuous Wound: Milton and Theism," *Milton Studies* 44 [2005]: 179–203; and Poole, "Two Early Readers," 76–99).

54. On early reception of *Paradise Lost*, see von Maltzahn, "First Reception," 479–99; Poole, "Two Early Readers," "Early Reception," and *Milton and the Idea of the Fall* (Cambridge: Cambridge University Press, 2005); and Stoll, "Discontinuous Wound."

55. Guillaume de Saluste du Bartas, *The Divine Weeks and Works of Guillaume de Saluste, Sieur du Bartas*, ed. Susan Snyder, trans. Josuah Sylvester (Oxford: Clarendon Press, 1979), 1:119–26.

56. Du Bartas, *La sepmaine*, 1:373; 2:781ff.

57. "Pria che facesse Dio la terra e 'l cielo, / non eran molti dei, né molti regi / discordi al fabricar del novo mondo. / Né solitario in un silenzio eterno / in tenebre viveasi il sommo Padre, / ma col suo Figlio e col divin suo Spirito / in se medesmo avea la sede e 'l regno, / de' suoi pensati mondi alto monarca: / perch'opera fu il pensier divina, interna." *Le sette giorni del mondo creato*, 1, lines 78–86. Cited from Torquato Tasso, *Opere*, ed. Bruno Maier (Milan: Rizzoli, 1963).

58. On Milton's use of the hexameral literature see J. Martin Evans, *Paradise Lost and the Genesis Tradition* (Oxford: Clarendon Press, 1968); Mary I. Corcoran, *Milton's Paradise with Reference to the Hexameral Background* (Washington, D.C.: Catholic University of America Press, 1945); and Robbins, *Hexaemeral Literature*.

59. On the interpretation of the two versions of Genesis in Renaissance literature in the context of spontaneous generation, see Jeanneret, *Perpetual Motion*, chap. 1.

60. On differences between the P and J texts, see Robert Alter, trans., *Genesis* (New York: W. W. Norton, 1996), and John J. Collins, *Introduction to the Hebrew Bible* (Minneapolis: Fortress Press, 2004). For analyses of the two texts in relation to *Paradise Lost*, see Schwartz, *Remembering and Repeating*; Acosta, *Reading Genesis*; and Mary Nyquist, "The Genesis of Gendered Subjectivity in the Divorce Tracts and Paradise Lost," in *Re-Membering Milton: Essays on the Texts and Traditions*, ed. Mary Nyquist and Margaret Ferguson (New York: Methuen, 1987), 99–127.

61. Augustine, *On Genesis: Two Books on Genesis Against the Manichees; And, On the Literal Interpretation of Genesis, an Unfinished Book*, trans. Roland J. Teske (Washington, D.C: Catholic University of America Press, 1991).

62. The competing voices are also noted by Elizabeth Sauer, *Barbarous Dissonance and Images of Voice in Milton's Epics* (Montreal: McGill-Queen's Press, 1996), 62; see also 174, note 1.

63. Despite these differences, there remains a tendency in Milton studies to treat these accounts as extensions or elaborations of the others. Milton has also been rescued many times over from these charges of inconsistency either by appeals to the "poetic" nature of the text or by ingenious explanations; see Evans, *Genesis Tradition;* Dennis R. Danielson, *Milton's Good God: A Study in Literary Theodicy* (Cambridge: Cambridge University Press, 1982); Michael Lieb, *The Dialectics of Creation: Patterns of Birth and Regeneration in Paradise Lost* (Amherst: University of Massachusetts Press, 1970); and Schwartz, *Remembering and Repeating.*

64. Sauer, *Barbarous Dissonance*, 62–63.

65. That Milton read and knew Lucretius thoroughly is not in question. Lucretius is part of the curriculum outlined in *Of Education;* Lucretian ideas underlie his early Latin poem, *Naturam non pati senium,* and the Epicurean challenge to theodicy is alluded to in *Of True Religion.* Milton's animist materialism and mortalism also resemble *De rerum natura* in important ways. Several recent studies have begun to highlight the Roman poet's importance for *Paradise Lost;* useful studies include: John Leonard, "Milton, Lucretius, and 'the Void Profound of Unessential Night,'" in *Living Texts: Interpreting Milton*, ed. Kristin A. Pruitt and Charles W. Durham (Selinsgrove, Penn.: Susquehanna University Press, 2000), 198–217; David Quint, "Fear of Falling: Icarus, Phaethon, and Lucretius in *Paradise Lost,*" *Renaissance Quarterly* 57, no. 3 (2004): 847–81; William Kerrigan, *The Sacred Complex: On the Psychogenesis of Paradise Lost* (Cambridge, Mass.: Harvard University Press, 1983), 195–98; Paul Hammond, "Dryden, Milton, and Lucretius," *The Seventeenth Century* 16, no. 1 (2001): 158–76; and Philip Hardie, "The Presence of Lucretius in Paradise Lost," *Milton Quarterly* 29, no. 1 (1995): 13–24.

66. John Leonard also notes this allusion (see "Milton, Lucretius").

67. See also Teskey, *Delirious Milton.*

68. Influential readings of this passage include: Barbara K. Lewalski, *Paradise Lost and the Rhetoric of Literary Forms* (Princeton: Princeton University Press, 1985); Rogers, *Matter of Revolution*, 122–25; and Schwartz, *Remembering and Repeating*, 21–22.

69. John Rogers, however, describes Satan in book 5 as a "vulgar empiricist" and argues that Abdiel articulates the orthodox position with its emphasis on first causes (Rogers, *Matter of Revolution*, 122–24).

70. Schwartz, *Remembering and Repeating*, 22.

71. See Lucretius *De rerum natura* 5.154–86.

72. See the discussion in Jeanneret, *Perpetual Motion*, 29–32. There is also suggestion that the stability of God's creation is itself subject to cyclical change: the Son's elevation coincides with the Platonic great year, implying that it is the beginning of a new cycle. Similarly, the cycles of creation and destruction ordained by God parallel a materialist, Epicurean notion of the cyclical dissolution and recreation of worlds.

73. Translators into English typically translate *religio* as "superstition," underplaying the range of Lucretius's term, which encompasses religion, ignorant belief, superstition, myth, manipulative theories, and narratives that constrain individual action.

74. Lactantius *De ira dei* 13; see Alexander Roberts et al., eds., *Ante-Nicene Fathers* (Peabody: Hendrickson Publishers, 1994).

75. Jaroslav Pelikan, *What Has Athens to Do with Jerusalem? Timaeus and Genesis in Counterpoint* (Ann Arbor: University of Michigan Press, 1997), 14. Pelikan speculates that Lucretius may well have been familiar with Plato's *Timaeus* and even Genesis. David Furley separates the question of cosmology in Greek philosophy into the Platonic/Aristotelian and Atomist camps—since the Platonic account became so much a part of Genesis interpretation, it was hardly surprising that Lucretius was the designated contrary; see *The Greek Cosmologists* (Cambridge: Cambridge University Press, 1987).

76. On the challenge to notions of divine agency, see Osler, *Divine Will*, and Ullrich Langer,

Divine and Poetic Freedom in the Renaissance: Nominalist Theology and Literature in France and Italy (Princeton: Princeton University Press, 1990). For a suggestive discussion of the increasing interest in Lucretius and the rise of modernity, see Blumenberg, *Legitimacy of the Modern Age*. Resurgence of theodicial texts in the late seventeenth century, ranging from Descartes's *Meditationes* (1641) to Liebniz's *Essais de Théodicée* (1710), attests to a broader cultural concern with rethinking the nature of deity itself.

77. It is worth noting the ambiguity in Job 42.2–6 that has made for problems of interpretation (translations suggest that Job speaks of himself ["I know . . ."] when the Hebrew suggests that Job speaks *to* God ["you know . . ."]), while the final lines of Job's speech (42.5–6) have no clear object, leaving open the possibility that Job may be voicing his anger against God. Milton may well be drawing on these tensions for Satan's speech. I owe this insight to an email exchange with Richard Allen (November 20, 2014).

78. Paul Ricœur, *The Symbolism of Evil*, trans. Emerson Buchanan (New York: Harper & Row, 1967), 319.

79. As Harold Fisch points out, the Book of Job is rarely seen as a great Creation-poem, even though Talmudic tradition (with which Milton was familiar) associated it closely with Genesis; Fisch, "Creation in Reverse: The Book of Job and Paradise Lost," in *Milton and Scriptural Tradition: The Bible into Poetry*, ed. James H. Sims and Lelan Ryken (Columbia: University of Missouri Press, 1984), 104–16. Parallels between Job and Genesis are numerous and established, particularly in the Buxtorf Rabbinic Bible (which included the Aramaic paraphrases that Milton drew on).

80. See Carol A. Newsom, *The Book of Job: A Contest of Moral Imaginations* (Oxford: Oxford University Press, 2003), 243.

81. See Gillies, "Space and Place," 27–32.

82. Antonella Piazza, "Milton and Galileo," 241.

83. Gillies also notes this ("Space and Place," 38).

84. Immanuel Kant, *Practical Philosophy*, ed. and trans. Mary J. Gregor, The Cambridge Edition of the Works of Immanuel Kant (New York: Cambridge University Press, 1996), 269.

85. For details of the changes between editions, see Stephen B Dobranski, "Editing Milton: The Case Against Modernisation," *Review of English Studies* 30 (2007): 5.

86. Reasons for Milton's re-division of the poem without substantive changes to the text are discussed in John K. Hale, "*Paradise Lost*: A Poem in Twelve Books or Ten?" *Philological Quarterly* 74, no. 2 (1995).

87. Also noted by Gillies, "Space and Place," 35. The trope echoes Seneca in *Quaestiones naturales* 1, a passage cited by Ortelius on his world map, *Typus orbis terrarum* (see Besse, *Les grandeurs de la terre*, 334).

88. On the debates over diurnal rotation and its relevance for *Paradise Lost*, see Grant McColley, "The Astronomy of *Paradise Lost*," *Studies in Philology* 34, no. 2 (April 1, 1937): 209–47.

89. For analysis of this problem in the philosophy of science, see André Mercier, "The Problem of the Imperfection of a World, Itself Created by a Perfect God," *Foundations of Physics (Historical Archive)* 22, no. 2 (1992): 205–19. Raphael's response is a twist on the traditional hexameral model where a lesson on the structure of the cosmos follows an Angelic account of creation; Raphael offers cosmic hypotheses instead of a clear lesson in cosmic order; see Gillies, "Space and Place," 34; McColley, "Astronomy of 'Paradise Lost,'" 223–24.

90. While the Fall was often invoked as a reason for cosmic imperfection, Milton blocks that defense by making Adam ask the question before the Fall. However, Milton also describes God's intentional disturbance of the cosmos after the Fall (10.650–715), enabling the deity to exercise a "special providence" that brings imperfect matter closer to perfection.

91. See Robert Hugh Kargon, *Atomism in England from Hariot to Newton* (Oxford: Clarendon Press, 1966), and W. R. Johnson, *Lucretius and the Modern World* (London: Duckworth, 2000).

92. *De rerum natura* 2.180–81.

93. The only other appearance of "atom" in *Paradise Lost* is in the Lucretian Chaos of 2.898–

900 ("For Hot, Cold, Moist and Dry, four champions fierce / Strive here for mastery, and to battle bring / Their embryo atoms . . .").

94. "Atom" was not necessarily synonymous with Lucretius, but atomism in early modern Europe was most closely associated with Lucretius and Epicurus, even though the philosophies of Democritus, Heraclitus, and Empedocles were known at second-hand. See David A. Hedrich Hirsch, "Donne's Atomies and Anatomies: Deconstructed Bodies and the Resurrection of Atomic Theory," *Studies in English Literature, 1500–1900* 31, no. 1 (Winter 1991): 69–94.

95. Considerable debate remains over Milton's science and whether he was a Copernican. See especially the arguments in Martin, "'What If the Sun'"; Gillies, "Space and Place"; and Quint, "'Things Invisible.'" Martin goes as far as to argue that Raphael paraphrases Galileo's heliocentric observations of the planets at 8.126–30 (245).

96. Barbara Lewalski connects Raphael and Adam's dialogue to the genre of the cosmological dialogue, made famous by Galileo's *Dialogo sopra i due massimi sistemi del mondo*; see *Rhetoric of Literary Forms*, 46–50. Dante's *Paradiso* is also an important source: Raphael's comment at 8.90–99 directly contradicts Beatrice in *Paradiso* 28.64–68.

97. For analysis of the clash of world-systems in the passage, see Martin, "'What If the Sun,'" who shows how Raphael's speech fits into scientific rhetoric of the period and is not anti-empirical but takes a stance that most of the virtuosi would have recognized and supported. David Masson recognized Milton's Copernican leanings early on, but attempted to explain this apparent anomaly away; see *The Life of John Milton: Narrated in Connexion with the Political, Ecclesiastical, and Literary History of His Time*, 6 vols. (Cambridge: Macmillan and Co., 1858–80), 6:39.

98. The OED gives this line in *Paradise Lost* as the first instance of the verbal use of "model" as meaning "to construct a model or theory of the structure of" something.

99. Milton, *Paradise Lost*, 2.566–67.

100. Martin, "'What If the Sun,'" 240, connects this to "the 'profoundly anthropocentric' perspective [Fernand] Hallyn associates with the Copernicans." Similarly, Kerrigan insists on Milton's bodily focus in *Paradise Lost* over "thinking in astronomy" and claims that the poem's focus is "unmistakably terracentric" (Kerrigan, *Sacred Complex*, 196), though he still identifies Milton's cosmology with the Ptolemaic world.

101. See the related argument about ontology in Henry Weinfield, "Skepticism and Poetry in Milton's Infernal Conclave," *Studies in English Literature 1500–1900* 45, no. 1 (2005): 191–214.

102. Michel de Montaigne, *The Complete Essays of Michel de Montaigne*, trans. Donald Frame (Stanford: Stanford University Press, 1958), 457.

103. "Des opinions de la philosophie, j'embrasse plus volontiers celles qui sont les plus solides, c'est à dire les plus humaines et nostres. . . . Elle faict bien l'enfant, à mon gré, quand elle se met sur ses ergots pour nous prescher. . . . C'est une absolue perfection, et comme divine, de scavoyr jouyr loiallement de son estre. Nous cherchons d'autres conditions, pour n'entendre l'usage des nostres, et sortons hors de nous, pour ne sçavoir quel il y fait." Montaigne, 3.13; 1113, 1115/855–56.

104. Gillies, "Space and Place," 37–38.

EPILOGUE

1. Key essays on the Fool's Cap Map include: Rodney Shirley, "Epichthonius Cosmopolites: Who Was He?" *The Map Collector* 18 (March 1982): 39–40 and *Mapping of the World*; Richard Helgerson, "Epilogue: The Folly of Maps and Modernity," in *Literature, Mapping and the Politics of Space in Early Modern Britain*, ed. Andrew Gordon and Bernhard Klein (Cambridge: Cambridge University Press, 2001), 241–62; Anne S. Chapple, "Robert Burton's Geography of Melancholy," *Studies in English Literature, 1500–1900* 33, no. 1 (Winter 1993): 99–130.

2. See Helgerson, "Epilogue." The map appears as the cover illustration for a number of re-

cent books: for example, William Shakespeare, *The Norton Shakespeare*, ed. Stephen Greenblatt et al. (New York: W. W. Norton, 2008) and David Turnbull, *Masons, Tricksters and Cartographers* (Routledge, 2000).

3. The cartouche reads: "Democritus Abderites deridebat, Heraclitus Ephesius deflebat, Epichthonius Cosmopolites deformabat" (Democritus of Abdera mocked it, Heraclitus of Ephesus wept for it, Epichthonius Cosmopolites deformed it). "Cosmopolite"—the Latinized form of the Greek *kosmopolites* seems a sixteenth-century invention, possibly linked to Guillaume Postel, who describes himself as a "cosmopolite" on the title page of his *De la république des Turcs* (1560). On earliest usages of the word, see Hazard, "Cosmopolite," 1:354–64.

4. I owe this insight to an email exchange with Douglas Pfeiffer (October 29, 2013). For *kosmopolites*, see Diogenes Laertius, *Lives of Eminent Philosophers*, trans. Robert Drew Hicks (Cambridge, Mass.: Harvard University Press, 1979), 6.63.

5. Immanuel Kant, *Kant: Political Writings*, ed. Hans Siegbert Reiss, trans. H. B. Nisbet (Cambridge: Cambridge University Press, 1991). See the comprehensive discussion of Kant's legacy in Pauline Kleingeld, *Kant and Cosmopolitanism: The Philosophical Ideal of World Citizenship* (Cambridge: Cambridge University Press, 2012). Anthony Pagden cautions against the implicit imperialism of Kant's view ("Stoicism, Cosmopolitanism, and the Legacy of European Imperialism," *Constellations* 7, no. 1 [2000]: 3–22). For discussion of the main issues in contemporary cosmopolitan thought, see Garrett Wallace Brown and David Held, *The Cosmopolitanism Reader* (Cambridge: Polity, 2010).

6. Appiah, *Cosmopolitanism*, xix.

7. The idea of a "project of mediation" is Seyla Benhabib's; see *Another Cosmopolitanism*.

8. The passage draws on Pliny, *Natural History* 2.72.

9. Davide Stimilli, "Aby Warburg's Pentimento," *The Yearbook of Comparative Literature* 56, no. 1 (2010), 175; citing Dante, *Paradiso*, 33.85–87.

10. From "Introduction" to Georges Didi-Huberman, *Atlas: How to Carry the World on One's Back?* (Madrid: TF Editores, 2010).

11. See Joshua Jelly-Schapiro, "All over the Map: A Revolution in Cartography," *Harper's Magazine*, September 2012, 79.

12. See Georges Didi-Huberman, *Atlas ou Le gai savoir inquiet* (Paris: Minuit, 2011).

Index